Sustainable Economy Models in the Age of Industry 5.0

Rakesh Kumar · Sachi Nandan Mohanty
Editors

Sustainable Economy Models in the Age of Industry 5.0

Springer

Editors
Rakesh Kumar
Department of Management Studies
Graphic Era Deemed to be University
Dehradun, Uttarakhand, India

Sachi Nandan Mohanty
VIT-AP University
Amaravati, Andhra Pradesh, India

ISBN 978-981-96-4103-1 ISBN 978-981-96-4104-8 (eBook)
https://doi.org/10.1007/978-981-96-4104-8

This Springer imprint is published by the registered company Springer Nature Singapore Pte Ltd.
The registered company address is: 152 Beach Road, #21-01/04 Gateway East, Singapore 189721, Singapore

Preface

The book "Sustainable economy growth in the age of industry 5.0" explores the integration of the economy and industry 5.0, focusing on sustainable growth and the Sustainable Development Goals of the United Nations. This book promotes different ideas of long-term, inclusive growth without major resource depletion, environmental damage, or social inequality. It focuses on renewable energy promotion, carbon emission reduction, and inclusive opportunities. The book highlights the importance of adopting green technologies, encouraging the circular economy, and switching to cleaner industrial techniques. Industry 5.0, the next stage of industrial change, is centred on resilient, sustainable, and human-centric manufacturing. It emphasizes human-machine cooperation, reviving human creativity and moral judgement. The book promotes the circular economy by reducing resource use and environmental effects, enabling precision manufacturing, predictive maintenance, and better supply chains. The book provides a comprehensive guide to the integration of the circular economy into Industry 5.0, addressing challenges and proposing solutions.

This volume explores the integration of Industry 5.0 and the green economy in promoting sustainable manufacturing practices. It highlights the use of AI, robotics, and IoT in Industry 5.0, focusing on efficient resource utilization and waste minimization. The text also discusses the link between Sustainable Development Goals and Industry 5.0, highlighting human-centred innovation and alignment with these goals. It delves the role of digital technologies in the circular economy, focusing on AI, IoT, and blockchain in overcoming environmental and resource depletion challenges. It discusses the circular economy concept for small and medium enterprises, highlighting the interdependence of Industry 5.0 and the circular economy. It covers green energy, inclusive growth, e-waste management, smart technology, and government policies. The book is applicable across various disciplines and offers case studies for in-depth research.

Dehradun, India — Rakesh Kumar
Amaravati, India — Sachi Nandan Mohanty

Introduction

The book title "Sustainable economy growth in age of industry 5.0 explore integration of economy and industry 5.0. This book provide comprehensive knowledge about sustainable economy growth. Suitable development gaol(SDG-2030) is as flagship agenda of united nation. The book delved about traditional economy and its challenges in sustainable growth. This provide how smart technology become essential part of sustainable economic growth.

Sustainable economy growth refers to long term inclusive growth of society. The long-term expansion of an economy free from major resource depletion, environmental damage, or social inequality is known as sustainable economic growth. Its main objectives are biodiversity conservation, renewable energy promotion, carbon emission reduction, and inclusive opportunity. Adopting green technologies, encouraging the circular economy, and switching to cleaner industrial techniques are important elements. Communities, corporations, and governments all have important responsibilities to play in this shift. To ensure resilience against environmental issues and to create a fairer society, sustainable growth is in line with the Sustainable Development Goals of the United Nations.

This volume examine sustainable economy growth with inclusion of advanced smart technology tools. The next stage of industrial change, known as Industry 5.0, is centred on resilient, sustainable, and human-centric manufacturing. It places an emphasis on human-machine cooperation, reviving human creativity and moral judgement. By utilising cutting-edge technology like artificial intelligence (AI) and 3D printing to develop more effective, ecologically friendly production techniques, Industry 5.0 incorporates the concepts of the circular economy. It also highlights adaptation and resilience to global issues, encouraging a comprehensive strategy where companies put profitability and societal well-being first. This change guarantees that sectors are more ethical, inclusive, efficient, and in line with equitable and sustainable future principles.

Through the integration of human-centric technology with environmental and social sustainability, Industry 5.0 aims to promote sustainable economic growth. To promote creativity and efficiency, it moves away from automation and towards a collaborative strategy that incorporates AI, robotics, and IoT. This book promotes the ideas of the circular economy by reducing resource use and environmental effect. Precision manufacturing, predictive maintenance, and better supply chains are made possible by advanced technologies. Industry 5.0 encourages equitable growth, chances for reskilling and upskilling, and the development of resilient economies to meet global problems.

This book provides a comprehensive guide to the integration of the circular economy into Industry 5.0, addressing challenges and proposing solutions. It covers green energy, inclusive growth, e-waste management, smart technology, and government policies. This delved about international regulation with focus on inclusive economic growth The book is applicable across various disciplines and offers case studies for in-depth research.

Contents

Editors and Contributors

About the Editors

Dr. Rakesh Kumar is an eminent educationist and decorated army officer with 25 years of experience. He is a classic example of courage, commitment and intimate relationships. His areas of expertise are Finance, Business law, Economic, Labor Law, Technology, and Sustainability. He is expertise in Structural Equation Modelling (SEM), ISM and Fuzzy TOPSIS research Methodology. He is an interdisciplinary researcher. He is also involved in the course and syllabus designing and developing on various Management topics, including analytics, and management curriculum. He has published more than 72 research papers in leading databases. He is editor of reputed books including Tylor & Francis, Nova Science, etc. He has published 33 Indian Patents. He has been part of the board of Study, board of examination, Evaluation department in government organization. He has been part of many international conferences. He also acts to lead and liaise between the student community and the industry delegates, with a keen interest in training around emotions and spirituality.

Dr. Sachi Nandan Mohanty was recognized as Top 2% World Scientists Ranking by Stanford University and Elsevier for years 2023. He received his PostDoc from Indian Institute of Technology Kanpur, India in the year 2019 and Ph.D. from Indian Institute of Technology Kharagpur, India in the year 2015, with MHRD scholarship from Govt of India. He has authored/edited Forty two books, published by IEEE-Wiley, Springer, Wiley, CRC Press, NOVA and DeGruyter. His research areas include Data mining, Big Data Analysis, Cognitive Science, Fuzzy Decision Making, Brain-Computer Interface, Cognition, and Computational Intelligence. Prof. S N Mohanty has received four Best Paper Awards during his Ph.D. from International Conference at Beijing, China, and the other at International Conference on Soft Computing Applications organized by IIT Rookee in the year 2013. He has awarded Best thesis award first prize by Computer Society of India in the year 2015. He has guided 9 Ph.D. Scholar, twenty three post-graduate student. He has published 241 International Journals of International repute. He has received 4th time International Travel support from Government of India Department of Science and Technology, SERB Funding for International travel for presenting research paper and Keynote speaker. He has awarded by many awards and fellowships during his career and to name a few are and has been elected as FELLOW of Institute of Engineers, IETE, Ambassador European Alliance Innovation (EAI), and Senior member of IEEE Computer Society Hyderabad chapter. He has been awarded by many awards and fellowship during his career and to name a few are Prof. Ganesh Mishra Memorial Award on 61st Annual Technical Session & 19th Prof. Bhubaneswar Behera Lecturer on 23-24th June 2020 from The Institution of Engineers (India) Odisha State Center, Bhubaneswar.

He is a member of professional bodies like IETE, CSI, IE, IEEE (Hyderabad section). He also the reviewer of Journal of Robotics and Autonomous Systems (Elsevier), Computational and Structural Biotechnology Journal (Elsevier), Artificial Intelligence Review (Springer), Spatial Information Research (Springer). He is leading as General chair of 2 international conferences such as ICISML, AIHC, and editor in chief of international journal EAI Transaction on Intelligent System and Machine learning Application. Dr. Mohanty has gone for academic assignment to New York, Atlanta, Orlondo, Georgia, Paris, Slovakia, Singapore, Abu Dhabi, Saraja, Dubai, Malaysia, Istanbul, Viena, and Germany for delivering Key note talk and chair the sessions in International conferences with travel support from Department of Science & Technology, Government of India, New Delhi, India.

Contributors

Kamal Ahmad HNB Garhwal University, Srinagar, Garhwal, Uttarakhand, India

Mirian Ngozi Alike Department of Philosophy, Nnamdi Azikiwe University Awka, Awka, Nigeria

Aditi Anand Christ University, Bengaluru, India

N. Arul Department of CSE (AIML), St. Peter's Engineering College, Hyderabad, Telangana, India

Yashmita Awasthi Christ University, Bengaluru, India

Rohit Bansal Vaish College of Engineering, Rohtak, India

Sukanta Kumar Baral Department of Commerce, Indira Gandhi National Tribal University, Amarkantak, Anuppur, Madhya Pradesh, India

Prashanth Beleya INTI International University, Nilai, Malaysia

Monu Bhardwaj School of Management Graphic, Era Hill University Dehradun, Dehradun, Uttarakhand, India

Tanupriya Choudhury Graphic Era Deemed to Be University, Dehradun, Uttarakhand, India

Ramesh Babu Damarla SR University, Warangal, Telangana, India

Smita Dash School of Management, Driems University, Bhubaneswar, Odisha, India

Sathya Swaroop Debasish Department of Business Administration, Utkal University, Bhubaneswar, Odisha, India

P. Deepan Department of CSE (AIML), St. Peter's Engineering College, Hyderabad, Telangana, India

K. Devender School of Business SR University, Warangal, India

Sunita Dwivedi IILM University, Gurugram, Haryana, India

Richa Goel Graphic Era University, Dehradun, India;
Symbiosis Centre for Management Studies, Noida, India;
Symbiosis International Deemed University, Pune, India

Gurunadham Goli School of Business SR University, Warangal, India

Neha Gupta IBCS SOA Deemed to Be University, Bhubaneswar, Odisha, India

V. Harish PSG Institute of Management, PSG College of Technology, Coimbatore, India

Mohammad Helal Hossain CEO & MD, Daystar HRM, Market Research Centre, Dhaka, Bangladesh

Ghousia Jabeen School of Business, SR University, Warangal, Telangana, 506371 India

Kafila School of Business SR University, Warangal, India

Kalyani Christ University, Bengaluru, India

Samta Kathuria Parul Institute of Law, Parul University, Vadodara, India

Jyotika Kaur Graphic Era Deemed to Be University, Dehradun, Uttarakhand, India

Rupa Khanna Graphic Era University, Dehradun, India

Tanaraj Krishna Quest International University, Ipoh, Malaysia

Chitra Krishnan Symbiosis Centre for Management Studies, Noida, India;
Symbiosis International Deemed University, Pune, India

Rakesh Kumar Graphic Era Deemed to Be University, Dehradun, India

Vandana Madaan Maharishi Markandeshwar (Deemed University), Mullana-Ambala, India

Rupa Khanna Mahlotra School of Commerce Graphic Era Deemed to Be University Dehradun, Dehradun, Uttarakhand, India

Chunchu Nagaraju School of Business, SR University, Warangal, India

Bhakti Parashar VIT Business School, VIT Bhopal University, Bhopal, India

Sumit Gouraba Patra Research Scholar, ASBM School of Business, ASBM University, Bhubaneswar, India

Pradeep Pillai AlRawabi Trading Co (LIFCO Group), Dubai, UAE

Namrata Prakash School of Management Graphic, Era Hill University Dehradun, Dehradun, Uttarakhand, India

Rita Prusty IBCS SOA Deemed to Be University, Bhubaneswar, Odisha, India

Fazla Rabby Stanford Institute of Management & Technology, Sydney, NSW, Australia

Mantena Siva Pavan Kumar Raju Sasi Institute of Technology and Engineering, Tadepalligudem, AP, India;
SR University, Warangal, Telangana, India

Geeta Rana Himalayan School of Management Studies, Swami Rama Himalayan University, Jolly Grant, Dehradun, India

Gayatri Rathi Faculty of Law, Swami Vivekanand Subharti University, Meerut, India

Sumit Roy Balaji Institute of Marketing Management, Sri Balaji University, Pune, India

Manmath Nath Samantaray Research Scholar, ASBM School of Business, ASBM University, Bhubaneswar, India

Ravindra Sharma Himalayan School of Management Studies, Swami Rama Himalayan University, Jolly Grant, Dehradun, India

M. Shravan School of Business SR University, Warangal, India

Pushpam Singh Department of Commerce, Indira Gandhi National Tribal University, Amarkantak, Anuppur, Madhya Pradesh, India

Ram Singh Symbiosis University of Applied Sciences, Indore, India

Shanu Singh Christ University, Bengaluru, India

Vinay Pal Singh Quantum University, Roorkee, Uttarakhand, India

Rachoru Himani Srihita School of Business, SR University, Warangal, India

Tilagavati Subramaniam Quest International University, Ipoh, Malaysia

R. Vidya Department of CSE (AIML), St. Peter's Engineering College, Hyderabad, Telangana, India

Historical Perspective of Circular Economy Growth from Industry 1.0 to 5.0

Pushpam Singh and Sukanta Kumar Baral

Abstract There has been a major transformation in economic and industrial practices that are increasingly centred on the concept of circular economy. It seeks to optimise resource utilisation by minimising its wastage and promote sustainable growth. This transformation is due to the historical revolutions that emerged from the late eighteenth century through Industry 1.0 to the continuing Fifth Industrial Revolution, which revolves around smart technologies. Every new era of industrialisation brought remarkable innovations and along with some challenges. The study reviews the historical progression of the circular economy by looking into every phase of industrial revolutions from Industry 1.0 to Industry 5.0. It offers a conceptual framework that addresses how technological innovations and industrial regulations have played an active part in shaping circular practices. It also discussed the challenges encountered during each phase of industrialisation. Lastly, it provides future solutions by drawing inferences from historical events for deriving new perspective on the circular economy in future. As this study is conceptual in nature, the historical events, progression, and challenges have been explained with the help of academic literature reviews, while stating the need of integrating industrial activities with circular economy practices.

Keywords Circular economy · Industrial revolutions · Industry 5.0 · Sustainable development · Technological innovations

P. Singh (✉) · S. K. Baral
Department of Commerce, Indira Gandhi National Tribal University, Amarkantak, Anuppur, Madhya Pradesh, India
e-mail: pushpamsingh4@gmail.com

S. K. Baral
e-mail: sukanta.baral@igntu.ac.in

R. Kumar and S. N. Mohanty (eds.), *Sustainable Economy Models in the Age of Industry 5.0*, https://doi.org/10.1007/978-981-96-4104-8_1

1 Introduction

Over centuries, the industrial landscape has evolved remarkably due to shifts in societal needs and technological breakthroughs. These shifts are classified into five distinct phases, termed as Industry 1.0 to Industry 5.0 (Leng et al., 2022). With each phase brings an innovative method of conceptualising technology, production process, and the way humans and machines interact. It aimed at restructuring industries and the global economy to promote personalised and sustainable solutions. The beginning of Industry 1.0, which brought mechanised production processes in the late eighteenth century (Mohajan, 2019), marked a turning point in the trajectory of industrialisation. During this phase, mining and textile industries were replaced by machines using steam engines and water, which predominantly took over conventional human labour. The adoption of steam engines, mechanised looms, and spinning machines together added an exemplary output, which in turn led to the onset of urbanisation as people migrated to industrial hubs in search of work opportunities (Groumpos, 2021). Despite being discovered that it was followed by obstructions like environmental deterioration and inadequate working conditions, this foundational phase was quite important in setting the framework for future improvements in industrial operations. The formation of Industry 2.0 in the late nineteenth century signalled the arrival of another new opportunity that was highlighted by electrification and large-scale production (Mokyr & Strotz, 1998). The invention of assembly line was made possible because of the use of electricity in industrial activities. Henry Ford's innovations in the automotive production are one of the prominent instances of this (Ferrigno et al., 2024). Not only did this period maximise the pace of production, but it also made simpler for large population to obtain goods, which spurred both consumer spending and economic development. Moreover, the creation of labour union movements happened during the years of Industry 2.0. This arose when workers tried to fight back against unfair labour practices that were evident during this time frame (Bosch, 1999). Electrified machineries and tools were produced as a result of mechanisation, which contributed to the transformation of industries and opened the path for massive production that remains to be the cornerstone of present manufacturing practices. Manufacturing was revolutionised with the inception of computerisation and automation in the twentieth century which signalled the starting point of Industry 3.0 (Greenwood, 1997). Programmable Logic Controllers (PLCs) and robotics were only made practicable through the integration of semiconductors, electronics, and information technology into industrial processes (Mohanty et al., 2024). These technological breakthroughs helped in achieving reproducibility, accuracy, and scalability, hence minimising the need for manual labour involvement. Computing power made it easy to optimise and simulate production workflows and helped in transforming industries. Moving towards the twenty-first century, the Fourth Industrial Revolution, which is popularly known as Industry 4.0, is a mix of physical and digital technologies that culminated an era of smart factories and systems that are linked globally (Philbeck & Davis, 2018). It led to an advancement in industrial operations by implementing automated maintenance and real-time monitoring. This paradigm shift was pushed through artificial intelligence (AI), Internet of Things (IoT), big data analytics, and cloud computing (Domenteanu et al., 2024). The capacity of digital twins to carry out computerised operations made manufacturing easier and more productive. According to the European Commission 2021 report, the modern world is currently experiencing a shift towards Industry 5.0 that focuses human-oriented innovation. Industry 5.0 puts an emphasis on the fostering interaction between smart machines and humans unlike its previous phases, which prioritised automation and productivity. Additionally, it introduces collaborative robots, also known as cobots that operates along with humans to strengthen innovation and creativity. Furthermore, it emphasises on social wellness and sustainability with an intent of syncing industrial developments with worldwide goals such as alleviating carbon emissions and promoting circular economy.

In recent times, the term circular economy has become increasingly popular everywhere across the nations because of its critical approach for attaining sustainability in industrial and economic contexts (Ghisellini et al., 2016). The circular economy places a primary focus on waste reduction, resource optimisation, and regenerative use of resources. As stated by the Ellen MacArthur Foundation, the circular economy has the possibility to elevate resource efficiency in Europe by 3% every year if digitalisation is used as a catalyst alongside circular practices. It is projected a gain of USD671.9 billion every year from primary resources by 2030 along with an incremental of USD1.35 trillion as additional benefits (MacArthur, 2015). The basic concept of circular economy can be found in the past industrial revolutions (Reike et al., 2018), despite it being considered as modern solutions to societal and environmental concerns. The application of this concept has gone through great progressive changes throughout every phase of industrial transformation, starting with resource intensive method used in Industry 1.0 to the technology-based innovations in Industry 5.0.

This research reviews the gradual evolution of circular economy throughout these five different industrial revolutions. It addresses how economic, social, and technological transformations have contributed towards sustainable development. The study revolves around key objectives, which are (a) to explore how new technologies help in the formation of ideas for circular economy, (b) to identify the challenges encountered during each phase of industrialisation, and (c) to draw inferences from historical events for deriving new perspective on the circular economy in future.

2 Technological Innovations Driving Circular Economy During Industrial Revolutions

Historically, the industrial revolution has been viewed as the major shift towards economic development. Since then, the world has gone through many significant changes. An era of advanced economy driven by high income per capita began with the industrial revolution (Kuznets, 1973). The very first industrial nation 'Britain'

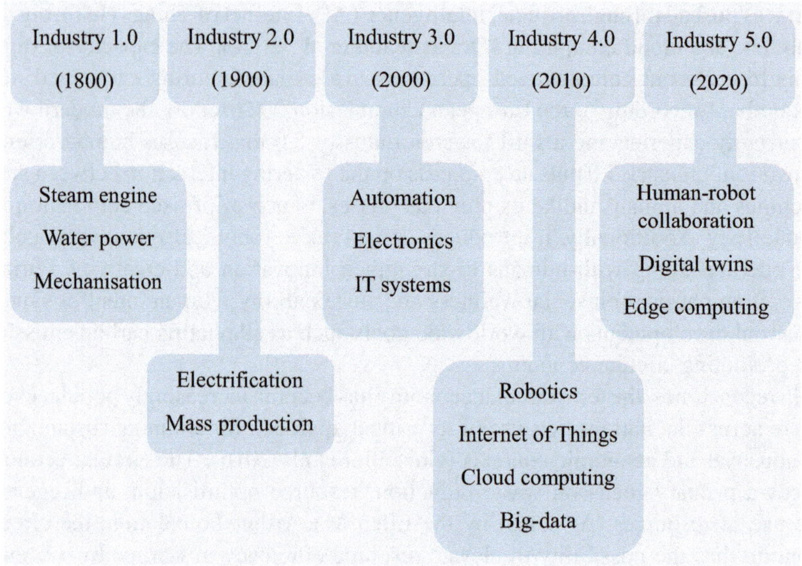

Fig. 1 Technological innovations from Industry 1.0 to Industry 5.0

was viewed as the foundation for subsequent industrial breakthroughs (O'Brien, 2006). Figure 1 shows technological innovations from Industry 1.0 to Industry 5.0. There was a wave of inventions that happened during the eighteenth century which eventually culminated industrial boom and the formation of manufacturing plants (MacLeod, 2009). Subsequently, other phases of industrial revolutions gave momentum to the circular economy concept by working on the flaws of linear approach. Each phase of industrial revolution has shaped the understanding of sustainability, waste management, and resource optimisation, which also brought transformations that are strongly connected to the formation of circular economy.

The evolution from linear to circular economy approach is explained in separate subsections through reviewing the technological innovations and major developments of each industrialisation phase.

2.1 Industry 1.0

The First Industrial Revolution not only brought remarkable economic progress, but it also laid the path for forming strategies towards circular economy. Although the linear "take-make-dispose" approach was prevalent at this stage, the structural changes and new technologies of Industry 1.0 proved to be the groundwork for effective utilisation of resources (Wyes, 2018). The production processes were completely transformed during the first industrialisation phase with the advent of mechanisation that brought in water and steam-powered machines and minimised manual labour. On the other side, demand for resources like iron, coal, and textiles increased due to these developments. The issues of scarcity and waste resulting from excessive resource utilisation forced the industrialists to search for alternative ways to use resources. An example of this would be the constant enhancement of machine productiveness like steam engine, which was developed by James Watt so as to lower the fuel consumption (Kanefsky & Robey, 1980). The ongoing upgradation of resource efficacy showed circular approach, wherein it emphasised on making the most out of resources. Moreover, the mechanised manufacturing of textiles helped in making further use of waste materials in the weaving process and that ultimately spurred the adoption of resource recycling practices (McDonough & Braungart, 2017). The introduction of new techniques for regenerating old clothing and scraps into tissue for paper manufacturing was an immediate solution to the growing demand for clothing. An early recognition of the opportunity to reproduce waste products gained momentum due to certain constraints which were put on the resource availability during the First Industrial Revolution. Waste management became more complex due to the rapid urbanisation that began with industrialisation, and it was evident in quick-growing industrial regions like Glasgow, Birmingham, and Manchester (Gibb & Lee, 2002). An early effort to mitigate harmful emissions and extract useful material from household and residential waste resulted to formation of municipal waste management practices. The residuals of coal combustion like ash particles were gathered and reused for cultivation as a fertiliser and also for construction. The steam-powered railway network also became the part of main technological development in this phase where it not only facilitated cost-effective transportation but also helped in recycling efforts for transporting waste materials (Barles, 2014). The recycling and reuse of waste materials that are considered to be essential aspects of circular economy were driven by growing urbanisation during the time of Industry 1.0 as shown through these actions. When looking back at the enduring impact of Industry 1.0, it reminds that despite linear economy being prevalent at that time, the roots of circular economy can be found while seeking productivity and inventions.

2.2 *Industry 2.0*

Industry made the revolutionary shift from steam power to electric power with the invention of electrification that happened during the Second Industrial Revolution. It enabled widespread power generation which reduced fuel wastage. Manufacturing processes turned energy efficient because of improvements in machines (Geels, 2006). The development of power grids and transformers made it easy for energy distribution, which minimised wastage of resource in industries (Kumar, R., et.al., 2024). The less expensive, efficient, and quick way of manufacturing steel came with the development of open-hearth furnace and Bessemer technique (Agarwal & Agarwal, 2017). The setup of steel factories that used scrap from blast furnace for cement and road constructions are pioneering instances of industrial synergy. This is an example of how waste from one product can be a resource for making another product. With the production of synthetic substances, petrochemicals, fertilisers, and dyes, the chemical manufacturing entered an era of quick growth during this period. A lot of these achievements were attained by recognising how to turn waste into valuable output. In this regard, one instance of an early understanding of waste as an asset is the recycling of coal tar, which is a byproduct of coke production. This coal tar was utilised to make colours and other chemicals (Tarr, 1985). Another sustainable solution to bio-fertilisers was the Haber–Bosch process that synthesises ammonia from hydrogen and nitrogen (Rouwenhorst et al., 2022). The assembly line model originally applied linear consumption behaviour, but Ford's unified products and equipments supported remanufacturing. Interchangeable components made it easy for repair, disassembly, replacement, which ultimately reduced product wastage. During this time, the initially developed waste-to-energy mechanisms had come to an effect. These mechanisms comprised combustion furnaces that were being utilised to generate electricity from municipal waste materials and this managed to lower landfills and created renewable source of energy (Hughes, 1993). Industries were reliant on recycled materials as a way of sustaining production during wartime and economic recessions. Although these initiatives were driven by economic necessities instead of ecological concerns, it provided crucial resources and ideas that were needed for shaping circular economy in future.

2.3 Industry 3.0

This era known by digital revolution brought disruptive innovations in the fields of computing, electronics, and telecommunications. It changed production techniques from mechanisation to automation. Even though, the linear "take-make-dispose" economy kept going at that time and was prevalent within the industry, but technological developments during Industry 3.0 moved towards circular economy practices by promoting improved resource administration, increased waste utilisation and fostering innovations in managing product life cycle. The manufacturing industry witnessed a transformation due to this digital infrastructure brought about by automation through an application of computer numerical control (CNC) devices. For instance, precise cutting of raw materials was done by CNC machines that reduced extra scraps (Naboni et al., 2015). The adoption of just-in-time manufacturing assisted in the reduction of excessive production and surplus inventory through automated machines. It was initially believed that items like electronics, plastics, and composite materials could not be recycled. But new discoveries of manufacturing technologies made recycling easy. The rise of information technology (IT) systems aided industrial optimisation and real-time monitoring. In order to safeguard economy and eliminate resource exploiation, alternative energy sources like wind and solar power were introduced that lowered dependency on energy from fossil fuels. Lithium-ion batteries ensured energy storage and grid stabilisation, reducing carbon emissions (Heinonen et al., 2015). Devices like computers and cellphones were easily upgraded, repaired, and recycled through 3D modelling and computer-aided design (CAD), reflecting the transformation from linear to circular model product. The advent of 3D printing changed the production techniques wherein it used only required materials so as to stop material wastage and transport emissions (Ward, 2015). The development of these technologies paved the way for subsequent industries to one step further towards regenerative economies that are consistent with circular economy practices.

2.4 Industry 4.0

This phase involved an integration of Internet of Things, blockchain, robotics, artificial intelligence, and big data analytics that have reshaped the industries with remarkable efficiency, connectivity, and automation. In order to speed up real-time monitoring and information gathering, IoT connects various devices and products for better operations. Like, industrial equipment that is fitted with sensors can detect damage and make automated maintenance that increases the lifetime of machineries. There are companies such as H&M that make use of IoT to retrieve rejected clothes from customers and thereafter send it to resale or recycling agencies. IoT is also employed in smart grids to control power consumption to encourage renewable energy sources like wind and solar. The use of robotic systems for recycling operations that are powered by AI possess the capacity to separate metals, plastics, and fabrics with great precision than conventional methods, leading to high recycling practices. It is crucial for circular supply chain to maintain records of transactions that are indestructible and transparent, and this function can be carried out with the help of blockchain technology (Esmaeilian et al., 2020). It allows ethical sourcing, supports recycling, and removes waste through monitoring resources from their initial extraction until their end life cycle stage. IBM and other organisations use blockchain technology for monitoring e-waste. The creation of robotics has proved to be a significant contributor towards improvement of recycling method. Robots that are installed with advanced vision technology and AI software have the power to dissect complex objects like electronic equipment and vehicles with high precision (Sarc et al., 2019). Sharing platforms have become popular with an effect of digitalisation in Industry 4.0, which promotes renting, sharing, or leasing of resources instead of having ownership of those resources. By making the best use of available resources, services like Airbnb, Uber, and OLIO lowered the excessive production. Businesses and industries are making a move to quit linear methods and started adopting sustainable and regenerative actions by including these technologies in their manufacturing operations.

2.5 Industry 5.0

The Fifth Industrial Revolution puts particular emphasis on building collaboration between human and emerging technologies to establish systems that are not only technically sophisticated but also human-centric and environmentally viable. Because of this synergy, industries are able to make use of creativeness and problem-solving skills of humans along with the scalability and proficiency of devices. Beyond automation, it integrates revolutionary innovations cobots, bioengineered materials, blockchain, AI, and improved connectivity. This phase of industrialisation brings new form of production and consumption that seamlessly merge environmental and social elements with technologies and create path for future where productivity, customisation, and sustainability are all intertwined. For example, Cobots perform together with humans to maximise task efficiency and it alleviate human exhaustion and provide high-quality assurance (Rahman et al., 2024). AI has moved beyond workflow optimisation and has evolved as catalyst of personalisation and decision making. Business serves customers with customised solutions by using databases of new fashion trends. Talking about fashion industry, AI is a game changer in terms of designing clothes that respond to user preferences while simultaneously achieving efficiency and waste reduction. In context of healthcare, AI-driven technologies devise personalised treatment services, which enhance patient's experience. One of the important features of Industry 5.0 is implementation of 5G network for sharing data instantaneously that helps in quick decision making. It not only enables remote operations but it is also feasible to collaborate globally, thereby reducing travels and its alongside environmental impact (Shehab et al., 2021). Blockchain technology makes sure that production process is in line with regulations and they are acquired in ethical way. Industry 5.0 also features digital twins that monitor and replicate system performance for preventive care and effectiveness. All these innovations have opened the avenues for ethical, efficient, and environment-friendly solutions. It is a huge step towards the development of industry and society by integrating industrial expansion with sustainable development.

3 Challenges Across Industrial Revolutions

Human society and economy have witnessed many changes throughout the industrial revolutions and these developments spanned over two centuries and are still continued. Each subsequent revolution has culminated in novel inventions that boosted developments. However, it also came with different challenges that gradually transformed the economies. These challenges were environmental deterioration, labour exploitation, technological disruptions, social inequality, ethical disputes, and inadequate resources. To tackle these issues, government interventions, societal change, and innovation were required. These challenges also caused long-lasting impressions on the countries.

When production was mechanised, it caused a major spike in low-cost labour. Many people including women and children were employed in risky jobs where they worked for long hours for nominal wages. There were textile plants in Manchester that engaged young children to do labour work for many long hours in harmful surroundings with inadequate ventilation (Kirby, 2013). It was common for accidents to happen and workers did not have choices to fight against detrimental activities. This caused severe medical issues and high death rate among labourers and especially children. The absence of legal rights for them caused a social instability. Fast urbanisation occurred due to huge migration from rural to urban industrial centres in search of work, leading to overcrowded slums and miserable living conditions. As a consequence of polluted water supplies and inadequate sewage systems, cholera epidemic was widespread in London during those times (Smith, 2002). In response to this, movements and reforms in healthcare like the Public Health Act in Britain enacted clean water supply and sewer system, and it succeeded in bringing down illness cases while improving people's living conditions. An extensive deployment of steam engines and dump of industrial waste into rivers were major factors of huge pollution. These challenges laid groundwork for large-scale environmental concerns in future, highlighting the need for environmentally conscious production.

Mass production intensified the extraction of resource that exploited local economies and environment, creating geopolitical conflicts. The advent of automation and computing technologies caused an inevitable decrease in demand for manual labour which subsequently results in job displacement in the industries (Pizzolato, 2004). Both industrialised and developing regions experienced uneven accessibility of technology and this proliferation of new technologies even introduced privacy and security risks. AI and robots also pose risk to job opportunities in areas like transportation, service, and manufacturing industry (Table 1).

Table 1 Major drawbacks of industrial revolutions

Phases	Drawbacks
Industry 1.0	• Factory labourers suffered long working hours in hazardous conditions with low wages • Slum areas became overcrowded due to mass migration • Industrial waste caused environmental degradation
Industry 2.0	• Mass production led to resource exploitation • Increase in monopolies caused market dominance and stiff competition • Conventional industries suffered from technological transition
Industry 3.0 & Industry 4.0	• Automation replaced manual labours • Uneven technology access caused digital divide • Rise in cybercrime • E-waste and energy consumption increased due to electrification • Robotics and other computing technologies replace human jobs • Sustainability

Source Authors' compilation

All these phases of transformation have necessitated the creation of adaptable measures and remedial approaches, starting from the conflicts in labour workforce during the First Industrial Revolution to the challenges of sustainability encountered at present. Industrial revolutions have emerged as a driving motivation behind global development where the challenges at each phase emphasised the need of strategies and regulations that maintain balance between economic prosperity and sustainability. Experiences from historical periods give critical perspectives that can potentially overcome the challenges at present economic environment.

4 Future Solutions for Restructuring Circular Economy

An increasing worldwide focus has been paid to the transforming power of circular economy in addressing social and environmental issues. Rising concern in this subject matter has been driven by prevailing sustainable development vision put forward by the United Nations (Khajuria et al., 2022). It is imperative to create strategies that are either global or regional in the task of drafting and executing circular economy plans. To foster businesses that promote circular economy and discourage unlawful waste disposal, changes need to be made in laws and policies. To make global impact, businesses need to bring creative ideas and implement customised strategies at individual level through the production of products and services that can be recycled, long-lasting, and are environment friendly. In context to climate change adaptions and lowering carbon emissions, many nations are incorporating circular economy in their nationally determined contributions (NDCs) as part of their planned actions (den Elzen et al., 2022). 79 countries have announced an explicit promise to promote circular economy in the year 2022 under the Paris Agreement (Nascimento, 2024).

Researchers found that both residential and commercial activities were accountable for huge percentage of carbon emissions (Xi et al., 2011). If cities are committed to respond to issues of extreme disconnection between surroundings and economic development, then they must necessarily find ways to new consumption practices. The entire surrounding environment of cities has to be designed tactically in order to minimise resource along with wealth generation. The waste hierarchy which has previously been applied by the waste management regulation of the European Union, presented an easy approach for tackling uncertainty issues that is similar to the 3R principles. The primary objective of waste hierarchy is to place importance to waste reduction and process minimisation that can create cost-saving prospects while extracting raw materials and disposing waste (Joensuu et al., 2020; Ribić et al., 2017; Yong, 2007). The global society has the responsibility to not only make a move towards equitable and inclusive circular economy but also be persistent to its continuous sustainable development. This can be achieved by giving importance to the needs of workers and societies that have been ignored in the past at great level. It has the potential to restructure development while safeguarding ecosystem for next generations by taking lessons from historical revolutions and adapting to changes.

5 Conclusion

The study of historical revolutions has consistently been a topic of interest for academicians and experts in different fields because of its transforming events that marked major turning points in human development. These revolutions stood as historic milestones that not only transformed societies and economies but also left with lasting impressions of innovations and adaptability. This study began by comprehensively reviewing the technological advancements made in each phase of industrial revolution and how these advancements gave path to the world towards circular economy. It has also discussed the challenges that were encountered during those times. At last, the future solutions for restructuring circular economies to have positive long-lasting progress have been presented. This conceptual study provides a base for further research and assessment by taking insights of historical events that outline technological changes and its contribution towards making an economy viable. It would help researchers and experts to understand past and modern industrial changes for an improvement in the current industrial practices.

References

Agarwal, H., & Agarwal, R. (2017). First industrial revolution and second industrial revolution: Technological differences and the differences in banking and financing of the firms. *Saudi Journal of Humanities and Social Sciences, 2*(11), 1062–1066.

Barles, S. (2014). History of waste management and the social and cultural representations of waste. In *The basic environmental history* (pp. 199–226). Springer.

Bosch, G. (1999). Working time: Tendencies and emerging issues. *International Labour Review, 138*(2), 131–149.

den Elzen, M. G., Dafnomilis, I., Forsell, N., Fragkos, P., Fragkiadakis, K., Höhne, N., Kuramochi, T., Nascimento, L., Roelfsema, H. V. S., & Sperling, F. (2022). Updated nationally determined contributions collectively raise ambition levels but need strengthening further to keep Paris goals within reach. *Mitigation and Adaptation Strategies for Global Change, 27*(5), 1–29.

Domenteanu, A., Cibu, B., & Delcea, C. (2024). Mapping the research landscape of Industry 5.0 from a Machine Learning and Big Data analytics perspective: A bibliometric approach. *Sustainability, 16*(7), 1–33.

Esmaeilian, B., Sarkis, J., Lewis, K., & Behdad, S. (2020). Blockchain for the future of sustainable supply chain management in Industry 4.0. *Resources, Conservation and Recycling, 163*, 1–16.

European Commission. (2021). *Industry 5.0: Towards more sustainable, resilient and human-centric industry*. Directorate-General for Research and Innovation. https://ec.europa.eu/info/news/industry-50-towards-more-sustainable-resilient-and-human-centric-industry-2021-jan-07_en

Ferrigno, G., Zordan, A., & Di Minin, A. (2024). The emergence of dominant design in the early automotive industry: An historical analysis of Ford's technological experimentation from 1896 to 1906. *Technology Analysis & Strategic Management, 36*(6), 1124–1135.

Geels, F. W. (2006). Major system change through stepwise reconfiguration: A multi-level analysis of the transformation of American factory production (1850–1930). *Technology in Society, 28*(4), 445–476.

Ghisellini, P., Cialani, C., & Ulgiati, S. (2016). A review on circular economy: The expected transition to a balanced interplay of environmental and economic systems. *Journal of Cleaner Production, 114*, 1–22.

Gibb, A., & Lee, R. (2002). Industrialization and demographic change: A case study of Glasgow, 1801–1914. *Population and Society in Western European Port Cities*, 1650–1939.
Greenwood, J. (1997). The third industrial revolution: Technology, productivity, and income inequality. *Economic Review, 35*(2), 1–23.
Groumpos, P. P. (2021). A critical historical and scientific overview of all industrial revolutions. *IFAC-PapersOnLine, 54*(13), 464–471.
Heinonen, S., Karjalainen, J., & Ruotsalainen, J. (2015). Towards the third industrial revolution. *Finland Futures Research Centre, 6*, 1–75.
Hughes, T. P. (1993). *Networks of power: Electrification in Western society, 1880–1930*. JHU Press.
Joensuu, T., Edelman, H., & Saari, A. (2020). Circular economy practices in the built environment. *Journal of Cleaner Production, 276*, 1–45.
Khajuria, A., Atienza, V. A., Chavanich, S., Henning, W., Islam, I., Kral, U., Liu, M., Liu, X., Murthy, I. K., Oyedotun, T. D. T., Verma, P., Xu, G., Zeng, X., & Li, J. (2022). Accelerating circular economy solutions to achieve the 2030 agenda for sustainable development goals. *Circular Economy, 1*(1), 1–9.
Kanefsky, J., & Robey, J. (1980). Steam engines in 18th-century Britain: A quantitative assessment. *Technology and Culture, 21*(2), 161–186.
Kirby, P. (2013). *Child workers and industrial health in Britain, 1780–1850* (Vol. 2). Boydell & Brewer Ltd.
Kumar, R., Goel, R., Singh, T., Mohanty, S. M., Gupta, D., Alkhayyat, A., & Khanna, R. (2024). Sustainable finance factors in indian economy: Analysis on policy of climate change and energy sector. *Fluctuation and Noise Letters, 23*(2), 2440004–339.
Kuznets, S. (1973). Modern economic growth: Findings and reflections. *The American Economic Review, 63*(3), 247–258.
Leng, J., Sha, W., Wang, B., Zheng, P., Zhuang, C., Liu, Q., Wuest, T., Mourtzis, D., & Wang, L. (2022). Industry 5.0: Prospect and retrospect. *Journal of Manufacturing Systems, 65*, 279–295.
Li, G., Hou, Y., & Wu, A. (2017). Fourth Industrial Revolution: Technological drivers, impacts and coping methods. *Chinese Geographical Science, 27*(4), 626–637.
MacArthur, E. (2015). Towards a circular economy: Business rationale for an accelerated transition. *Greener Manag International, 20*, 1–20.
MacLeod, C. (2009). Patents for invention: Setting the stage for the British industrial revolution. *Empiria: Revista De Metodología De Ciencias Sociales, 18*, 37–58.
Matuszak, J. (2022). Is your business ready for Industry 5.0? https://knowhow.distrelec.com/manufacturing/is-your-business-ready-for-industry-5-0/
McDonough, W., & Braungart, M. (2017). The next industrial revolution. In *Sustainable solutions* (pp. 139–150). Routledge.
Mohajan, H. (2019). The first industrial revolution: Creation of a new global human era. *Journal of Social Sciences and Humanities, 5*(4), 377–387.
Mohanty, R., Samal, A., & Kumar, R. (2024). Green finance supports sustainable environment: Empirical approach towards applicability in tourism sector. *Theoretical and Practical Research in Economic Fields, 15*(1), 114–126.
Mokyr, J., & Strotz, R. H. (1998). The second industrial revolution, 1870–1914. *Storia Dell'economia Mondiale, 21945*(1), 219–245.
Naboni, R., Paoletti, I., Naboni, R., & Paoletti, I. (2015). The third industrial revolution. In *Advanced customization in architectural design and construction* (pp. 7–27). Springer.
Nascimento, L. (2024). *Enabling future climate policy: An analysis of policy gaps and progress under the Paris Agreement* (Doctoral dissertation, Wageningen University).
O'Brien, P. (2006). *Provincializing the first industrial revolution*. Working Papers of the Global Economic History Network (GEHN) (17/06). Department of Economic History, London School of Economics and Political Science, London, pp. 1–39.
Philbeck, T., & Davis, N. (2018). The fourth industrial revolution. *Journal of International Affairs, 72*(1), 17–22.

Pizzolato, N. (2004). Workers and revolutionaries at the twilight of Fordism: The breakdown of industrial relations in the automobile plants of Detroit and Turin, 1967–1973. *Labor History, 45*(4), 419–443.

Rahman, M. M., Khatun, F., Jahan, I., Devnath, R., & Bhuiyan, M. A. A. (2024). Cobotics: The evolving roles and prospects of next-generation collaborative robots in Industry 5.0. *Journal of Robotics, 2024*(1), 1–22.

Reike, D., Vermeulen, W. J., & Witjes, S. (2018). The circular economy: New or refurbished as CE 3.0?—exploring controversies in the conceptualization of the circular economy through a focus on history and resource value retention options. *Resources, Conservation and Recycling, 135*, 246–264.

Ribić, B., Voća, N., & Ilakovac, B. (2017). Concept of sustainable waste management in the city of Zagreb: Towards the implementation of circular economy approach. *Journal of the Air & Waste Management Association, 67*(2), 241–259.

Rouwenhorst, K. H., Travis, A. S., & Lefferts, L. (2022). 1921–2021: A century of renewable ammonia synthesis. *Sustainable Chemistry, 3*(2), 149–171.

Sarc, R., Curtis, A., Kandlbauer, L., Khodier, K., Lorber, K. E., & Pomberger, R. (2019). Digitalisation and intelligent robotics in value chain of circular economy oriented waste management–A review. *Waste Management, 95*, 476–492.

Shehab, M. J., Kassem, I., Kutty, A. A., Kucukvar, M., Onat, N., & Khattab, T. (2021). 5G networks towards smart and sustainable cities: A review of recent developments, applications and future perspectives. *IEEE Access, 10*, 2987–3006.

Smith, G. D. (2002). Commentary: Behind the Broad Street pump: Aetiology, epidemiology and prevention of cholera in mid-19th century Britain. *International Journal of Epidemiology, 31*(5), 920–932.

Tarr, J. A. (1985). Industrial wastes and public health: Some historical notes, Part I, 1876–1932. *American Journal of Public Health, 75*(9), 1059–1067.

Ward, M. (2015). Tool-less manufacture: Digital fabrication, 3D printing and the third industrial revolution. In *Handbook of Manufacturing Industries in the World Economy* (pp. 100–108). Edward Elgar Publishing.

Wyes, H. (2018). Connecting sustainable lifestyles, Industry 4.0, and the circular economy. *Industry, 4*, 36–66.

Xi, F., Geng, Y., Chen, X., Zhang, Y., Wang, X., Xue, B., Dong, H., Liu, Z., Ren, W., Fujita, T., & Zhu, Q. (2011). Contributing to local policy making on GHG emission reduction through inventorying and attribution: A case study of Shenyang, China. *Energy Policy, 39*(10), 5999–6010.

Yong, R. (2007). The circular economy in China. *Journal of Material Cycles and Waste Management, 9*, 121–129.

Industry 5.0: Pioneering Sustainable Manufacturing in the Green Economy

Richa Goel, Rupa Khanna, Chitra Krishnan, and Rakesh Kumar

Abstract This chapter explores the integration of Industry 5.0 and the green economy in promoting sustainable manufacturing practices. Industry 5.0 takes the concepts of Industry 4.0 a step further through more emphasis on AI, robotics, and IoT integration with the human brain to increase efficiency while putting more emphasis on utilizing resources efficiently. By focusing on collaborative technologies, human–robot relations, and the sustainable business models of circular economy for waste minimization and energy conservation, as well as sustainable product development. This chapter has also used Tesla, Siemens, and Unilever examples to demonstrate how Industry 5.0 technologies support sustainable manufacturing and sustainability goals at large. As industries evolve, the synergy between technological advancements and sustainability will be vital for fostering resilient and green production systems. The chapter calls on industry leaders, policymakers, and innovators to embrace Industry 5.0 as a pathway to achieving sustainable development in manufacturing.

Keywords Green economy · Human–robot collaboration · Sustainable manufacturing · Circular economy · Renewable energy · Smart technologies

R. Goel (✉) · R. Khanna
Graphic Era University, Dehradun, India
e-mail: richasgoel@gmail.com

R. Goel · C. Krishnan
Symbiosis Centre for Management Studies, Noida, India

Symbiosis International Deemed University, Pune, India

C. Krishnan
e-mail: chitra@scmsnoida.ac.in

R. Kumar
Graphic Era Deemed to Be University, Dehradun, India

R. Kumar and S. N. Mohanty (eds.), *Sustainable Economy Models in the Age of Industry 5.0*, https://doi.org/10.1007/978-981-96-4104-8_2

1 Introduction

The green economy is an economic development model that integrates economic, social, and ecological imperatives for sustainable growth. It calls for the shift from business as usual, which involves massive exploitation of resources, pollution, and hindrance to the use of natural resources, excessive emission of greenhouse gases, and poor exploitation of renewable energy. Green economy has been described by the United Nations Environment Programme (UNEP) as "low carbon, resource efficient, and inclusive economy" improving human well-being while addressing environmental challenges and degradation (UNEP, 2011). The concept of the green economy is now recognized as part of the global response to climate change and sustainable development as defined in SDG 12 (Responsible Consumption and Production) and SDG 13 (Climate Action). Such change pertains to transformations toward a green economy where sectors like manufacturing, which utilize substantial amounts of resources, emissions, and wastes, are critical in degrading the environment.

The green economy in the manufacturing sector mainly focuses on the environmentally friendly ways of production, effective utilization of resources, minimization of wastes, and adoption of circular economy. Businesses are now embracing the use of green technologies to feed the market demand of consumers who are increasingly becoming conscious about their impacts on the environment and the general calls for high level of Corporate Social Responsibility. For example, sustainability in manufacturing enlists renewable energy, resource efficiency, and life cycle assessment to enhance the sustainability of goods (Sánchez-García et al., 2024). However, Industry 5.0 goes beyond automation, focusing on human-centric innovation, where collaboration between humans and smart technologies enhances creativity, precision, and sustainability. Industry 5.0 is described as the next revolution in manufacturing that emphasizes synergy between human intelligence and machine efficiency (Javaid et al., 2024).

An important part of Industry 5.0 is the focus on sustainability. Whereas Industry 4.0 aimed at making production systems more efficient, Industry 5.0 focuses on making these systems more efficient and sustainable. This ranges from the use of fewer resources, less utilization of resources, and customizing products to fit specific markets with negligible effect on the environment. In Industry 5.0, augmented with AI, big data, and IoT technologies, manufacturers can control resources in real time, thereby furthering sustainability goals (Tóth et al., 2023).

Furthermore, Industry 5.0 also advocates the concept of cobots, which integrate with human labor to perform various operations. Cobots that exist in industries are meant to aid in sustainable manufacturing through minimizing rubbing, wastage of material, and power consumption. This is because the idea of human creativity and artificial intelligence working together enables new and diversified thinking in production, which is close to the green economy concept (Alves et al., 2023).

The goal of this chapter is to discuss the integration of the concept of green economy and the technological revolution of industry 5.0 toward embracing sustainable manufacturing. Analyzing such factors as human-oriented technologies, environmentally friendly production, and efficient use of resources, this chapter attempts to explore the ways to reach sustainability in manufacturing industries while remaining competitive. The discussion will focus on the following key aspects:

- How Industry 5.0 enhances sustainable manufacturing through collaborative technologies.
- The role of AI, IoT, and robotics in minimizing waste and optimizing resource efficiency.
- Case studies of industries that have successfully integrated green economy principles and Industry 5.0 technologies to drive sustainability.

At the end of this chapter, the reader would be equipped with knowledge on how integrating the green economy that centers on sustainability with the newer industrial revolution known as Industry 5.0 to take manufacturing industries backward to being the major driver toward sustainable development.

2 Understanding the Green Economy in Manufacturing

2.1 Key Principles of the Green Economy

The principles behind the green economy framework are designed to make economic development environmentally sustainable. They help the manufacturing industries to implement practices that not only facilitate the manufacturing business, but also enhance the environmental and social sustainability on the long run.

Sustainability: The green economy encompasses sustainability as its core theme since it seeks to lessen the impacts of production on the environment. It includes lowering the use of finite resources, lowering emissions, and avoiding utilizing up resources in such a manner that they cannot be replenished in the future (Evans, 2023). Sustainable manufacturing involves energy-efficient production, the minimization of pollution, and; water utilization, contributing to efforts that make industries more environmentally friendly and still competitive. Switching from high carbon sources, decreasing energy intensity, and utilizing clean technologies of energy generation like solar or wind power.

Resource Efficiency: The last of the principles inherent in the concept of green economy is resource efficiency, which implies using resources sparingly and, if possible, reusing, recycling, and repairing end-products. The circular economy model is a component of resource efficiency that differs from the linear model by focusing on a circular model that prolongs product life, and renews natural resources (Ghisellini et al., 2024). For the manufacturing industry, this implies when designing the product to consider the end of life recovery, input of various resources, and output of waste at each step. For instance, companies such as Patagonia and Apple have of late incorporated circular economy whereby reused materials such as scrap are utilized in the making of new products to cut on the usage of raw materials.

Innovation in Green Technologies: Most industries consider innovation as an important factor in manufacturing since it can foster environmentally sustainable practices on the production line. Such technologies stem from renewable energy technology (solar energy, wind energy, geothermal energy), environment-friendly materials (biodegradable plastic, recycled metal), and sustainability engineering in products (Kiefer et al., 2024). The advanced manufacturing processes, including three-dimensional printing and effective and efficient manufacturing processes, also make the manufacturing process sustainable by minimizing waste and an optimal use of materials. They aid in the separation of economic development from resource consumption and negative impact on the environment, which is a central idea of the green economy.

2.2 *Impact of Manufacturing on Environmental Sustainability*

Manufacturing has traditionally been acknowledged as a major source of pollution as it frequently forms its business model based on fossil fuels, intense mining of resources, and production of waste. As stated by the International Energy Agency (IEA), CO2 emissions in the industrial segment make up approximately 24% around the globe, explaining the environmental impact of the conventional industrial processes (Dechamps, 2023).

Environmental Footprint of Traditional Manufacturing: Historical manufacturing systems have been characterized by energy-intense, non-sustainable practices, and the generation of waste products such as greenhouse gases, poisonous chemicals, and solids. To illustrate, steel and cement production requires large amounts of energy and produces large volumes of carbon dioxide emissions which affects climate change. Secondly, many industries such as textiles and electronics produce significant hazardous wastes that are commonly dumped on land or on bodies of waters, this has serious negative effects on the environment in the long run (Bocken et al., 2023).

The Necessity for Transforming Industrial Practices: This means that the whole manufacturing processes have to change in order to meet the objectives of green economy. It relates to strategies like energy conservation, cogeneration, efficient use of energy, and emission reduction measures. Ongoing green manufacturing is vital to improving the sustainability of the sector and support the country's commitments to combating climate change. Business organizations are gradually beginning to understand that adopting environmentally friendly/sustainable processes brings about efficiencies, better image, and compliance with recent stringent environmental laws. For instance, Siemens recently launched what it has termed as its "green factory" whereby new technologies and green energy are incorporated in the production process in an attempt to demonstrate that even industries can strike the right balance between production and sustainability.

2.3 Sustainable Development Goals (SDGs) and Manufacturing

Framework for the creation of sustainability came with the formulation of what is called the Sustainable Development Goals (SDGs) in 2015. Manufacturing being one of the significant economic sectors has various contributions to fulfillment of the various SDGs.

The goal fosters efficiency by pushing industries to prevent wastage of resources and practice conservation. Companies have been encouraged to come up with products with longer service lives, minimize their impacts on the environment, and manage material recycling and use responsibly. This can be done through the implementation of green technology and following the circular economy (Arora & Mishra, 2023).

Climate Action (SDG 13): Greenhouse gas emissions have been observed to be paramount especially in organizations within the manufacturing industry must therefore take active measures toward climate change. Implementing efficient technologies, enhancing energy use proficiency, and boosting the use of renewable energy demonstrate a perfect way that the manufacturing sector can reduce its carbon emission in line with the eighth target of SDG 13 (Yadav et al., 2023). For instance, the automotive industry has embarked on the manufacturing of electric cars and follows sustainable supply chain strategies.

Affordable and Clean Energy (SDG 7): Energy consumption is a common feature in manufacturing systems due to the processes used. SDG 7 aims at promoting the use of renewable energy sources in the industries through the use of wind, solar, and bioenergy for the industrial processes. The use of renewable energy sources by the manufacturing industries plays a significant role in reinventing the capability of the fossil fuel economy while at the same time having a positive impact on tackling climate change effects. Several firms like Tesla and IKEA are already in the process of transitioning to entirely source energy from renewables, which leaves manufacturers with no reason not to be a part of the energy transition toward sustainability (Akaev & Davydova, 2023).

In conclusion, having an understanding and application of green economy in manufacturing is crucial so that that many environmental disasters are prevented and that manufacturing performs more efficiently in resource usage and fulfilling the sustainability goals in the current world. Linking the manufacturing processes with the SDGs ensures that industries are longevity, sustainable, and socially just.

3 Journey from Industry 4.0 to Industry 5.0

3.1 Brief Recap of Industry 4.0

Industry 4.0 also known as the Fourth Industrial Revolution brought innovative advances such as automation, Industrial IoT, AI, and data analysis into production line. With the help of new innovative technology, the cyber physical systems advanced, generating smart factories where devices interacted with each other, worked more efficiently minimizing the need of intervention from humans, and provided real-time decision-making (Lasi et al., 2014). This combination has also allowed the possibilities of predictive maintenance, supply chain visualization, and customization for high-quantity production (Kagermann et al., 2016).

However, Industry 4.0 paradigm had some drawbacks. There was also a question of job displacement since the narrative about automation dimmed the demand for workers' physical labor. This together with concerns of job insecurity and increase of de-humanization of work, provoked criticism of uses of machines in Industry 4.0. Also, in the Industry 4.0 question of efficiency and mass production has dominated over the ecological question. As Industry 4.0 grew common, issues such as overconsumption of resources that contribute to the generation of electronic waste, energy consumption also emerged as major issues that this technology could only partly overcome (Stock & Seliger, 2016).

3.2 Introduction to Industry 5.0

Industry 5.0 is an improvement on Industry 4.0 as it combines human imagination with the precision of technology. While Industry 4.0 meant complete automation of production processes, Industry 5.0 is based on the interaction between man and machine. This synergy leverages the strengths of both: concerning competency, machines are good at tasks that require handling of large data sets and repetitiveness while humans are good at tasks that require creativity, emotional intelligence, and decision-making (Nahavandi, 2019). AI continues to be strategic in Industry 5.0 manufacturing but is different from seeking complete automation of production processes; AI is employed to enhance humans' work in production lines by allowing real-time collaboration between AI and human workers simultaneously. This collaboration coupled with personalization are ideals that are core to Industry 5.0. It enables the provision of mass customization for manufacturers since they can supply products that meet the needs of certain customers more effectively. By moving from mass production to mass personalization, manufacturers can quickly adapt to the needs of a changing consumer while avoiding the cost of excess inventory. For example, with an increasing focus on sustainability as a notion that is promoted among the consumers, the provision of solutions that include environmentally friendly and individually tailored goods is a valuable asset (Adel, 2022).

The last fundamental concept of Industry 5.0 is sustainability. While Industry 4.0 focuses on aspects of productivity and revenues, Industry 5.0 incorporates environmental and social concerns directly into the production line. AI and IoT systems assist in the provision of efficient usage of resources in production, thereby lessening the exploitation of natural resources in the manufacturing processes of products (Lom et al., 2016). Accordingly, the integration of sustainability into AI processes within industries that seek carbon neutrality is becoming increasingly relevant.

3.3 Human-Centric Approach

Human-Centered Smart Automation means that human creativity returns from the periphery to the center of industrial production. Typically, in Industry 4.0. human intervention was reduced to the barest minimum through the use of automated systems. Nevertheless, the Industry 5.0 recognizes that while machines are very fast, accurate, and good at repetitive work, humans are unique in creativity, problem-solving skills and emotional intelligence (Nahavandi, 2019). Therefore, the Industry 5.0 creates a workplace environment where both human and technology work together enhancing the capabilities of each other.

Ethical considerations must also be integrated. Industry 5.0 recognizes the emergent imperative for addressing the issues that AI generates and the prejudice of AI-ambiguities it brings to decision-making. That is why they still maintain control over activities of the AI system and guarantee that the identified ethical challenges, including transparency, fairness, and accountability, are considered in all stages of product development (Longo et al., 2020). This will help to ensure that people develop confidence in use of artificial intelligence in their daily lives besides making sure that the application of new technology is for the benefit of society and does not violate basic human rights.

Sustainability is a guiding principle fundamental to Industry 5.0. So, this approach involves social and environmental factors combined with technology use with the consideration of the manufacturing environmentally friendly functions. When used in conjunction with manufacturing innovation, the use of creativity, combined with artificial intelligence, allows for the creation of products that are as bespoke as they are efficient and environmentally friendly. The aim is to foster sustainable forms of business that can generate long-term revenues for organization and not at the expense of exploiting the natural resources or polluting the environment. Therefore, Industry 5.0 embraces technology and innovation for the global sustainable development, leading to improved responsible production.

4 Industry 5.0's Role in Driving Sustainable Manufacturing

4.1 Enhancing Production Efficiency While Reducing Waste

Thus, the key contribution of Industry 5.0 in manufacturing sustainability is the improvement of productivity, or output per employment, without growing inefficiency, or the creation of waste. Unlike Industry 4.0 that emphasized on automation in mass manufacturing, Industry 5.0 combines the human ingenuity with the intelligence of advanced AI and robotics to create a better method of manufacturing. The real-time data helps the company to reduce the time required at each step of the production process and also reduce material and energy consumption. This shift eliminates overproduction and stockpiling, and in the process, minimizes wastage of resources.

Many industries also find predictive maintenance to be instrumental in increasing production capacity where maintenance and repair are conducted based on analysis provided by artificial intelligence systems. These systems monitor potential problems developing in the attached equipment that would otherwise lead to halting of production, frequent equipment breakdowns, or calls for repair or replacement of some of its parts (Bonilla et al., 2018). It not only makes equipment durable but also helps the environment by disposing of wreckage from breakages and overuse of resources.

4.2 Using AI and Robotics to Optimize Resource Utilization and Improve Energy Efficiency

Industry 5.0 primarily focuses on the way the concept of using artificial intelligence and robotics in manufacturing to enhance the resource utilization and energy efficiency. Through AI-driven smart energy management systems, manufacturers are able to detect energy consumption in real-time possible energy optimization of production operations to avoid the usage of energy during the peak periods (Tallat et al., 2023). For instance, analytical capabilities of the AI algorithms can find the wasteful patterns prevailing in energy consumption, and control the systems to run only when required, thus minimizing the carbonization of the factory premise (Xu et al., 2021).

However, they can be used to efficiently allocate resources which reduces wastage most of the time. For instance, through data checkpoints and original inputs the AI systems can guarantee that supply chain and production of raw materials do not undergo wastage. Robotics also occupies its importance in making precision, thereby manufacturing, cutting, molding, or assembling material in the least possible manner. The connection helps the manufacturers to minimize wastage of material while still producing quality products (Stock & Seliger, 2016).

4.3 Customizing Production for Sustainable Outcomes, Including Personalized, Low-Waste Product Designs

A notable characteristic of the Industry 5.0 model is the emphasis on individualized production to create sustainable solutions. On the same note of AI, robotics, and human ingenuity, it is now possible for manufacturers to design and create products to suit unique consumer needs, without having to make numerous identical items; thus, little to no wastage. Also, unlike Industry 4.0, which focused on mass production, Industry 5.0 provides mass personalization when a customer can order a product made in a way that suits their individuality and not have unwanted products that cannot be sold (Demir, 2023). Moreover, the concept of Industry 5.0 also touches upon design process as more and more products are designed with sustainable materials and can be easily recycled. AI has been noted to influence design software used in manufacturing thus creating products that suit the customers' needs while at the same time using environmentally friendly or recyclable materials (Nahavandi, 2019). This form of consumption is consistent with the increased demand for environmentally friendly products and allows businesses to practice sustainable production that supports the achievement of sustainable development goals.

5 Green Economy and Circular Manufacturing

Industry 5.0 supports a circular economy due to its importance as a sustainable economy model that seeks to reduce waste and make the best use of resources. This model stands in stark difference to the conventional linear economy or "take-make-dispose" model. Rather, the circular economy focuses on the concepts of reuse, remanufacturing, and recycling which are extended with the help of industry 5.0 intelligent technologies. The Internet of Things (IoT) and artificial intelligence (AI) systems in Industry 5.0 enable real-time tracking of materials and products across their life cycle. These technologies allow organizations to gather information on the usage, wear, and disposal of products and then provide insights into the opportunities to prolong product lives and enhance resource recycling. For example, by using the information from predictive analytics, producers can find out when it is most efficient to restore or redistribute products, thus using them as long as possible and reducing their waste. Further, digital technologies of smart manufacturing including advanced robotics and automation promote the cost-effective remanufacturing. These technologies can dematerialize products in a selective way, reclaim components, and enable recycling of decomposed material thus minimizing the use of raw materials. Furthermore, digital platforms prove productive in the construction of a circular supply chain and direct consumers to recycles and remanufacturing industries to increase product returns and reduce wastage.

The Industry 5.0 technologies are core to the establishment of circular economies that involve recirculation of waste products in production cycles. A textbook example is the textile sector, which has included circular approaches by employing IoT of clothes' life cycle by several brands such as Patagonia. They also conduct repair operations and accept worn-out clothes to reuse or reclaim them and reuse them in other products. This not only cuts the use of textiles and clothing but also ensures the adoption of sustainable consumerism (Niinimäki et al., 2020).

Another example is the same as that in the electronics industry where the firms like Dell and HP have integrated closed-loop systems into the silver value chain by recycling plastics to offer in the production of new goods. For instance, Dell has created a program known as Closed Loop Recycling where it takes used electronics and turns them into new computer products (Aggarwal et al., 2023). Sorting and processing of electronic waste require utilization of AI and machine learning algorithms to enhance the efficiency of the processes which enable proper extraction of materials that can be reused or recycled.

BMW in the automotive industry has adopted circular manufacturing through using recycled materials on cars. To the extent that the company leverages artificial intelligence to assess and oversee data and data about the lifecycle of parts that produce cars, they are able to consolidate and recycle used parts (Gomez, 2024). Indeed, this approach saves on material costs of manufacturing while at the same time protecting the environment.

In general, Industry 5.0 innovations define new forms of circular economy that imply balancing resource consumption, effectiveness, and reduction of post-consumption waste using cutting-edge technologies and systems for circularity.

6 Human–Robot Collaboration for Sustainable Practices

Industry 5.0 has made the combination of human imagination and robot efficiency as the bedrock for promoting sustainable business practices. While industry 4.0 relied heavily in automation, industry 5.0 reintroduces the human element to the process allowing for more innovation, creativity, and the much-needed ethical considerations. It is possible to maximize tangible business values with cross-pollinated human creativity and the motionless perfection of machines. Human values intuition, innovation, and moral conduct while robots rely on precision, efficiency, and dependability for routine or risky operations. Such a balance results in increased efficiency in manufacturing processes especially in areas where human designers can only contribute toward development of sustainable products but are overridden by robot that can perform complex, detailed work with zero percent error (Nahavandi, 2019). This kind of collaboration facilitates the use of resources in the best way possible which minimizes cases of producing excess, producing substandard products, or even draining too much energy (Brettel et al., 2014). Moreover, advanced today's technologies such as ML and AI can help to study a huge amount of data to identify successful uses of energy and materials, as well as the effectiveness of working processes. These data insights can be understood by humans, adapted to the requirements of the current production, and calibrated to optimize the functioning of machines. For example, while AI platforms can forecast load demand, human operators decide on operation changes to achieve sustainability goals or the likelihood of changing the rate at which machines operate to decrease power usage during peak periods (Lu et al., 2022).

6.1 Examples of Industry 5.0 Applications in Reducing Waste, Energy Consumption, and Environmental Impact

Here is the list of Industry 5.0 applications that showed the usefulness of human–robot collaboration in decreasing waste, energy, and environmental effects. One of the most prominent roles, which has been assigned to key performance indicators, is in automotive industry. Manufacturers such as Tesla and BMW have incorporated robotic structures that function alongside employees in the development of more efficient cars. For example, robots are used in the formation of battery packs with accurate welding avoiding any wastage of material while human beings deal with the design and direction that the company needs to use recycle materials. The synergism of these activities allows for the proper consumption of resources and a reduction in unnecessary energy expenditure in production lines. Another good example is in the food and beverage industry where the use of humans overseeing robots has transformed the production process. For instance, Nestlé has adopted the automated use of robots for packaging since it optimizes the use of materials by at least 30%. In the same way, human workers work on coming up with more environmentally friendly packaging solutions, which adds to the sustainable efforts (Butt, 2024). Several robots used in Nestlé facilities have incorporated the use of AI to manage energy where they monitor peak energy usage and plan for functioning during other off-peak hours of energy usage.

As for the construction industry, 3D printing robotic aides assist architects and engineers in designing and building environment-friendly shelters for the needy. Such robots are capable of operating under the direction of human designers where they can print structures using recycled materials and other forms of green composite materials; the structures designed are sustainable and have been optimized to fit into the local climate. This form of human–robot collaboration minimizes waste production, energy usage, and the negatives that come with conventional construction. The healthcare manufacturing sector also offers examples of how Industry 5.0 might help decrease waste and overall energy utilize. In the current world, pharmaceutical companies have incorporated human–robot collaboration in drug production, whereby robots perform delicate activities such as the mixing and filling processes while people oversee the issue of quality and morality. This results in less material waste and energy consumption while attaining the highest production quality standards.

In these examples, Industry 5.0 harnesses the strengths of both people and technology to build efficient structures for production. When the creative operations of humans are combined with analytical capabilities of machines, corporations can significantly decrease expenses and inefficiency within enterprises.

7 Challenges in Implementing Industry 5.0 for Sustainable Manufacturing

7.1 Technological Integration

Several factors impede the transition to Industry 5.0, including the challenge of interoperability between legacy systems of production and the use of AI, robotics, and IoT. The classical manufacturing systems were built for the sequential order of processes while the Industry 5.0 encompasses integrated systems for which the improvements are required in terms of technology. Integrating new components with the existing system is usually a challenge since it could require significant readjustments of the system or complete reinvention, which could take a lot of time or money (Nicora et al., 2021). Furthermore, it is crucial to integrate these systems into a single network where data can be easily transferred to optimize the value derived from co-operating with robots and employing AI-reliant workflows.

Another challenge is the relatively long training time required for workers to learn about tools of Industry 5.0. The employment of robots and humans entails not only technical know-how but also organizational culture transforms. Employees must be taught how best they can work with robots, that is, the co-functioning of human creativity and knowledge, coupled with the accuracy of robots in completing their tasks (Nahavandi, 2019). This transition calls for substantial investments in training and development of human resources' skills in the use of these new technologies. The adoption of technology in the Industry 5.0 also has its weaknesses, mainly caused by the reluctance of the workers who are used to the customary approaches to manufacturing.

7.2 Cost Implications

Industry 5.0 technologies require high capital investment at the initial stage of their integration. Businesses have to capitalize on modern technology by acquiring superior tools, sensors, artificial intelligence, and application software to support symbiotic relationships between humans and robots besides practicing sustainability in production. In addition, adopting Industry 4.0 technologies requires a modification of the existing manufacturing facilities; a process that is expensive and unaffordable for many SMEs who may need to borrow capital to finance the new technology (Rüßmann et al., 2015). The additional cost, including the integration of the sustainability aspect in manufacturing, like the use of renewable resources, recycling technologies, and more, makes such costs even higher.

Even though Industry 5.0 provides such advantages as improved productivity, minimal material loss, and lower energy consumption in the long run, the transition cost is expensive for numerous businesses. Any company should consider the costs of implementing any changes over the benefits that can be gained from more efficient and sustainable processes in the long run (Di Marino et al., 2022). The benefits and the cost of these investments can sometimes be hard to determine as seen from the next section especially in fluctuating economic conditions where the returns might be unpredictable.

7.3 Regulatory and Policy Barriers

The shift to Industry 5.0 also has hurdles that relate to the political environment that are worth discussing. Consumer demands increased to have supportive government policies and regulation that enables incorporation of green technologies and sustainable manufacturing. At the moment, few regions have policy guidelines that would encourage the use of Industry 5.0, especially in the context of the circular economy and the use of sustainable materials (Gmelin & Seuring, 2014). If there is not adequate policy backing, businesses may not be willing to integrate the appropriate technologies because of lacking confidence in the market as well as license issues.

These issues are important not only on a domestic level, but also relating to the global accessibility of supply chains there are international standards for sustainability. Internationalization of sustainability standards is challenging because resolutions differ from one country to another based on laws, environmental concerns, and innovation adaptation (Shahbaz et al., 2018). Governments also have the responsibility of spelling out policies that corporations must implement, encouraging corporations to invest in green projects, and generally encouraging corporations in all sectors to meet sustainability objectives. This is due to the inconvenience resulting from the absence of universally approved sustainability standards, which would ease the green policies' application across different countries.

8 Supply Chain Disruptions

Industry 5.0 seeks to incorporate sustainability into supply chains: doing so in a globalized economy is not without its difficulties. Most of the supply chains are strongly decoupled with many suppliers at multiple locations, creating a lack of commitment toward sustainable SCM. The enhancement of green manufacturing principles means its reflection in all supply chain segments; therefore, it is vital to provide strict coordination between the members. Additionally, factors such as politics, economy, and changes on the market that affect supply chain in one country disrupt the careful balance required in the implementation of sustainable measures (Ivanov & Dolgui, 2020). Another problem associated with supply chain management is the lack of orientation on using long-distance shipping and resource-intensive logistics also does not correspond to the goal of reducing CO2 emissions and resource consumption.

Nonetheless, technologies of Industry 5.0 provide the solution to design a supply chain that is more robust and sustainable. Technologies such as blockchain may also be employed to promote traceability and transparency throughout the supply chain system in order to better track sustainability aspects of the materials all the way down to the constituent product. AI and data analysis can assist the supply chain management in evaluating both risks of disruption and allocation of resources more efficiently, while minimizing associated losses and energy consumption. Moreover, the using of collaborative robots known also as cobots and autonomous logistics systems will minimize human interferences in transport jobs and, therefore, optimize the carbon footprint of logistics.

9 Opportunities for the Future

9.1 Sustainable Product Innovation

Industry 5.0 can bring a guideline to create an eco-friendly innovation through relationship centric on human and machine and through integration of the modern technologies like artificial intelligence and robotics. These technologies result in more detailed and variable production techniques that lowers resource usage and waste production during product development and assembly (Nahavandi, 2019). Through the use of AI in data analytics the designers are able to determine the environmental impacts of every stage from raw material procurement, to product use and disposal to design products that are not only effective but environmentally friendly. Industry 5.0 is a new model of green innovation by the promotion of sustainable ideas, circulating designs, and more importantly materials are biodegradable or recycled.

The rising global concern for the environment creates a clearance for manufactures who come across a big market in these products. This demand is well supported by Industry 5.0 since it creates products that are unique, environmentally friendly and consumer friendly products in a market. Through the use of advanced technologies like 3D printing and IoT, it is possible for the manufacture to produce goods that meet the specific need of the customer while at the same time minimizing on the rate of production. This approach conforms with the green economy development agenda that is forged on sustainable use of resources, efficiency, and social equity. By these innovations, firms are in a position to manage and satisfy market requirements on sustainability while meeting their competitive edge.

9.2 Green Jobs and Workforce Development

Industry 5.0, with its focus on sustainability and integration of human beings into the production process, also opens up new opportunities for the creation of new green jobs in the areas like renewable power, environment-friendly production processes, and sustainable supply chains. These jobs are mainly aimed at minimizing the effects on the environment, conserving resources, and encouraging the use of clean energy sources. For example, robotics engineers, AI specialists, or sustainability analysts are emerging as common occupations in sustainable manufacturing industries (Demir et al., 2019). Industry 5.0 consecrates worker engagement in green innovation, in the face of cooperatively designing sustainable products and optimizing operations.

Consequently, as Industry 5.0 beckons it becomes necessary to not just develop new supply chains, but also upskill the employees to align with the set pace and sustainability standards. These are the training of workers in how to work alongside cobots or collaborative robots and making them appreciate systems that use AI in decision-making as noted by Prinz et al. Managers need to ensure proper training programs which are aimed at developing the necessary skills of their employees working in tomorrow's smart but human-oriented manufacturing industries. Moreover, it involves the formulation of policies as well as programs by governments and educational institutions to promote lifelong learning and skills training for workers hence preparing them for the Industry 5.0 mandate.

10 Real-Life Examples of Green Economy and Industry 5.0 in Action

10.1 Tesla's Gigafactories

Tesla's Gigafactories can be considered as benchmarking examples of Industry 5.0 where sustainable manufacturing goals are achieved. At gigafactories, artificial intelligence and robotics work hand in hand with humans to produce electric vehicles (EVs) and battery storage systems. Such factories switch to the use of renewable energy including solar as a way of implementing green economy principles because the use of renewable energy fosters the reduction in the use of non-renewable energy (Cooke, 2020). There's also waste minimization that is depicted in Tesla's closed-loop manufacturing that takes lithium and cobalt from used batteries and reuses it. It also conserves resources hence optimizing resources which is the concept of Industry 5.0. Gigafactories represent the vision of a sustainable source of production; or, how sustainable technologies are suitable to fight environmental issues.

10.2 Siemens' Digital Factory Solutions

Siemens, in particular, has undergone significant strides in pushing technology in Industry 5.0 through their Digital Factory Solutions. These solutions use artificial intelligence, internet of things, and big data to enhance production procedures, decrease energy, and material usage. Through regular tracking of the assembly lines in manufacturing, Siemens achieves efficient use of resources besides minimizing on emission and energy consumption. Some of the key offerings include, MindSphere which is a cloud-based open IoT operating system that the company uses in capturing relevant data for analysis toward operational improvement (Stock & Seliger, 2016). Moreover, sustainability management is a significant aspect of Siemens' supply chain that requires every supply chain node to meet environmental sustainability. This demonstrates that Industry 5.0 holds the core of bringing innovation in the tradition manufacturing processes through the intelligent and sustainable solutions.

10.3 Unilever's Sustainable Manufacturing Initiatives

Unilever has been a pioneer company in integrating sustainability into production processes with help of Industry 5.0 concepts. Their manufacturing plants have integrated efficient knowledge processing tools such as artificial intelligence and automation that enhance production potential, reduce energy consumption, or reduce waste or how some of its accomplishments are the "Zero manufacturing waste to landfill" where all of Unilever's factories globally avoid sending waste to landfill by minimizing, repurposing, or recycling material. With its emphasis on the sustainable production of its products and energy, Unilever is able to locate opportunities for growth the firm can capitalize on through regular innovation, particularly in the area of circular business models. Explaining how Industry 5.0 initiatives can be applied to manufacture sustainably at scale, the company's innovative approaches to cutting greenhouse gas emissions as well as improving water efficiency were also described in detail.

11 Conclusion

In summary, Industry 5.0 represents a significant advancement in manufacturing paradigms, emphasizing the integration of human creativity and advanced technologies to promote sustainable practices. This new era prioritizes collaboration between humans and machines, facilitating innovations that enhance efficiency while minimizing waste and environmental impact. As demonstrated through case studies of leading companies like Tesla, Siemens, and Unilever, Industry 5.0 fosters sustainable manufacturing by leveraging smart technologies, optimizing resource utilization, and adopting circular economy principles. It bridges the gap between technological advancement and environmental stewardship, laying the groundwork for a greener economy.

Looking ahead, the evolution of manufacturing is poised to become even more intertwined with sustainability goals. As we move toward a more interconnected and intelligent manufacturing ecosystem, Industry 5.0 will continue to emphasize human-centric processes that prioritize ethical considerations and environmental responsibility. Innovations in AI, robotics, and digital twin technologies will further enhance the ability of manufacturers to tailor their processes for sustainability. Future developments will likely focus on creating resilient supply chains, promoting eco-friendly product designs, and fostering collaboration across industries and sectors to meet global sustainability objectives.

To realize the full potential of Industry 5.0 in advancing sustainable manufacturing, it is crucial for industry leaders, policymakers, and innovators to embrace this paradigm shift. Collaborative efforts should focus on integrating sustainable practices into all levels of manufacturing processes and ensuring that technological advancements align with environmental goals. Policymakers must establish supportive regulations and incentives to facilitate the adoption of sustainable technologies. Meanwhile, innovators are encouraged to explore the synergy between Industry 5.0 and the green economy, developing solutions that not only meet market demands but also contribute positively to the environment. By working together, we can drive a more sustainable future that harnesses the power of human ingenuity and technological innovation.

Competing Interests There is no conflict of interest for this research.

References

Adel, A. (2022). Future of industry 5.0 in society: human-centric solutions, challenges and prospective research areas. *Journal of Cloud Computing, 11*(1), 40.

Aggarwal, R., Singh, H., & Garg, P. (2023). Dell Technologies Inc.: Innovating toward circularity. *Journal of Information Technology Teaching Cases*, 20438869231215089.

Akaev, A., & Davydova, O. (2023). Climate and energy: Energy transition scenarios and global temperature changes based on current technologies and trends. *Reconsidering the limits to growth: A report to the Russian Association of the Club of Rome* (pp. 53–70). Springer International Publishing.

Alves, J., Lima, T. M., & Gaspar, P. D. (2023). Is industry 5.0 a human-centred approach? a systematic review. *Processes, 11*(1), 193.

Arora, N. K., & Mishra, I. (2023). Responsible consumption and production: A roadmap to sustainable development. *Environmental Sustainability, 6*(1), 1–6.

Bocken, N., Baldassarre, B., Keskin, D., & Diehl, J. C. (2023). Design thinking tools to catalyse sustainable circular innovation. In *The Routledge handbook of catalysts for a sustainable circular economy* (pp. 359–387). Routledge.

Bonilla, S. H., Silva, H. R., Terra da Silva, M., Franco Gonçalves, R., & Sacomano, J. B. (2018). Industry 4.0 and sustainability implications: A scenario-based analysis of the impacts and challenges. *Sustainability, 10*(10), 3740.

Brettel, M., Friederichsen, N., Keller, M., & Rosenberg, M. (2014). How virtualization, decentralization and network building change the manufacturing landscape: An Industry 4.0 Perspective. *International Journal of Information and Communication Engineering, 8*(1), 37–44.

Butt, U. I. (2024). Nestlé's role in global food security, climate change and consumer health. *International Journal of Social Sciences and English Literature, 8*, 7–15.

Cooke, P. (2020). Gigafactory logistics in space and time: Tesla's fourth gigafactory and its rivals. *Sustainability, 12*(5), 2044.

Dechamps, P. (2023). The IEA World Energy Outlook 2022—A brief analysis and implications. *European Energy & Climate Journal, 11*(3), 100–103.

Demir, K. A., Döven, G., & Sezen, B. (2019). Industry 5.0 and human-robot co-working. *Procedia Computer Science, 158*, 688–695.

Di Marino, C., Rega, A., Vitolo, F., & Patalano, S. (2022, June). Enhancing human-robot collaboration in the industry 5.0 context: Workplace layout prototyping. In *International joint conference on mechanics, design engineering & advanced manufacturing* (pp. 454–465). Springer International Publishing.

Evans, S. (2023). An integrated circular economy model for transformation towards sustainability. *Journal of Cleaner Production, 388*, 135950.

Ghisellini, P., Lazarevic, D., Passaro, R., & Liu, G. (2024). Editorial special issue who will benefit from the transition to the circular economy? *Journal of Cleaner Production*, 142137.

Gmelin, H., & Seuring, S. (2014). Achieving sustainable new product development by integrating product life-cycle management capabilities. *International Journal of Production Economics, 154*, 166–177.

Gomez, R. (2024). Driving sustainability and innovation through design: The BMW Group+ QUT design academy. In *Research Journeys to Net Zero* (pp. 43–56). Routledge.

Hazen, B. T., Boone, C. A., Ezell, J. D., & Jones-Farmer, L. A. (2014). Data quality for data science, predictive analytics, and big data in supply chain management: An introduction to the problem and suggestions for research and applications. *International Journal of Production Economics, 154*, 72–80.

Ivanov, D., & Dolgui, A. (2020). Viability of intertwined supply networks: extending the supply chain resilience angles towards survivability. A position paper motivated by COVID-19 outbreak. *International Journal of Production Research, 58*(10), 2904–2915.

Javaid, M., Haleem, A., Khan, I. H., Singh, R. P., & Khan, A. A. (2024). Industry 4.0 and circular economy for bolstering healthcare sector: A comprehensive view on challenges, implementation, and futuristic aspects. *Biomedical Analysis, 1*(2), 174–198.

Kagermann, H., Anderl, R., Gausemeier, J., Schuh, G., & Wahlster, W. (Eds.). (2016). *Industrie 4.0 in a Global Context: Strategies for cooperating with international partners*. Herbert Utz Verlag.

Khan, S. A., Al Rashid, A., Muhammad, J., Ali, F., & Koç, M. (2024). 3D printing technology for rapid response to climate change: Challenges and emergency needs. *Intelligent and Sustainable Manufacturing, 1*(1), 10004.

Kiefer, C. P., Carrillo-Hermosilla, J., & del Río, P. (2024). How does corporate environmental culture enable the eco-innovation transition of firms towards the circular economy? *Corporate Social Responsibility and Environmental Management, 31*, 5911–5937.

Lasi, H., Fettke, P., Kemper, H. G., Feld, T., & Hoffmann, M. (2014). Industry 4.0. *Business & Information Systems Engineering, 6*, 239–242.

Lom, M., Pribyl, O., & Svitek, M. (2016, May). Industry 4.0 as a part of smart cities. In *2016 Smart Cities Symposium Prague (SCSP)* (pp. 1–6). IEEE.

Longo, F., Padovano, A., & Umbrello, S. (2020). Value-oriented and ethical technology engineering in industry 5.0: A human-centric perspective for the design of the factory of the future. *Applied Sciences, 10*(12), 4182.

Lu, Y., Zheng, H., Chand, S., Xia, W., Liu, Z., Xu, X., ... & Bao, J. (2022). Outlook on human-centric manufacturing towards Industry 5.0. *Journal of Manufacturing Systems, 62*, 612–627.

Mathew, R., McGee, R., Roche, K., Warreth, S., & Papakostas, N. (2022). Introducing mobile collaborative robots into bioprocessing environments: Personalised drug manufacturing and environmental monitoring. *Applied Sciences, 12*(21), 10895.

Nahavandi, S. (2019). Industry 5.0—A human-centric solution. *Sustainability, 11*(16), 4371.

Nicora, M. L., Ambrosetti, R., Wiens, G. J., & Fassi, I. (2021). Human–robot collaboration in smart manufacturing: Robot reactive behavior intelligence. *Journal of Manufacturing Science and Engineering, 143*(3), 031009.

Niinimäki, K., Peters, G., Dahlbo, H., Perry, P., Rissanen, T., & Gwilt, A. (2020). The environmental price of fast fashion. *Nature Reviews Earth & Environment, 1*(4), 189–200.

Rüßmann, M., Lorenz, M., Gerbert, P., Waldner, M., Justus, J., Engel, P., & Harnisch, M. (2015). Industry 4.0: The future of productivity and growth in manufacturing industries. *Boston Consulting Group, 9*(1), 54–89.

Sánchez-García, E., Martínez-Falcó, J., Marco-Lajara, B., & Manresa-Marhuenda, E. (2024). Revolutionizing the circular economy through new technologies: A new era of sustainable progress. *Environmental Technology & Innovation, 33*, 103509.

Shahbaz, M. S., Rasi, R. Z. R., Ahmad, M. F. B., & Sohu, S. (2018). The impact of supply chain collaboration on operational performance: Empirical evidence from manufacturing of Malaysia. *International Journal of Advanced and Applied Sciences, 5*(8), 64–71.

Stock, T., & Seliger, G. (2016). Opportunities of sustainable manufacturing in industry 4.0. *Procedia CIRP, 40*, 536–541.

Tallat, R., Hawbani, A., Wang, X., Al-Dubai, A., Zhao, L., Liu, Z., ... & Alsamhi, S. H. (2023). Navigating industry 5.0: A survey of key enabling technologies, trends, challenges, and opportunities. *IEEE Communications Surveys & Tutorials.*

Tóth, A., Nagy, L., Kennedy, R., Bohuš, B., Abonyi, J., & Ruppert, T. (2023). The human-centric Industry 5.0 collaboration architecture. *MethodsX, 11*, 102260.

UNEP. (2011). *Towards a green economy: Pathways to sustainable development and poverty eradication. A synthesis for policymakers.*

Xu, M., David, J. M., & Kim, S. H. (2018). The fourth industrial revolution: Opportunities and challenges. *International Journal of Financial Research, 9*(2), 90–95.

Yadav, S., Samadhiya, A., Kumar, A., Majumdar, A., Garza-Reyes, J. A., & Luthra, S. (2023). Achieving the sustainable development goals through net zero emissions: Innovation-driven strategies for transitioning from incremental to radical lean, green and digital technologies. *Resources, Conservation and Recycling, 197*, 107094.

Sustainable Development Goal and Industry 5.0

Ghousia Jabeen, Gurunadham Goli, and Kafila

Abstract Industry 5.0 can promote SDGs and reveal a future with enhanced technologies, humanity, and environmental responsibility. This chapter discusses the link between Sustainable Development Goals (SDGs) and Industry 5.0. Facilitating human-centric and environmentally sustainable industrial development was the focus of this study. We initiated Industry 4.0 concentrated on automating industries, Industry 5.0 emphasizes human-technology interaction, innovation, and environmental thinking. The human-centered approach thus reiterates most of the dimension keys Social, Economic, and Environmental challenges. However, limitations include high implementation costs, workforce skills gap and cybersecurity challenges. In this study, we used a qualitative research methodology to explore Sustainable Development Goals (SDGs) in relation to Industry 5.0, focusing primarily on an analysis of pertinent literature. The chapter highlights an industry-supported innovation and alignment with the most sustainable development goals. In this chapter, research findings show Industry 5.0 promotes SDG 9 by combining human-centered innovation with AI, robotics, and the IoT, this fosters the goal of development. The generator conception of new jobs that prioritize human contributions in industrial operations promotes sustainable development by contributing to SDG 8 (Decent Work and Economic Growth). Industry 5.0 supports SDG 12 on responsible consumption and in the same respect SDG 13 on Climate Action by reducing waste, using resources efficiently, and using low-carbon production methods. Additionally Industry 5.0 fosters social justice toward SDG 10 by ensuring that technological growth benefits all stakeholders. The chapter also examines how Industry 5.0 technologies support SDG 11-compliant towns and communities.

G. Jabeen (✉) · G. Goli · Kafila
School of Business, SR University, Warangal, Telangana 506371, India
e-mail: ghousiajabeen786.gj@gmail.com

G. Goli
e-mail: mbagurunath@gmail.com

Kafila
e-mail: kafila@sru.edu.in

R. Kumar and S. N. Mohanty (eds.), *Sustainable Economy Models in the Age of Industry 5.0*, https://doi.org/10.1007/978-981-96-4104-8_3

Keywords Sustainable development goal · Incorporating human-centered · Decent work and economic growth · Industry 5.0 · Industry 4.0 · Innovation

1 Introduction

Sustainable Development Goals (SDGs) are an interdependent outline consisting of environmental, social, and human developmental goals (Moyer & Hedden, 2020). This structure has 17 goals and 169 targets. From the overview provided in (Giller et al., 2018), the foregoing were the goals of the world. The key to the SD of corporate entities is the SDG, which is an important subject highlighted in the discussion. In the study performed by (Bennich et al., 2020), the SDGs covered a broad spectrum of issues, including poverty eradication (SDG 1), value education (SDG 4), gender equality (SDG 5), clean water and sanitation (SDG 6), affordable and clean energy (SDG 7), and climate action (SDG 13) (Mahanayak, 2024). Regarding the targets of each of the development goals to some extent, there are correlations between them. Such interrelationships can be either synergic (an Incan goal's improvement may lead to the improvement of other interference effects (the accomplishment of one goal has a negative impact on the accomplishment of the other defined goal) or a conflicting effect where progress in one objective may hinder advancement in another objective. The idea under discussion was further explored by (Elavarasan et al., 2022). Pradhan et al. (2017) define synergistic effects as positive correlations between indicator pairs. These positive correlations were found to be more common than the negative relationships (trade-offs) observed among the SDGs.

1.1 Industry 5.0: A New Paradigm

Industry 4.0 represents a collaborative effort among various technologies and trends aimed at transforming conventional practices through the implementation of digital tools (Aceto et al., 2020). Industry 5.0 is a brand-new context when man and machine cooperate and enhance productivity efficiency in different ways. Industry 4.0, has been proven in lifecycle manufacturing. Chain in the industry has changed from when it was in young adulthood during the industrial revolution to when it is now in the digital age. A general outline of the revolution from the 1970s to the twenty-first century is illustrated in Fig. 1. Each signifies the experience of revolution within the making process and its way, revolutionized the way the industry operates. Industry 5.0 emphasizes the collaborative connection among people with machineries. While the primary focus is on automation and cutting-edge manufacturing techniques, it is essential to address another crucial aspect in this context: the interaction between human workers and robotic systems. This element plays a significant role in shaping the future of industrial processes and deserves careful consideration. Consequently, Industry 5.0 can perfectly be described by realizing a system's efficiency through

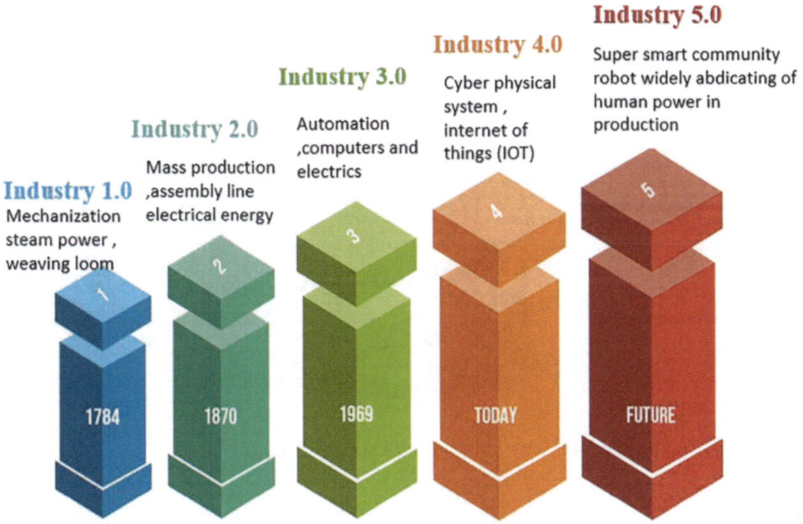

Fig. 1 Evaluation industry sector 1784 to 21 centuries (Author created)

automated and effective concepts with adherence to conventional and belief-specific humane elements. This type of revolution was completely encoded. Automation processes and creates new manufacturing prospects for participants in the medical world. Major disruptions to conventional IoS, such as the Internet of Things and Internet of Services, point to promising new directions for progress. Novel manufacturing technologies, software, and applications that interface with those technologies are exhibited and explained below: high technology such as robots, sensors, and others guarantees data exchange and connection.

When examining the difference between Industry 4.0 and Industry 5.0, it becomes apparent that the latter emphasizes interactions between humans and technology. The collaboration of human capabilities and machine intelligence has already been observed in various sectors. Industry 5.0 primarily concentrates on the interplay between people and automated systems. Hence, owing to the factors of automated machines adding strength to the workforce as well as advancing technologies that make all types of work accessible in one place owing to the incorporation of digital platforms, the innovation of co-working of such centers can be from anywhere. People with machines, and smart manufacturing facilities are already available. Industry 5.0 targets to elevate these device interactions to a more innovative stage, building upon existing connections. The main topics are human–machine interfaces (Janssen et al., 2019). Automated Industry 5.0 is expected to incorporate smart technologies and big information, new technology policies, accountable implementation science, and three-dimensional reflection symmetry, safety measures for innovative ecosystem design, in addition to development plans (Sabourin, 1999). The integration will be enhanced, faster, and better automated in accordance with human power (Fig. 2).

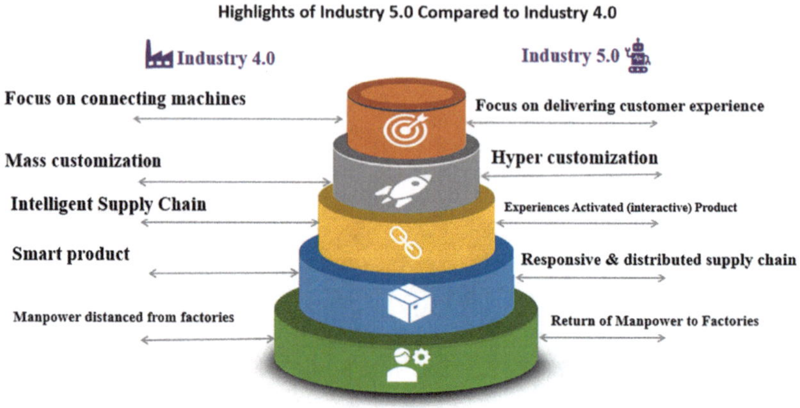

Fig. 2 Comparing Industry 4.0 with Industry 5.0 (Author created)

Therefore, human resources and computational intelligence are complementary to the production of value-added products and good of high standards. In the performance sectors, in the future, automation will not replace human beings' cognitive and critical thinking processes. Thus, while in the previous technological revolution, the use of technology-impersonalized workers, Industry 5.0 enhances different roles performed by them in manufacturing (Leng et al., 2022). The application of repetitive maneuvers carried out via mechanical means, like drilling or data input, leads to a reduction of human supervision that is a component of cooperatively performed work. Staff members can be tasked with blood pressure-related responsibilities, including monitoring systems, making quick decisions, and conducting thorough assessments to improve the comprehension of quality and production processes. The integration of mechanical performance optimization with cognitive processing is anticipated in the near future. In Industry 5.0, data gathering is automated, with manufacturing facilities, machines, and various devices connected to the internet, allowing for increased data input. The core aspect of Industry 5.0 is the continuous availability of real-time information. Transparency for the integration of one system with another as well as interaction (Xu et al., 2021). A simple example is the improvement in the periods of machine-tool availability. Real-time monitoring of actual conditions is possible using sensors integrated into machine tools. Tool lifetime according to the amount of wear it undergoes during machining process. For instance, the machinery can autonomously send a requisition for a replacement cutting tool to the inventory management system, thereby minimizing downtime and boosting productivity.

2 Introduction to Industry 5.0

2.1 From Automation to Human-Centric Innovations

Different from Industry 4.0, Industry 5.0 takes another position and is centered on the integration of intelligent production systems and human workers in future manufacturing systems. This emergent framework seems to lay a much higher premium on the need to integrate AI into the retained human imperatives in manufacturing systems. This new industrial revolution has more to do with human–robot partnership an amalgamation of man and machine, the end product of which is utility, economy, and welfare. Industry 5.0 refers to the fifth industrial revolution which superimposes advanced technologies such as AI, robotics, and big data alongside promoting the human employees to do imaginative thinking and interpersonal skills (Fig. 3).

2.1.1 Human-Centric Collaboration with Industry 5.0

Industry 5.0 is centered on the integration of robotics in the production process with human workers in a bid to increase competency (Yitmen & Almusaed, 2024). The implementation of cobots and the issues of robot learning allowed for safe and efficient collaboration between humans and robots. S. Seenuvasamurthi et al. (2024) emphasize that soft skills and wise learning will be necessary for staff training for Industry 5.0 focuses on human requirements and merges man-centric applications

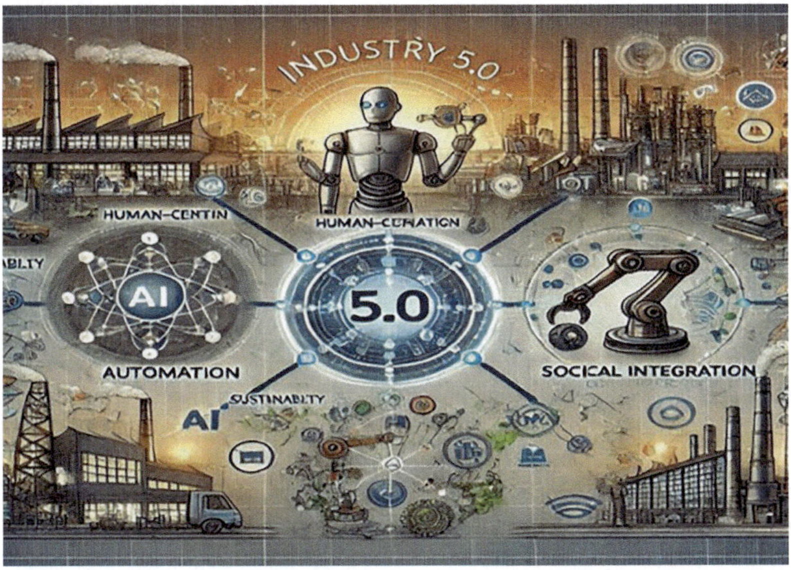

Fig. 3 Industry 5.0

into bio-inspired methodologies as a way of working with artificial intelligence. This approach is vital in smart cities and communities fulfilling SDGs 3, 4, 9, and 11 with aligned frameworks (Chenary et al., 2024).

Industry 5.0 builds on Industry 4.0 by adding a human component to the otherwise purely technological model of automation, digitalization, and artificial intelligence (AI; Bregar, 2024). Many industries, including production, health care, depend on such innovations as artificial intelligence, internet of things, and machine learning. The idea of having a digital human being proved resourceful, to have organizations work according to the capabilities and preferences of the personnel. Transformative technologies like CDTs and AI play a crucial role in modifying and enhancing manufacturing techniques, responding to market forces and the pull factors for sustainable industrialization as highlighted by R. Sharma and H. Gupta (2024). Disruptive technologies support smart cities and villages, and the enabler for numerous SDGs, with particular regard to the ninth one, as indicated by P. Kasinathan et al. (2022).

2.1.2 Societal and Ethical Considerations

Industry 5.0 includes a reversible approach to sustainable development, risk management, and restriction of undesirable phenomena (Kulkarni et al., 2024). It covers issues of data protection, privacy, training, and skill development. The human-centricity concept covers calls for empowering people and providing their well-being along with an adequate supply of human demands regarding technological progress. Industry 5.0 presents a great promise in increasing productivity as well as improving the welfare of societies but comes with a number of issues, such as talent development in the workforce, ethical issues, and privacy and security issues. To embrace this new industrial age, society must adopt a model that encourages harmony between human capital and technological capital.

2.2 The Impact of Cutting-Edge Technologies in the Fifth Industrial Revolution: AI, Robotics, IoT, and Blockchain

Industry 5.0 marks a transformative shift from traditional industrial methods to a new paradigm that prioritizes human-centered values, customization, collaboration, and sustainability. In contrast to Industry 4.0, which aims to enhance corporate intelligence through digital technologies and automation, Industry 5.0 charts a unique course for industrial progress. This new approach diverges from its predecessor by placing greater emphasis on the human element and environmental considerations in industrial development. This shift can be explained by the incorporation of emotional intelligence and soft competencies into workforce training, as well as the correct combination of IT use of information technology (Patil et al., 2023) and human values and rights.

Industry 5.0 has become foundational for the advent of other advanced technologies that involve human-AI cooperation. Today, AI, robotics, IoT, and Blockchain are crucial aspects of this change and bring value to the industrial world. Practical applications of AI and ML include intelligent decision-making and control that adds more productivity to human–machine interfaces (Prasad et al., 2024). The synergy between robotics and AI means that robotization removes resource intensity from repetitive tasks, enabling human workers to be more innovative and efficient. The integration of artificial intelligence enables the utilization of vast amounts of data gathered from IoT devices to make informed decisions, manage, and monitor events in real-time. Blockchain technology provides a secure means of transactions that is essential when dealing with interdependent systems (Tariq et al., 2024). Together, these technologies facilitate the shift toward Industry 5.0, an improved state of industrial sustainability, and focus on human factors.

2.2.1 Artificial Intelligence (AI)

AI and machine learning are at the core of Industry 5.0 where machines actively learn from people's interactions (Shaolei & Heng, 2019). Data inference in edge computing fundamentally depends on AI-based algorithms for obtaining meaningful data from heterogeneous big data. Incorporating the principles of intelligence in the organization of work improves productivity through management by decisions and controls.

2.2.2 Robotics

Robotics integrated with AI can perform repetitive actions and enable workers to perform more creative and productive work. All of these amplify efficiency and production capacity in Industry 5.0 by promoting human–robot association (Raffik et al., 2024).

2.2.3 Internet of Things (IoT)

IoT devices create tremendous amounts of data, and the only way to manage and harness them is through the use of AI, the evolution of IoT devices, and wireless communication, IoT ecosystems have evolved, and users are able to access information in relatively isolated zones (Goswami & Rai, 2024).

2.2.4 Blockchain

Cryptographic mechanisms provide an immutable ledger base for trade with an important role in data integrity in computerized networks. Owing to the distributed nature of blockchains, more risks associated with data leakage and cheating are prevented. Although such technologies have been revolutionary in their respective fields, issues such as data security, ethical issues, and sustainability persist. Industry 5.0 seeks to fix these by applying human-centered robotics and the use of sustainable technology technologies, together with enhanced and continued development of technology.

2.3 Methodology

This study concerning the SDGs in the era of Industry 5.0 was performed utilizing a qualitative research methodology, with a specific focus on the contingency of literature. This chapter advocates for techno-optimism, asserting that economic and social resilience depends on Industry 5.0, defined by human centricity, inclusivity, environmental goals, and sophisticated technologies.

3 Understanding Sustainable Development Goal (SDG)

3.1 Definition and Historical Context

The United Nations in 2015 moved from the Millennium Development Goals (MDGs) as a new development framework the Sustainable Development Goals (SDGs) which have been integrated into its National Agenda 2030. These SDGs function as a worldwide initiative, promoting international cooperation toward common goals to improve developmental conditions across countries. The core principles of the SDGs are inclusiveness and equity, as reflected in their motto "leaving no one behind".

3.2 Key Objectives and Pillars of SDGs

As highlighted before, the United Nation's 2015 Sustainable Development Goals aimed at solving global problems. These SDGs are structured around three important facets, which are *Social, Economic, and Environmental* on which the concept of sustainable development was developed (Fig. 4).

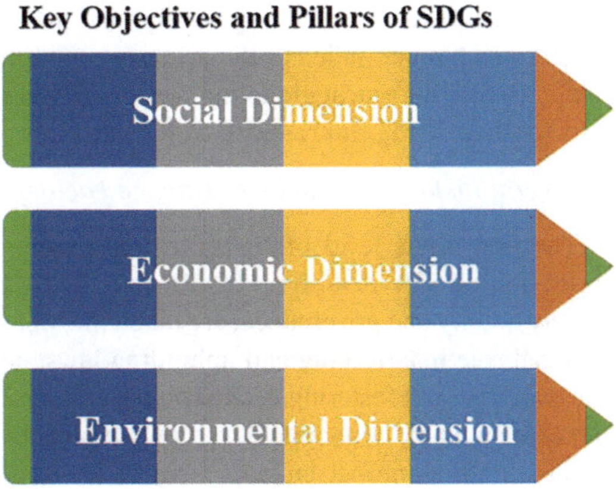

Fig. 4 Key objectives and pillars of SDGs (Author created)

3.2.1 Social Dimension

One of the pillars of the 2030 agenda is the social dimension, which not only reflects all three dimensions, including economic and environmental aspects, but also provides a fundamental principle: the importance of considering these aspects by considering their social equity, inclusion, and well-being perspectives to achieve sustainable development (SDGs). However, this dimension is also very important: it can be implied that sustainable development efforts are not only economically and environmentally viable but also socially efficient. It deals with the environment and myriad other issues related to poverty eradication, gender equality, and education on health and social justice that are imperative if an equitable society is to be established. However, the papers on which we commented include the following key elements of the social dimension of SDGs.

The social dimension of the SDGs focuses on improving human well-being, promoting inclusive societies, and ensuring that basic needs such as health, education, and gender equity are met. The primary goals centered on eliminating poverty (SDG 1), providing high-quality education (SDG 4), and enhancing health and wellness (SDG 3). The initiative also strives to attain gender equality and strengthen the position of women and girls across all sectors (Jabeen et al., 2024).

3.2.2 Economic Dimension

Economic foundations of SDGs as a key pillar: For the economic dimension, it is third and is critical to support growth. This dimension focuses on no poverty, zero hunger, and responsible consumption and production.

- *Everywhere End Poverty in All Forms*

Goals 1 and 2 distinguish the economic characteristics of the SDGs, which emphasize ending poverty and hunger. Similarly, Bogor Regency is one of the 25 target areas that has met the goals of ending poverty and improving health services. Focusing on poverty and hunger indicates the primary economic pillars of the SDGs, (Horemans et al., 2016) which struggle to build a world where everyone has what they need at least for their living time as well as business development.

- *Several Institutional and Governance Factors*

The first is the level of centralization of the education system. The economic dimension of SDGs can only be effective and efficient when three characteristics, namely quality and governance, are taken into consideration. Strong institutions play a crucial role in enforcing and upholding laws, addressing issues of corruption and transparency, and fostering economic growth, ultimately contributing to the realization of sustainable economic development objectives (Breque et al., 2021). In developing countries, proper enhancement of institutional foundations is of paramount importance to virtually eliminate most problems associated with governance and economic policy management.

- *Economic Thriving and Social Justice for Women*

Economic equity, especially female equity, is a critical area of concern for sustainable economic development goals. After the analysis, we find that African women experience systematic prejudice that denies them basic economic or resource rights, making a stronger call for policies that support the economic liberation of women. To achieve sustainable development objectives and enable all individuals to participate in and reap the benefits of economic progress, it is crucial to tackle gender inequalities.

- *Integration with ESG Factors*

It is possible to improve the effects of economic activity on the environment by implementing economic elements in ESG technologies. The proposed ECON-ESG model includes macroeconomic factors and supports the sustainability model to achieve SDGs and counteract negative externalities, thereby offering an appropriate approach to increase business sustainability. This is important in fashioning strategies that will sustain both economic growth and care for the environment, as well as respect social responsibilities.

While Sustainable Development Goals (SDGs) are crucial and primarily focus on economic growth for sustainable progress, it is essential to recognize the interconnectedness of economic, social, and environmental aspects. Understanding and addressing these three dimensions in tandem is vital for achieving comprehensive sustainable development. Successful outcomes of the economic dimension depend on functionality, social equity, and environmental sustainability. However, to accomplish the goal of the SDGs, there are several barriers, including cost restrictions, infrastructure development, and systematic corruption. The economic aspect, therefore, should have an ethical perspective and an institutional perspective, as well as a gender perspective, and should, therefore, achieve a broad development on its own.

3.2.3 Environmental Dimension

Environmental management is one of the strongest pillars of the SDG and considers the environmental aspects of the sustainable development agenda. The significance of this dimension to the elaboration of strategies and actions for the achievement of the SDGs is also understood in that it involves the proper use of natural resources, the definition of ecosystems, and the avoidance of environmental disasters. Environmental concerns are closely linked to social and economic concerns, demonstrating the importance of an integrated approach to the problem. This section explores the different elements of the environmental dimension of the SDGs based on enclosed research papers and seeks to combat climate change (SDG 13) and protect life below water (SDG 14) and land (SDG 15) (Bhardwaj et al., 2024). It is used to solve environmental and socioeconomic problems by contributing to making the right decisions when interacting with nature environmental management related courses at higher learning institutions is crucial in the integration of the SDG in learning institutions as this will enable future leaders in dealing with complex issues of sustainability. (Masri & Jaaron, 2017). SDGs are ambitious and require significant collaboration among countries, institutions, and stakeholders. Although they offer a comprehensive blueprint for sustainable development, progress can be hindered by various challenges, particularly socioeconomic and political barriers, especially in less developed areas. These are barriers for the delivery of quality healthcare services by the governments to their citizens that are complemented by the need of effective governance, structural changes, and international cooperation in order to include all the sustainable development goals (SDGs) by 2030.

4 Aligning Industry 5.0 with SDGs

The Sustainable Development Goals (SDGs) and Industry 5.0 are aligned in their pursuit of a sustainable industrial future (Mariappanadar, 2003). Industry 5.0 is the next level up from Industry 4.0 and focuses on sustainability, reliability, and personnel aligned with SDG 9 that treats innovation, industrialization, and facilities. This is found well aligned especially in the adoption of innovative technologies like AI, ML, and CDTs for effective resource utilization and organization developmental factors for sustainable business growth.

The subsequent sections examine key areas where this integration, grounded in the interplay between IT and human factors, acts as a vital enabler of the SDGs. It supports SDG 8's objective of economic growth and decent work by integrating digital technologies and work synergy to stimulate activity and drive new innovations. Furthermore, it reinforces infrastructure for SDG 9 and advocates for more sustainable production and consumption practices within SDG 12. Moreover, Industry 5.0 aids in sustainable urban development for SDG 11 and backs climate action for SDG 13 through technological advancements (Fig. 5).

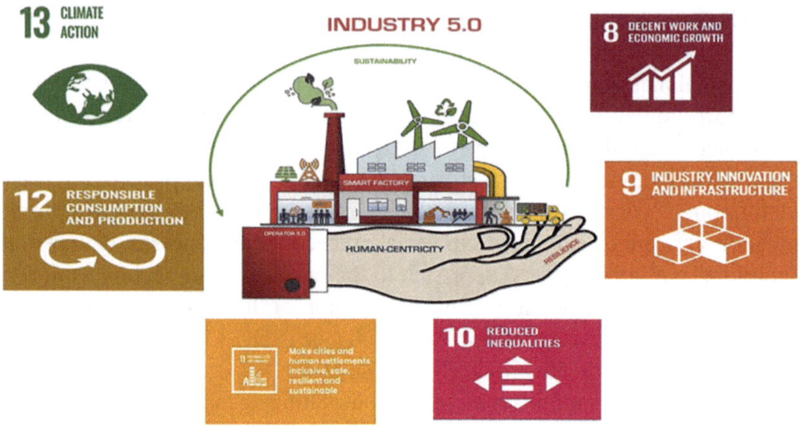

Fig. 5 Aligning Industry 5.0 with SDGs (Author created)

4.1 Enhancing SDG 8: Working Environment and Economic Development

Collaboration is an important feature of Industry 5.0 it creates jobs rather than destroying them and improves skill, further promoting decent work. Accurate use of AI and automation increases resource utilization, economic development, and sound systems of production (Ahmad, 2015). By focusing on human well-being and social-centric solutions, Industry 5.0 fosters a competitive and resilient economy, aligning with economic growth objectives.

4.2 Contribution to SDG 9: Industry, Innovation, and Infrastructure

CDTs are critical to Industry 5.0 because they provide instant optimal control and fast data processing while promoting an efficient and sustainable industrial evolution and new developments. The technological advancements associated with Industry 5.0 serve as a foundation for the creation of intelligent urban areas and communities, thereby driving progress in infrastructure and fostering innovation. Industry 5.0, representing the subsequent stage in digital transformation and environmentally conscious business operations, brings improvements to industrial processes and frameworks.

4.3 SDG 10: Reduce Inequalities

Industry 5.0 brings humans and artificial intelligence together, which opens new possibilities and risks for advancing SDG 10, which focuses on inequality. If not handled well, then we are likely to deepen inequalities with the introduction of Industry 5.0, because advancement will always support those who have capital and education. However, it offers ways to overcome this disadvantage through positive policies. The following sections discuss how Industry 5.0 can help achieve SDG 10 to lessen inequalities. The increasing technologies in Industry 5.0 can create value to lessen inequalities by enabling inclusive business models. Such technologies are being implemented to achieve SDG10 where benefits within organizations cut across all segments of society, especially marginalized (Kira, 2003). Sustainable banking is a core of Industry 5.0, which tackles income inequality due to improved economic prospects in areas of low institutional endowment. It revealed that while economic development like India shows progression, it is also socially unfavorable because it augments inequality. Industry 5.0 should focus on reducing intragroup gaps in a society and demography to enhance equal distribution of gains across these groups, for instance between urban and rural settings (Tooranloo et al., 2017). On the one hand, Industry 5.0 proclaims future opportunities to lessen inequalities, but on the other hand, it is crucial to recognize the possibility of deepening the existing gaps in the case of unsuitable technological development management. However, policy points should concentrate on developing acceptable customs that would ensure the full participation of disadvantaged groups in the opportunities offered by Industry 5.0, thereby enhancing the cause of SDG 10 in general.

4.4 SDG 11: Sustainable Cities and Communities

Society 5.0 was realized by the Fifth Science and Technology Basic Plan of Japan as the future model for a Human-Centered Society in harmony with advancement and responsibility. This concept can be broadly understood as a new approach to doing business and using technology seen as the force behind social change. Any continuing changes in these domains become the technological aspects that are incorporated into this concept at the moment. The primary concern should be citizenry well-being and nurturing of an über-smart population of the nation. Society 5.0 is thus pictured as a kind of framework, which in fact allows to stimulate economic development and social responsibility.

4.5 SDG 12: Responsible Consumption and Production

The implementation of Industry 5.0 enables manufacturing sectors to enhance their production processes and resource efficiency, resulting in minimal to zero waste production. It is believed that the circular economy encourages sustainable production and consumption practices through the extensive use of digital technologies. As a result, Industry 5.0, which encompasses the entire product lifecycle, plays a crucial role in promoting sustainable methods of production and consumption.

4.6 SDG 13: Climate Change

Industry 5.0 is another advancement in industries making use of people, as well as being environmentally sustainable in line with Sustainable Development Goal 13 on Climate Action. While Industry 4.0 mostly focuses on technology, Industry 5.0 incorporates humans into the new system to create better sustainability. This integration is special for responding to climate change and fostering environmental responsibility by retaining the accurate calculation ability of machines, while also retaining people's imagination and moral judgment. Industry 5.0 goal of sustainability becomes apparent based on impacts on different sectors, such as manufacturing, agriculture, and smart cities, among others, where the reduction of resources and emissions would be key aspects toward the achievement of SDG 13.

5 Challenges

Industry 5.0 poses some challenges that are not observable in the other versions as explained below, such as, the alternative type involves diversity in society, encompassing both worth and recognition. Determining optimal method for accessing the creation of environment and social value in corporating clients across the entire supply chain into SMEs and faster in the Faster in the cross disciplinary integration of research domains and systems integration all with a lean and traget result focus. Productivity must be achieved and at the same time huge investments are desirable. Like Industry 4.0 and Industry 5.0, achieving success will necessitate substantial financial support from governing bodies. The core principles of Industry 5.0, namely human-centered design and sustainability, remain crucial regardless of its future trajectory. These values have emerged as key drivers of societal advancement, fostering qualities such as creativity and resilience, rather than focusing solely on GDP-driven prosperity. This shift is evident in recent governmental initiatives to integrate these principles into broader strategies, including the Paris Agreement, the United Nations' SDGs, the Well-being of Future Generations Act, Genuine Progress

Indicator 2.0, the Economy of Well-being, National Ecclesia (Lu, 2017), and the OECD better life index.

6 Discussion: Toward a Sustainable and Human-Centric Industrial Revolution

Industry 5.0's future perspectives depend on its capability to implement sustainable and human-centered industrial change to answer the SDGs. Therefore, Industry 5.0 brings forward improvements over Industry 4.0, and it focuses on human–robot interaction, sustainability, and survival. They envision that it occupies a central position toward the fulfillment of long-term development goals in the world through the fashionable pairing of technology with human resources and sustainable processes. There are some trends and predictions that define the shift toward Industry 5.0 and show the potential of this future industry type.

6.1 The Future of Industry 5.0 in SDG Achievements

Industry 5.0 relates most closely to the goal that addresses industrialization sustainably and innovations, which is SDG 9. This focuses on the utilization of resources and integration of humans and machines to improve operations while achieving higher efficiency and profitability (Turkyilmaz et al., 2019). This paper finds it imperative for emotional intelligence and soft skills to be incorporated in workforce training for employees in Industry 5.0 for human-centered achievement of the SDGs. Industry 5.0 transformative attribute to intelligence urban cities and communities in creating Sustainable Development Goals 3, 4, 9, and 11; good health, quality education, innovation, and sustainable urban development.

6.2 Predictions and Emerging Trends in Sustainable Technology

Industry 5.0, depending on digital human modeling (DHM), brings success to design and optimize the manufacturing, textile, robotics, and energy sectors in terms of efficiency, safety, and sustainability. Blockchain is expected to deliver the value of Industry 5.0, with solid theoretical references: satisfying the need of industry for access to secure, decentralized information while fostering mutual understanding and trust in industrial uses. Augmented reality and other human-centric technologies are key components of smart manufacturing that enhance worker aid and digitalization-derived instruments.

6.3 Industry 5.0's Role in Long-Term Global Development Goals

Industry 5.0 seeks to balance the practices of human and robot workers to enhance sustainability and ethical performance as organized by strategic planning and innovation. Industry 5.0 is more suitable for the development of developing nations because it fosters sustainable service provision and consumption across service sectors in nations to include health services, education, and governance. The European Commission embraces Industry 5.0, supported by the Industry 4.0 paradigms as an opportunity to build a sustainable human-oriented resilient industry for future development.

7 Conclusion

This chapter, sustainable development goal in Industry 5.0, has showcased how Industry 5.0 helps the achievement of the SDGs and glimpses into a future where, once again, man and machine are intertwined to appreciate the environment, as well as to affect sustainable development. If compared to the focus on mere automation observed in Industry 4.0, Industry 5.0 is strongly connected with several critical SDGs because of the human-oriented approach and environmentally friendly concept. It fundamentally contributes to industrial innovation (SDG 9), growth of decent work (SDG 8), and responsible consumption (SDG 12), as well as acting on climate change (SDG 13). In addition, it advocates for equality and reduced inequalities (SDG 10) that will ensure that technologies are developed for the benefit of all the relevant stakeholders and in the process of improving the quality of cities and communities (SDG 11). Industry 5.0 focuses on people and seeks to incorporate AI & Cobots in a way that can make utilization of resources more effective in operation. This approach cuts across the United Nations Sustainable Development (Goal 9), which deals with the provision of resilient infrastructure and support for sustainable industrialization and innovation. That's why partnerships play a significant role in this context, as they help share information and solutions and bring about changes at the national level.

Acknowledgements The editors and anonymous reviewers are gratefully acknowledged by the author for their valuable feedback and recommendations, which contributed to enhancing the quality of the manuscript.

Funding This chapter didn't receive any financial support from public, private, not profit entities.

Competing Interests The author declares that they have no competing interest.

References

Aceto, G., Persico, V., & Pescapé, A. (2020). Industry 4.0 and health: Internet of things, big data, and cloud computing for healthcare 4.0. *Journal of Industrial Information Integration, 18.*

Ahmad, S. (2015). Green human resource management: Policies and practices. *Cogent Business & Management, 2*(1), 1030817.

Bennich, T., Weitz, N., & Carlsen, H. (2020). Deciphering the scientific literature on SDG interactions: A review and reading guide. *Science of the Total Environment, 728.*

Bhardwaj, V., Bhardwaj, R., & Sharma, A. C. (2024). ACHIEVING THE SDGs: Physics propelled multidisciplinary emerging areas. *International Journal of Innovation in Multidisciplinary Scientific Research.*

Bregar, A. (2024). Decision process for industry 5.0. *Journal of Decision System*, 1–15. https://doi.org/10.1080/12460125.2024.2359158

Breque, M., De Nul, L., & Petridis, A. (2021). *Industry 5.0: Towards a sustainable, human-centric and resilient European industry*. Luxembourg, LU: European Commission, Directorate-General for Research and Innovation

Chenary, K., Pirian Kalat, O., & Sharifi, A. (2024). Forecasting sustainable development goals scores by 2030 using machine learning models. *Sustainable Development.* https://doi.org/10.1002/sd.3037

Elavarasan, R. M., Pugazhendhi, R., Shafiullah, G. M., Kumar, N. M., Arif, M. T., Jamal, T., & Dyduch. (2022). Impacts of COVID-19 on sustainable development goals and effective approaches to maneuver them in the post-pandemic environment. *Environmental Science and Pollution Research, 29*(23), 33957–33987.

Europa.Eu. Retrieved November 11, 2024, from https://ec.europa.eu/info/research-and-innovation/research-area/industr

Giller, K. E., Drupady, I. M., Fontana, L. B., & Oldekop, J. A. (2018). Editorial overview: The SDGs-aspirations or inspirations for global sustainability. *Current Opinion in Environmental Sustainability, 34*, 1–2.

Goswami, S., & Rai, A. K. (2024). An assessment of prospects of geothermal energy in India for energy sustainability. *Renewable Energy.*

Horemans, J., Marx, I., & Nolan, B. (2016). Hanging in, but only just: Part-time employment and in-work poverty throughout the crisis. *IZA Journal of European Labor Studies, 5*, 1–19.

Jabeen, G., Goli, G., Kafila, & Gobinath, R. (2024). A bibliometric review on gender equity in human resource management. *Future Business Journal, 10*(1). https://doi.org/10.1186/s43093-024-00381-x

Janssen, C. P., Donker, S. F., Brumby, D. P., & Kun, A. L. (2019). History and future of human-automation interaction. *International Journal of Human-Computer Studies, 131*, 99–107.

Kasinathan, P., Pugazhendhi, R., Elavarasan, R. M., Ramachandaramurthy, V. K., Ramanathan, V., Subramanian, S., ... & Alsharif, M. H. (2022). Realization of sustainable development goals with disruptive technologies by integrating industry 5.0, society 5.0, smart cities and villages. *Sustainability, 14*(22), 15258.

Kira, M. (2003). *From good work to sustainable development-Human resources consumption and regeneration in the post-bureaucratic working life (Doctoral dissertation).*

Kulkarni, A. V., Joseph, S., & Patil, K. P. (2024). Artificial intelligence technology readiness for social sustainability and business ethics: Evidence from MSMEs in developing nations. *International Journal of Information Management Data Insights, 4*(2).

Leng, J., Sha, W., Wang, B., Zheng, P., Zhuang, C., Liu, Q., Wuest, T., Mourtzis, D., & Wang, L. (2022). Industry 5.0: Prospect and retrospect. *Journal of Manufacturing Systems, 65*, 279–295. https://doi.org/10.1016/j.jmsy.2022.09.017

Lu, Y. (2017). Industry 4.0: A survey on technologies, applications and open research issues. *Journal of Industrial Information Integration, 6*, 1–10.

Mahanayak, B. (2024). Meaning and concept of sustainable development for the protection of environment and the role of India. *World Journal of Advanced Research and Reviews, 22*(3), 1460–1465.

Mariappanadar, S. (2003). Sustainable human resource strategy: The sustainable and unsustainable dilemmas of retrenchment. *International Journal of Social Economics, 30*(8), 906–923.

Masri, H. A., & Jaaron, A. A. (2017). Assessing green human resources management practices in Palestinian manufacturing context: An empirical study. *Journal of Cleaner Production, 143*, 474–489.

Moyer, J. D., & Hedden, S. (2020). Are we on the right path to achieve the sustainable development goals? *World Development, 127*.

Patil, R., Kulkarni, P., Kulkarni, S., & Bhatt, A. (2023). A hybrid model for rain prediction using machine learning algorithm. In *International conference on soft computing for problem-solving* (pp. 855–869). Springer Nature.

Pradhan, P., Costa, L., Rybski, D., Lucht, W., & Kropp, J. P. (2017). A systematic study of sustainable development goal (SDG) interactions. *Earth's Future, 5*(11), 1169–1179.

Prasad, S., Yadav, K. K., Kumar, S., Pandita, P., Bhutto, J. K., Alreshidi, M. A., & Cabral-Pinto. (2024). Review on biofuel production: Sustainable development scenario, environment, and climate change perspectives—A sustainable approach. *Journal of Environmental Chemical Engineering, 12*(2).

Raffik, R., Swetha, M. M., Sathya, R. R., Vaishali, V., Adithya, B. M., & Balavedhaa, S. (2024). Integration of Unmanned Aerial Vehicle Systems With Machine Learning Algorithms for Wildlife Monitoring and Conservation. In *Applications of machine learning in UAV networks* (pp. 121–159). IGI Global.

Sabourin, V. (1999). Technological revolutions and the formation of strategic groups. *Journal of Engineering and Technology Management, 16*(3–4), 271–293. https://doi.org/10.1016/s0923-4748(99)00012-0

Seenuvasamurthi, S., Ashok, S., Shankarlal, B., Abbas, A. M., & Vajravelu, A. (2024). Brain image compression and reconstruction system using deep learning. *International Journal of Medical Engineering and Informatics, 16*(5), 401–413. https://doi.org/10.1504/ijmei.2024.140799

Shaolei, F., & Heng, C. (2019). *Developing the far east and Chinese-Russian relations: New perceptions and new practices*. Valdai Papers.

Sharma, R., & Gupta, H. (2024). Harmonizing sustainability in industry 5.0 era: Transformative strategies for cleaner production and sustainable competitive advantage. *Journal of Cleaner Production, 445*(141118), 141118. https://doi.org/10.1016/j.jclepro.2024.141118

Tariq, M., Xu, Y., Ullah, K., & Dong, B. (2024). Toward low-carbon emissions and green growth for sustainable development in emerging economies: Do green trade openness, eco-innovation, and carbon price matter. *Sustainable Development, 32*, 959–978.

Tooranloo, H. S., Azadi, M. H., & Sayyahpoor, A. (2017). Analyzing factors affecting implementation success of sustainable human resource management (SHRM) using a hybrid approach of FAHP and Type-2 fuzzy DEMATEL. *Journal of Cleaner Production, 162*, 1252–1265.

Turkyilmaz, A., Guney, M., Karaca, F., Bagdatkyzy, Z., Sandybayeva, A., & Sirenova, G. (2019). A comprehensive construction and demolition waste management model using PESTEL and 3R for construction companies operating in Central Asia. *Sustainability, 11*(6), 1593.

Xu, X., Lu, Y., Vogel-Heuser, B., & Wang, L. (2021). Industry 4.0 and Industry 5.0—Inception, conception and perception. *Journal of Manufacturing Systems, 61*, 530–535. https://doi.org/10.1016/j.jmsy.2021.10.006

Yitmen, I., & Almusaed, A. (2024). Synopsis of Industry 5.0 paradigm for human-robot collaboration. In *Artificial Intelligence*. IntechOpen.

"Leveraging Digital Technologies for Circular Economy: Exploring the Role of AI, IoT, and Blockchain in Overcoming Future Challenges"

V. Harish, Ravindra Sharma, Geeta Rana, and Bhakti Parashar

Abstract The shift toward a circular economy offers a critical opportunity to address the environmental and resource depletion road-blocks driven by traditional linear economic models. The world of circular economy practices aims to design waste out of the system, extend product lifecycles, and enable closed-loop systems where resources are reprocessed and cast-off efficiently. However, transitioning to such a model at scale by firms and entities requires overcoming significant operational, logistical, and systemic barriers. Digital technologies, particularly Artificial Intelligence, the Internet of Things, and Blockchain, offer transformative potential to mitigate these road-blocks and allow the successful implementation of circular economy principles across industries. This study explores the part of the recent technologies in overcoming the key challenges closely connected with the achievement of circular economy. While earlier studies indicate that technologies such as Artificial Intelligence can optimize resource management by enabling predictive maintenance, improving manufacturing processes, and enhancing decision-making capabilities for sustainable product design. IoT by connecting physical assets through real-time monitoring and data collection across the entire product lifecycle enables precise tracking of materials and facilitating efficient resource use. Blockchain, with its decentralized system, augments transparency and accountability in circular supply chains, allowing for secure and trustworthy data on product provenance, recycling, and material reuse. The research adopts a mixed-methods approach, integrating case studies of industries that have successfully implemented these technologies in circular economy initiatives with a detailed analysis of the environmental and economic impacts.

V. Harish (✉)
PSG Institute of Management, PSG College of Technology, Coimbatore, India
e-mail: harish@psgim.ac.in

R. Sharma · G. Rana
Himalayan School of Management Studies, Swami Rama Himalayan University, Jolly Grant, Dehradun, India

B. Parashar
VIT Business School, VIT Bhopal University, Bhopal, India

R. Kumar and S. N. Mohanty (eds.), *Sustainable Economy Models in the Age of Industry 5.0*, https://doi.org/10.1007/978-981-96-4104-8_4

Keywords Digital technologies · Circular economy · Artificial intelligence · Internet of things · Blockchain

1 Introduction

1.1 Background

The mounting apprehensions over environmental damage and the widespread exhaustion of the resources have led governments, policymakers, and industries to venture out for sustainable economic models, whereas traditional "linear economic models" which take the "take, make, dispose" approach have significantly contributed to waste generation and unsustainable utilization of limited resources. In contrast, the new age "Circular Economy" (CE) model proposes a sustainable alternate that aims and targets on limiting waste, increasing the product shelf life and creating closed-loop systems, in which products are continuously reprocessed and cast-off, thereby aiming to create a sustainable economy (Kara et al., 2022). This shift from linear to circular is seen as critical, crucial, and vital for addressing climate change, reducing environmental impact, and conserving resources for future generations.

The amalgamation of new age digital technologies such as "Artificial Intelligence (AI), the Internet of Things (IoT), and Blockchain (BC)" has created unparalleled opportunities to support and fasten the much-required changeover toward creating a CE (Parashar et al., 2023).

1.2 Problem Statement

While the benefits of shifting to a CE are well-accepted, adopting this model across firms remains a huge challenge due to various operational, logistical, and systemic barriers and as cited, many firms struggle with issues such as inefficient resources, inability to track consignments in real time, lack of transparency in supply chains, and the escalating costs of implementing circular strategies. While digital technologies specifically AI, IoT, and BC offer potential answers, yet their adoption is not without its own challenges, such as high adoption costs, data privacy issues, and the need for robust digital infrastructure to name a few.

1.3 Objectives

The aim of the chapter is to understand the role of AI, IoT, and BC in overcoming key barriers in the adoption and CE principles and the aims are as follows:

- To recognize the role of Artificial Intelligence in achievement of CE.
- To comprehend the role of IoT in achievement of CE.
- To understand the role of Blockchain in achievement of CE.
- To do a comparison of AI, IoT, and BC in CE.
- To identify the barriers to incorporate Digital Technologies in CE.
- To present various Case Studies of Successful Integration of Digital Technologies in CE.

1.4 Scope and Significance

The scope of this article includes an in-depth examination of AI, IoT, and BC technologies and their application in various industries striving to implement CE practices. The study implements a mixed-method, combining a detailed analysis of the existing literature with specific cases of industries that have successfully utilized these technologies. The implication of this article is in its outcome to provide significant insights for various stakeholders and policymakers, helping to drive the sustainable transformation of economic systems.

1.5 Chapter Organization

The paper is structured as follows:

- An outline of the CE and its key principles.
- A detailed exploration of digital technologies, focusing on AI, IoT, and BC.
- An analysis of the applications and benefits of these technologies in CE initiatives.
- Identification of challenges and barriers to their adoption.
- Conclusions and recommendations for future research.

2 Circular Economy

2.1 Definition and Understanding

CE is designed in such a way as to lessen waste and make the most of existing products contrasting to the linear economy, which adopts raw material extraction to disposal (Suchek et al., 2021) while CE emphasizes keeping products in use for a prolonged time and some of the philosophies of CE are:

- **Reducing Waste**: Products and goods are designed and developed with the complete lifecycle of the product in mind rather than a one-time use, reducing waste and using sustainable materials (Sarkar et al., 2022).

Table 1 Comparison of LE and CE

Aspect	Linear economy (LE)	Circular economy (CE)
Resource use	Finite, single-use	Regenerative, continuous use
Waste	High, disposable	Minimal, reused, or recycled
Value Flow	One-way (take-make-dispose)	Closed-loop (reuse-recycle)
Environmental impact	High	Reduced through sustainability

- **Reuse Materials**: Emphasizes reusing and recycling materials to extend their usage and thus minimize the necessity for novel products (Foster, 2020).
- **Regenerating Natural Systems**: Ensures that economic activities enhance the natural environment, for instance, by using biodegradable materials that return nutrients to the ecosystem (Morseletto, 2020).

2.2 *Comparison with Linear Economy*

Table 1 indicates comparison of LE and CE. In a **linear economy**, goods are manufactured, utilized, and disposed of without considering the environmental impact or potential for reuse which leads to significant waste generation and higher resource exhaustion. On the other hand, a **CE** focuses on closing the loop, where waste is minimized, and the product value is maintained within the economic system (Neves & Marques, 2022).

2.3 *Key Elements of CE*

A detailed review of earlier studies on CE has indicated the following key elements, namely:

a. **Product Life Extension**: CE aims to design products that can be easily mended, renovated, or remanufactured to extend their useful life and not aim at one-time use of products, thereby extending their life (Milios, 2021).
b. **Resource Efficiency**: The focus of CE is to maximize the use of raw materials and goods and services aimed at reducing waste and thereby employing sustainable sourcing and supply chain practices leading to higher productivity and efficiency in terms of materials and resources (Cainelli et al., 2020).
c. **Closed-Loop Supply Chains**: CE aims at establishing robust and fool-proof systems that are aimed at collecting, combining, recycling, and reintegrating the products back into production methods and thus minimizing the wastages in the production and supply chain process (Harish et al., 2023a, 2023b; MahmoumGonbadi et al., 2021).

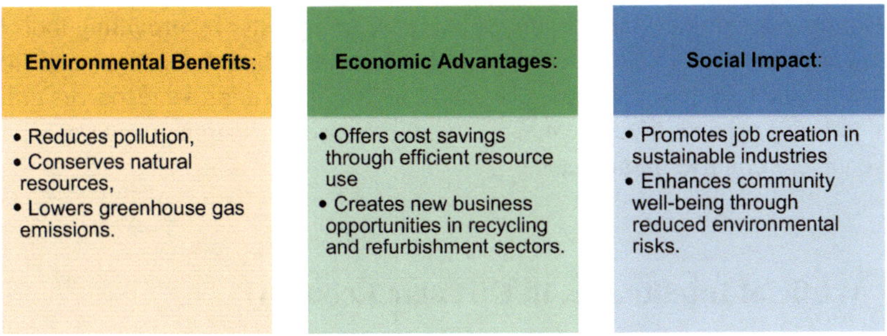

Fig. 1 Benefits of transitioning to a circular economy. *Source* Authors own

d. **Collaborative Consumption**: CE promotes the shared use of products and services across the supply chain entities, such as leasing or rental services, or sharing of resources materials and people in order to minimize waste and optimize the resources and save on the consumption of the resources (Minami et al., 2021).

2.4 Benefits of Transitioning to a CE

Various studies have indicated numerous benefits of circular which can be consolidated into the following three major ones. Figure 1 shows benefits of transitioning to a circular economy.

3 Overview of Digital Technologies

3.1 Introduction to Digital Transformation (DT)

DT is the incorporation of new age technologies into various domains, altering the way industries work. Technologies like "Artificial Intelligence (AI), the Internet of Things (IoT), and Blockchain (BC)" have become key tools in enabling businesses to innovate, improve operational efficiency, and respond to evolving market demands. These technologies are particularly relevant for firms not only to adopt CE principles but also to survive the competitive space, as they offer solutions for improved resource utilization, product lifecycle enhancement, and improved transparency in supply chains (Kraus et al., 2021). The union of DT with new age sustainable practices marks a radical shift toward an efficient and resilient economic system and by leveraging these technologies, companies can overcome traditional barriers to circularity, such as increased wastes, lack of real-time data, and challenges in tracking materials across supply chains (Pasqualino et al., 2021).

Digital technologies have revolutionized modern industry by providing tools that augment productivity, reduce waste, and facilitate data-based decisions and in the context of the CE, these technologies can be leveraged to improve resources utilization, improve product lifecycle management, increase transparency and traceability, and enable predictive maintenance.

4 Artificial Intelligence in Circular Economy

4.1 Introduction to AI

Artificial Intelligence (AI) in simple terms enables machines to do work that is ideally done by humans and these jobs include data analysis using natural language processing among others and its ability to analyze massive quantity of data and derive actionable insights makes it a valued technology for driving efficiency and sustainability in business processes (Zhang & Lu, 2021).

4.2 AI in Resource Optimization

AI contributes to optimizing resources in CE by:

- **Predictive Analytics**: AI algorithms can examine past data to forecast resource requirements in future, ensuring companies avoid overproduction and minimize waste and allow firms to forecast more accurately (Agbehadji et al., 2020).
- **Smart Manufacturing**: AI-enabled smart systems can constantly monitor production processes in real time, proactively identifying potential inefficiencies and suggesting adjustments to reduce resource consumption and avoiding wastage and defects in the manufacturing process (Sahoo & Lo, 2022; Sharma, Saini et al., 2021).
- **Enhanced Recycling Processes**: AI-powered tools using robots can automatically classify and isolate reusable resources, refining efficiency and accuracy of reprocessing process and at the same time enhancing the safety of the workers doing the tasks (Yu et al., 2021).

4.3 AI in Predictive Maintenance

Conventional machine maintenance processes are reactive in nature and often lead to unnecessary downtime and inefficient use of resources and non-utilization of expensive and critical machines, whereas AI through their effective forecasting can aid in predictive maintenance, where it analyzes from sensors as the machine runs to predict equipment failures even before they happen, allowing companies to perform maintenance only when needed and take a proactive approach instead of a reactive one (Nimmagadda, 2022). This approach:

- **Reduces Downtime**: By predicting equipment failures in advance, and the parts that are likely to fail, businesses can plan maintenance activities without disrupting production schedules and have inventory of spares so that the downtime is reduced to a significant level (Kumar et al., 2024).
- **Extends Equipment Lifespan**: Proactive maintenance reduces wear and tear, extending the useful life of machinery which often is very expensive and by improving the life span offers a more sustainable option for the firms and also reduces the need for replacements.
- **Minimizes Waste**: Efficient maintenance practices coupled with a proactive approach in maintenance recommended by AI systems reduce the waste generated from premature disposal of machinery and spare parts.

4.4 Enhancing Sustainable Product Design with AI

AI can significantly contribute to sustainable product design by providing insights into materials selection, product lifecycle analysis, and product end life management and by simulating different design scenarios (Liu et al., 2020), AI can help designers:

- **Choose Sustainable Materials**: AI can nalyse the environmental impact of various materials and suggest alternatives that are more sustainable in short term, long term, health factors, disposal factors, and much more and recommend better raw material alternatives.
- **Optimize Product Lifespan**: AI systems can predict how products will perform over time and over the years of usage and thereby help product designers to create goods more robust and easier to repair or recycle and thus remain sustainable.
- **Facilitate Design for Disassembly**: AI-enabled systems can assist in creating modular designs that make it easier to disassemble products for recycling or refurbishment over the life of the product and post usage of the product.

4.5 Industry Examples

1. **Manufacturing Sector**: A leading electronics manufacturer uses AI algorithms to predict defects in production, reducing waste and improving yield and productivity (Ghelani, 2024).
2. **Textile Industry**: AI is employed to analyze textile waste and identify patterns, helping companies design clothing that uses fewer raw materials and generates less waste and remain sustainable (Ahmad et al, 2020).

3. **Automotive Industry**: AI-driven predictive maintenance systems have been implemented by automotive manufacturers, reducing equipment downtime and extending the lifecycle of production tools (Robnik-Šikonja, 2023).

4.6 Challenges and Limitations of AI Integration

While AI offers significant potential for enhancing CE practices, several challenges must be addressed by firms (Wilson et al., 2022) such as:

1. **Data Adequacy**: AI systems for their functioning require massive high-quality data, which may not always be available or consistent across companies and industries and this becomes an important determinant for the validity of the AI inferences and most of the companies lack having data in terms of quality and quantity.
2. **High Implementation Costs**: The expenses of designing and deploying AI solutions can be very huge, particularly for SMEs and, in few cases, even big size firms and an added constraint is the lack of knowledge in these new age technologies.
3. **Ethical Dilemma:** The adoption of AI poses dangers related to privacy of the data and unless these ethical concerns are addressed, the adoption of the technologies might have an impact on the usage across the firms.

5 Internet of Things (IoT) in Circular Economy (CE)

5.1 Introduction to IoT

IoT denotes a system of interconnected gadgets that consolidate data with sensors and share across network and internet allowing real-time monitoring and in the framework of the CE, IoT plays a vital role in tracking products, monitoring resource flows, and enhancing the effectiveness of waste administration systems.

The variety of IoT devices that are available across platforms has enabled the conception of smart systems that can check on environmental conditions, look into the status of assets, and collect data throughout the product lifecycle and this data allows firms to optimize usage of materials, minimalize waste, and progress in sustainability of their operations (Voulgaridis et al., 2022).

5.2 Applications of IoT in Circular Economy

IoT can assist firms to transform and adopt CE practices by providing enhanced visibility and control over resources and few key uses of IoT in CE initiatives include:

1. **Real-Time Monitoring of Resource Flows**

IoT devices equipped with sensors can identify and disseminate the movement of materials across the supply chain and this capability helps companies gain insights into resource usage, identify inefficiencies, and optimize resources efficiently (Mboli et al., 2022; Prakash et al., 2021).

- In **manufacturing**, IOT devices can check the usage of raw materials, energy consumption, and production output, enabling real-time adjustments to minimize waste and improve efficiency in the production process.
- In **agriculture**, sensors can track soil conditions, crop growth, optimizing resource inputs and reducing waste, and optimize usage of resources and save on cost for farmers.

2. **Enhanced Product Lifecycle Management**

IoT technology enables businesses to collect data on product performance, usage, and status of the issues, errors, and much more allowing for better product lifecycle management and by checking the condition of products (Rusch et al., 2023), firms can:

- **Extend Product Lifespan**: Identify signs of wear and tear early, issues in the usage of the product, and have preventive schedule maintenance or repairs before failures occur, thereby extending life of the product.
- **Enable Product-as-a-Service Models**: IoT devices can facilitate leasing or subscription models, where products are returned for refurbishment or recycling after use, rather than being disposed of, resulting in a more sustainable business model.
- **Support Design for Circularity**: Feedback from IoT data from the products on and off use can provide significant insights and assist in the designing of future products, making them easier to repair, upgrade, or recycle.

3. **Optimizing Waste Management Systems**

IoT has revolutionized the waste management business by enabling smart waste management process that improves the efficiency of collection, sorting, and recycling processes where sensors can track the fill levels of waste containers, minimizing transportation costs and additionally, IoT-enabled sorting systems can automatically identify variety of waste, streamlining the recycling process and increasing the recovery rate of valuable materials (Jogarao et al., 2024).

5.3 IoT-Driven Supply Chain Transparency

Supply chain transparency is a crucial requirement for realizing CE practices, especially in complex global supply chains, and IoT improves visibility by providing data on location, status, and condition of items and few of the advantages are:

1. **Improved Traceability of Materials**

IoT sensors can track the movement of products across the supply chain, ensuring that companies have accurate information about the sources of their inputs and increase the traceability of materials which helps verify compliance with sustainability standards and reduces the risk of using non-recyclable or environmentally harmful materials (Santana & Ribeiro, 2022).

2. **Enhanced Inventory Management**

Instant data on inventory levels and product conditions, capable through IoT devices enable better inventory, mitigating the risk of overproduction and minimizing waste and optimize the inventory level and this capability is especially valuable for perishable goods, where timely monitoring can prevent spoilage and reduce losses (Akbari, 2024).

3. **Enabling Reverse Logistics**

IoT technology supports reverse logistics, a key component of the CE, by tracking products returned for refurbishment, remanufacturing, or recycling. By monitoring the condition and location and status of returned products, companies can streamline the reverse logistics process and ensure that materials are efficiently reintegrated into production cycles (Khan et al., 2022).

5.4 Case Studies and Industry Applications

1. **Smart Packaging in Retail**

In the retail sector, IoT-enabled smart packaging provides real-time data on product freshness, temperature, and location, for example, food retail stores use IoT sensors to track the temperature of perishable goods and in shelf ensuring that products remain fresh and reducing food waste (Trejo Beltran, 2023).

2. **Industrial IoT in Manufacturing**

Manufacturing companies are using Industrial IoT (IIoT) systems to monitor machinery performance and optimize production processes by collecting data on equipment status; IIoT devices can detect potential issues early, enabling predictive maintenance and reducing downtime, which in turn minimizes waste from defective products and work efficiency of machines (S.Nagpal et al., 2020; Haghnegahdar et al., 2022).

6 Blockchain (BC) Technology in Circular Economy (CE)

6.1 *Introduction to BC Technology*

Blockchain technology, initially conceptualized for financial transactions, "is a decentralized, distributed ledger system that archives transactions in a safe, clear, and absolute manner and each transaction, or 'block,' is added to a chain of previous transactions, creating a comprehensive, tamper-proof record that is accessible to all participants in the network" (Upadhyay et al, 2021).

6.2 *Key Features of Blockchain for Circular Economy*

The adoption of BC technology in CE initiatives is driven by its unique features:

1. **Decentralization**

Conventional databases are managed by a central authority, whereas BC works on a decentralized policy wherein data is distributed across multiple nodes and increases the safety of the data and ensures privacy (Kumar & Chopra, 2022).

2. **Incentivizing Sustainable Practices**

BC can be used to create incentive systems that reward companies and consumers for participating in sustainable practices, for instance, a BC-based token system could provide credits to consumers who return used products for recycling or to companies that use recycled materials in their production processes and these tokens can be traded or redeemed for rewards, encouraging broader participation in CE initiatives (De Giovanni, 2022).

6.3 *Challenges in Adopting Blockchain for Circular Economy*

Despite its potential, the adoption of BC technology in CE practices faces several barriers:

1. **Scalability Issues**

Scalability is often a major concern, particularly in industries with high transaction volumes, as it can affect the speed and cost of data processing and questions the sustainable aspect of the same (Kumar & Chopra, 2022).

2. **High Energy Consumption**

The process of validating transactions on a BC, especially in developing systems, requires high energy consumption and indirectly affects sustainability as power is mostly generated using conventional methods and this energy usage can conflict with the sustainability goals of CE initiatives (Basile et al., 2023).

3. **Legal Challenges**

The regulatory environment for BC is still evolving, with different jurisdictions implementing varying rules and standards and certain countries not having any legal framework regarding BC (Steenmans et al., 2021).

4. **Data Privacy Concerns**

While BC's transparency is a key advantage, it also raises concerns about data privacy as sensitive business information recorded on a public BC could be exposed to competitors, making it critical to maintain between transparency and confidentiality and as the technology is relatively new the data privacy needs to be evaluated and monitored.

7 Comparison of AI, IoT, and Blockchain in CE

A detailed analysis of the three digital technologies—AI, IoT and Blockchain—and their roles in enabling CE practices where the analysis is presented both in textual form and in a comprehensive table to provide clarity on the synergies, unique contributions, and limitations of each technology.

Textual Analysis

1. **Synergistic Potential of AI, IoT, and Blockchain**

The combined use of AI, IoT, and BC can significantly enhance CE initiatives which for example AI's predictive capabilities, when combined with data collected by IoT gadgets, can optimize decision-making across the entire product lifecycle, BC can provide a reliable platform for recording and sharing this data, ensuring accountability and trust among stakeholders.

2. **Unique Contributions of Each Technology**

Table 2 shows comparison of AI, IoT, and BC in CE. While AI, IoT, and BC can work together, each technology also has distinct strengths. AI excels in data analysis, IoT provides visibility in resource flows to monitor assets throughout the product lifecycle, and BC offers enhanced transparency and traceability.

Table 2 Comparison of AI, IoT, and BC in CE

Aspect	AI	IoT	BC
Primary Function	Predictive analytics, optimization, and decision-making	Real-time data collection and monitoring	Secure, transparent, and immutable ledger
Key Applications	Demands forecasting, predictive maintenance, sustainable product design	Resource tracking, asset monitoring, waste management	Traceability, product provenance, compliance
Strengths	Analyzes large datasets for insights, improves efficiency	Provides real-time visibility, connects physical assets	Ensures transparency, enhances data integrity
Challenges	High implementation costs, data privacy concerns	Requires robust infrastructure, potential data security risks	Scalability issues, high energy consumption
Integration Potential	Works well with IoT for enhanced data analysis	Complements AI for data generation and analysis	Provides a secure platform for AI and IoT data sharing
Impact on CE	Optimizes resource usage, reduces waste, enhances product lifecycle	Improves visibility of resource flows, supports reverse logistics	Increases transparency, ensures reliable data on material usage
Industry Use Cases	Smart manufacturing, sustainable supply chain management	Smart packaging, waste tracking, inventory management	Supply chain transparency, e-waste tracking, ethical sourcing
Future Trends	Integration with edge computing, AI for circular design	Expansion of IoT networks, advancements in sensor technology	Development of energy-efficient BCs, integration with AI and IoT

8 Barriers to Incorporate Digital Technologies in CE

While these technologies offer immense potential, there are significant barriers to their widespread adoption. There are key challenges faced by industries in implementing these digital solutions and the article attempts to provide strategies for overcoming these obstacles.

8.1 High Implementation Costs

Barrier: The initial investment required for adopting digital technologies is often substantial, like in the case of AI, it requires huge and expensive equipment's,

IoT devices require sensors and BC demands specialized technical expertise for deployment and these are vital concerns for small enterprises. (Trevisan et al., 2023).

Mitigation Strategy:

- **Government Subsidies and Grants:** Policymakers can offer monetary enticements such as grants and subsidies to lower the entry barriers for firms adopting these technologies.
- **Public–Private Partnerships:** Collaborations between private enterprises and government bodies can speed up with technology adoption, assisting them to go for adoption of these technologies particularly SMEs.
- **Scalable Solutions:** Businesses can start with pilot projects that focus on specific processes and gradually scale up as they gain experience and demonstrate value but for which they should have the necessary information and knowledge on where and how to start the implementation of these DT.

8.2 Nonexistence of Advanced Digital Infrastructure

Barrier: Effective deployment of AI, IoT, and BC technologies requires necessary facilities such as internet and reliable power supply, and many regions, particularly in developing and under developed countries, face infrastructural challenges that hinder the integration of these technologies (Kandasamy et al., 2023).

Mitigation Strategy:

- **Investment in Infrastructure Development:** Governments need to prioritize investments in digital infrastructure, focusing on expanding broadband internet access and supporting cloud-based services, and improve the necessary infrastructure for the digital technologies.
- **Edge Computing Solutions:** Utilizing edge computing can reduce the reliance on centralized infrastructure, minimizing latency and connectivity issues, and initiative should be taken to create awareness on such solutions.
- **Adoption of Hybrid Models:** Companies can use a combination of cloud and local (on-premise) solutions to balance infrastructure demands and ensure continuity in areas with limited connectivity and mitigate the challenges.

8.3 Data Security Concerns

Barrier: The extensive use of AI, IoT, and BC encompasses the gathering, storage, processing of data and this increases significant concerns about data privacy, particularly when personal or sensitive information is involved and cyber-attacks are possible, and BC's transparency can expose confidential business information (Bag et al., 2022; Kandasamy et al., 2023; Trevisan et al., 2023).

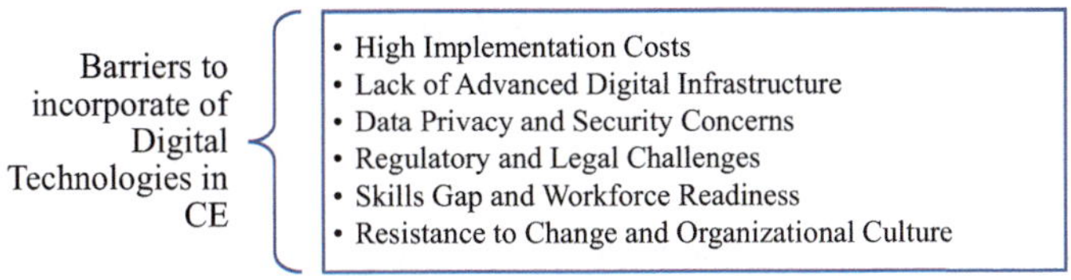

Fig. 2 Barriers to incorporate digital technologies in Circular Economy. *Source* Author's Own

Mitigation Strategy:

- **Enhanced Governance Frameworks:** Developing policies and frameworks for governance can help address privacy concerns, confirming that data is consolidated, stored, and shared in compliance with regulations of the country and ensuring safety of data. Figure 2 shows barriers to incorporate digital technologies in Circular Economy.
- **Private BCs:** Using private or permissioned BCs instead of public ones can help control access to sensitive data, reducing the risk of exposure to the general public and users in general.

8.4 *Regulatory and Legal Challenges*

Barrier: With evolving laws, varying standards, and compliance requirements across different jurisdictions, inconsistent regulations among different regions can create uncertainty for businesses, particularly those operating across international borders and additionally, the use of AI and BC raises ethical and legal concerns related to accountability, data ownership, and liability (Bag et al., 2022; Kandasamy et al., 2023; Trevisan et al., 2023).

Mitigation Strategy:

- **Harmonization of Regulations:** Governments and international bodies need to work toward standardizing regulations related to digital technologies, creating a consistent framework that facilitates cross-border operations that aim to facilitate firms in adopting new technologies.
- **Development of Ethical regulations:** Framing laws for the use of AI, IoT, and BC can help address concerns related to accountability and data ownership, ensuring responsible technology adoption and facilitating mutual benefits to all concerned stakeholders.
- **Legal Expertise:** Businesses should engage legal experts to navigate compliance requirements effectively and in order to achieve this, there have to be legal experts available to the firms and users.

8.5 *Skills Gap and Workforce Readiness*

Barrier: Implementing advanced technologies like AI, IoT, and BC requires specialized skills in data science, cybersecurity, and software development and many companies, particularly those in traditional industries, struggle to find qualified personnel, leading to a significant skills gap or the cost to hire the experts is high which is unaffordable to most of the firms (Bag et al., 2022; Kandasamy et al., 2023; Trevisan et al., 2023).

Mitigation Strategy:

- **Upskilling and Reskilling Programs:** Making staffs participating in relevant training programs can address the skills gap, giving necessary technical expertise and have continuous learning programs for employees to be efficient on the new technologies and their uses.
- **Academic-Industry Collaborations:** Partnering with academic institutions can help develop tailored curricula and training programs focused on emerging digital technologies, ensuring a steady pipeline of skilled professionals in the upcoming workforce and academicians can also be engaged in providing necessary training for the firms in these technologies and their applications.
- **Hiring Experts:** Companies can initially hire external consultants or technology experts to assist with the deployment and management of these technologies while building internal capabilities over time and the consultants can handhold the firms' employees in the initial phases and slowly aim at technology transfer.

8.6 *Resistance to Change and Organizational Culture*

Barrier: Adopting digital technologies often requires momentous changes to existing processes, which can encounter resistance from staff as most are resistant to changes and stakeholders accustomed to traditional methods hinder the embracing of new CE practices (Bag et al., 2022; Kandasamy et al., 2023; Rana & Sharma, 2017; Trevisan et al., 2023).

Mitigation Strategy:

- **Change Management Initiatives:** Implementing comprehensive change management programs that include training, stakeholder engagement, can help overcome resistance and taking things in a phased manner.
- **Leadership Support:** Strong top management support is key and leaders need to champion the adoption of new technologies and align them with the firms' goals and motivate their teams to go for these changes.
- **Pilot Projects for Demonstrating Value:** Starting with small, well-defined pilot projects can help demonstrate the tangible benefits of digital technologies, gaining buy-in from stakeholders and easing the transition rather than a complete change which will be over whelming for the employees.

8.7 Conclusion

The research article has aimed to provide proof that digital technologies, specifically AI, IoT, and BC, are key enablers in achieving Circular Economy to remain sustainable and they offer numerous benefits such as improved resource efficiency, reduced waste, and creation of sustainable and transparent supply chains. However, the successful implementation of these technologies requires mitigating several barriers such as high cost of deployment, making necessary infrastructure, data privacy, regulatory compliance, and sufficient knowledge. The role of collaboration between governments and firms is crucial for creating a conducive environment to achieve CE through digital technologies and governments should create favorable policies, while firms should spend in the required infrastructure and update required skills to adopt these technologies. Ultimately, the transition to a CE is not only a technological change but also a cultural shift toward sustainability and resource conservation and by adopting digital innovations and overcoming existing barriers, industries can work for a more sustainable future (Harish et al., 2023a), where goods and resources are endlessly reused, reducing environmental impact and contributing to long-term economic growth.

References

Agbehadji, I. E., Awuzie, B. O., Ngowi, A. B., & Millham, R. C. (2020). Review of big data analytics, artificial intelligence and nature-inspired computing models towards accurate detection of COVID-19 pandemic cases and contact tracing. *International Journal of Environmental Research and Public Health, 17*(15), 5330.

Ahmad, S., Miskon, S., Alabdan, R., & Tlili, I. (2020). Towards sustainable textile and apparel industry: Exploring the role of business intelligence systems in the era of industry 4.0. *Sustainability, 12*(7), 2632.

Akbari, M. (2024). Revolutionizing supply chain and circular economy with edge computing: Systematic review, research themes and future directions. *Management Decision, 62*(9), 2875–2899.

Bag, S., Sahu, A. K., Kilbourn, P., Pisa, N., Dhamija, P., & Sahu, A. K. (2022). Modeling barriers of digital manufacturing in a circular economy for enhancing sustainability. *International Journal of Productivity and Performance Management, 71*(3), 833–869.

Basile, D., D'Adamo, I., Goretti, V., & Rosa, P. (2023). Digitalizing circular economy through blockchains: The blockchain circular economy index. *Journal of Industrial and Production Engineering, 40*(4), 233–245.

Cainelli, G., D'Amato, A., & Mazzanti, M. (2020). Resource efficient eco-innovations for a circular economy: Evidence from EU firms. *Research Policy, 49*(1), 103827.

Centobelli, P., Cerchione, R., Del Vecchio, P., Oropallo, E., & Secundo, G. (2022). Blockchain technology for bridging trust, traceability and transparency in circular supply chain. *Information & Management, 59*(7), 103508.

De Giovanni, P. (2022). Leveraging the circular economy with a closed-loop supply chain and a reverse omnichannel using blockchain technology and incentives. *International Journal of Operations & Production Management, 42*(7), 959–994.

Foster, G. (2020). Circular economy strategies for adaptive reuse of cultural heritage buildings to reduce environmental impacts. *Resources, Conservation and Recycling, 152*, 104507.

Ghelani, H. (2024). AI-driven quality control in PCB manufacturing: Enhancing production efficiency and precision. *Valley International Journal Digital Library*, 1549–1564.

Haghnegahdar, L., Joshi, S. S., & Dahotre, N. B. (2022). From IoT-based cloud manufacturing approach to intelligent additive manufacturing: Industrial Internet of things—An overview. *The International Journal of Advanced Manufacturing Technology, 119*(3), 1461–1478.

Harish, V., & Sharma, R. (2024). 6 Uncovering the shifting landscape: A comprehensive analysis of emerging trends and best practices adopted in green supply chain management. In *Green innovations for industrial development and business sustainability: Models and implementation strategies*, 82.

Harish, V., Sharma, R., Rana, G., & Nayyar, A. (2023a). Artificial Intelligence in sustainable education: Benefits, applications, framework, and potential barriers. In *The role of sustainability and artificial intelligence in education improvement* (pp. 219–236). Chapman and Hall/CRC.

Harish, V., Sharma, R., Saini, R. K., & Negi, D. (2023b). Intelligent supply chain orchestration: A framework for seamless integration of Industry 4.0 technologies. In *2023 seventh international conference on image information processing (ICIIP)* (pp. 265–270). IEEE.

Jogarao, M., Lakshmanna, B. C., & Naidu, S. T. (2024). AI-enabled circular economy management for sustainable smart cities: Integrating Artificial intelligence in resource optimization and waste reduction. In *Smart cities and circular economy* (pp. 83–96). Emerald Publishing Limited.

Kandasamy, J., Venkat, V., & Mani, R. S. (2023). Barriers to the adoption of digital technologies in a functional circular economy network. *Operations Management Research, 16*(3), 1541–1561.

Kara, S., Hauschild, M., Sutherland, J., & McAloone, T. (2022). Closed-loop systems to circular economy: A pathway to environmental sustainability? *CIRP Annals, 71*(2), 505–528.

Khan, S. A., Laalaoui, W., Hokal, F., Tareq, M., & Ahmad, L. (2022). Connecting reverse logistics with circular economy in the context of Industry 4.0. *Kybernetes, 52*(12), 6279–6320.

Komyshova, L. N., Serebryakova, E. A., Kryuchkova, I. V., Belyantseva, O. M., Smorodina, E. P., & Gunina, M. A. (2020, August). Guidelines of digitalizing waste industry infrastructure. In *Russian Conference on Digital Economy and Knowledge Management* (RuDEcK 2020) (pp. 353–357). Atlantis Press.

Kraus, S., Jones, P., Kailer, N., Weinmann, A., Chaparro-Banegas, N., & Roig-Tierno, N. (2021). Digital transformation: An overview of the current state of the art of research. *SAGE Open, 11*(3), 21582440211047576.

Kumar, N. M., & Chopra, S. S. (2022). Leveraging blockchain and smart contract technologies to overcome circular economy implementation challenges. *Sustainability, 14*(15), 9492.

Kumar, R., Goel, R., Singh, T., Mohanty, S. M., Gupta, D., Alkhayyat, A., & Khanna, R. (2024). Sustainable finance factors in indian economy: Analysis on policy of climate change and energy sector. *Fluctuation and Noise Letters, 23*(2), 2440004–339.

Liu, Y., Zhang, Y., Ren, S., Yang, M., Wang, Y., & Huisingh, D. (2020). How can smart technologies contribute to sustainable product lifecycle management? *Journal of Cleaner Production, 249*, 119423.

MahmoumGonbadi, A., Genovese, A., & Sgalambro, A. (2021). Closed-loop supply chain design for the transition towards a circular economy: A systematic literature review of methods, applications and current gaps. *Journal of Cleaner Production, 323*, 129101.

Mboli, J. S., Thakker, D., & Mishra, J. L. (2022). An Internet of things-enabled decision support system for circular economy business model. *Software: Practice and Experience, 52*(3), 772–787.

Milios, L. (2021). Overarching policy framework for product life extension in a circular economy—A bottom-up business perspective. *Environmental Policy and Governance, 31*(4), 330–346.

Minami, A. L., Ramos, C., & Bortoluzzo, A. B. (2021). Sharing economy versus collaborative consumption: What drives consumers in the new forms of exchange? *Journal of Business Research, 128*, 124–137.

Morseletto, P. (2020). Restorative and regenerative: Exploring the concepts in the circular economy. *Journal of Industrial Ecology, 24*(4), 763–773.

Nandi, S., Sarkis, J., Hervani, A. A., & Helms, M. M. (2021). Redesigning supply chains using blockchain-enabled circular economy and COVID-19 experiences. *Sustainable Production and Consumption, 27*, 10–22.

Neves, S. A., & Marques, A. C. (2022). Drivers and barriers in the transition from a linear economy to a circular economy. *Journal of Cleaner Production, 341*, 130865.

Nimmagadda, V. S. P. (2022). Artificial intelligence for predictive maintenance of banking IT infrastructure: Advanced techniques, applications, and real-world case studies. *Journal of Deep Learning in Genomic Data Analysis, 2*(1), 86–122.

Nyman, T. (2019). *Increased transparency and prevention of unethical actions in the textile industry's supply chain through blockchain.*

Parashar, B., Sharma, R., Rana, G., Balaji, R.D. (2023). Foundation concepts for industry 4.0. In A. Nayyar, M. Naved, & R. Rameshwar (Eds.), *New horizons for industry 4.0 in modern business.* Contributions to environmental sciences & innovative business technology. Springer. https://doi.org/10.1007/978-3-031-20443-2_3.

Pasqualino, R., Demartini, M., & Bagheri, F. (2021). Digital transformation and sustainable oriented innovation: A system transition model for socio-economic scenario analysis. *Sustainability, 13*(21), 11564.

Prakash, C., Saini, R., & Sharma, R. (2021). Role of internet of things (IoT) in sustaining disruptive businesses. In *Internet of things and businesses in a disruptive economy.*

Qvist-Sørensen, P. (2020). Applying IIoT and AI–Opportunities, requirements and challenges for industrial machine and equipment manufacturers to expand their services. *Central European Business Review, 9*(2), 46–77.

Rakha, N. A. (2023). Artificial intelligence and sustainability. *International Journal of Cyber Law, 1*(3).

Rana, G., & Sharma, R. (2017). Organizational culture as a moderator of the human capital creation-effectiveness. *Global HRM Review, 7*(5), 31–37.

Rejeb, A., Zailani, S., Rejeb, K., Treiblmaier, H., & Keogh, J. G. (2022). Modeling enablers for blockchain adoption in the circular economy. *Sustainable Futures, 4*, 100095.

Robnik-Šikonja, M. (2023). AI-driven predictive maintenance for autonomous vehicle sensor systems. *Journal of Bioinformatics and Artificial Intelligence, 3*(2), 119–137.

Rusch, M., Schöggl, J. P., & Baumgartner, R. J. (2023). Application of digital technologies for sustainable product management in a circular economy: A review. *Business Strategy and the Environment, 32*(3), 1159–1174.

Nagpal, S., Athavale, V. A., Saini, A. K., & Sharma, R. (2020). Indian health care system is ready to fight against COVID-19 a machine learning tool for forecast the number of beds. 2020 Sixth International Conference on Parallel, Distributed and Grid Computing (PDGC), Waknaghat, India, pp. 61–65. https://doi.org/10.1109/PDGC50313.2020.9315825

Sahoo, S., & Lo, C. Y. (2022). Smart manufacturing powered by recent technological advancements: A review. *Journal of Manufacturing Systems, 64*, 236–250.

Santana, S., & Ribeiro, A. (2022). Traceability models and traceability systems to accelerate the transition to a Circular Economy: A systematic review. *Sustainability, 14*(9), 5469.

Sarkar, B., Debnath, A., Chiu, A. S., & Ahmed, W. (2022). Circular economy-driven two-stage supply chain management for nullifying waste. *Journal of Cleaner Production, 339*, 130513.

Sharma, R., Harish, V., & Rana, G. (2024). Navigating Risk In The Age Of Artificial Intelligence: Assessing And Identifying Risks With AI Strategies. *Educational Administration: Theory and Practice, 30*(4), 2130–2140.

Sharma, R., Rana, G., & Agarwal, S. (2021). 11 Techno innovative tools. *Employer Branding for Competitive Advantage: Models and Implementation Strategies*, 171.

Sharma, R., Saini, A. K., & Rana, G. (2021). Big data analytics and businesses in Industry 4.0. *Design Engineering, 2*, 238–252.

Shojaei, A., Ketabi, R., Razkenari, M., Hakim, H., & Wang, J. (2021). Enabling a circular economy in the built environment sector through blockchain technology. *Journal of Cleaner Production, 294*, 126352.

Steenmans, K., Taylor, P., & Steenmans, I. (2021, December). Regulatory opportunities and challenges for blockchain adoption for circular economies. In *2021 IEEE international conference on Blockchain (Blockchain)* (pp. 572–577). IEEE.

Suchek, N., Fernandes, C. I., Kraus, S., Filser, M., & Sjögrén, H. (2021). Innovation and the circular economy: A systematic literature review. *Business Strategy and the Environment, 30*(8), 3686–3702.

Trejo Beltran, D. Y. (2023). *Barriers and drivers to adoption of IoT-enhanced smart packaging in the food supply chain.*

Trevisan, A. H., Lobo, A., Guzzo, D., de Vasconcelos Gomes, L. A., & Mascarenhas, J. (2023). Barriers to employing digital technologies for a circular economy: A multi-level perspective. *Journal of Environmental Management, 332*, 117437.

Upadhyay, A., Mukhuty, S., Kumar, V., & Kazancoglu, Y. (2021). Blockchain technology and the circular economy: Implications for sustainability and social responsibility. *Journal of Cleaner Production, 293*, 126130.

Vazquez Melendez, E. I., Bergey, P., & Smith, B. (2024). Blockchain technology for supply chain provenance: Increasing supply chain efficiency and consumer trust. Supply Chain Management: *An International Journal.*

Voulgaridis, K., Lagkas, T., Angelopoulos, C. M., & Nikoletseas, S. E. (2022). IoT and digital circular economy: Principles, applications, and challenges. *Computer Networks, 219*, 109456.

Wilson, M., Paschen, J., & Pitt, L. (2022). The circular economy meets artificial intelligence (AI): Understanding the opportunities of AI for reverse logistics. *Management of Environmental Quality: An International Journal, 33*(1), 9–25.

Yu, K. H., Zhang, Y., Li, D., Montenegro-Marin, C. E., & Kumar, P. M. (2021). Environmental planning based on reduce, reuse, recycle and recover using artificial intelligence. *Environmental Impact Assessment Review, 86*, 106492.

Zhang, C., & Lu, Y. (2021). Study on artificial intelligence: The state of the art and future prospects. *Journal of Industrial Information Integration, 23*, 100224.

Green and Blue Economic Concept Toward Sustainable Society

Sumit Roy and Mirian Ngozi Alike

Abstract The blue and green are the colors that color the economy. The blue-green economy is now a subject of discussion as sustainability standards are now getting implemented in most of the companies. Each nation has a target to fulfill certain environmental objectives which align with UNEP and ultimately it helps all of us. With data analytics now in the forefront, the economy of our environment is under the subject of analysis, though many of its attributes are beyond economics. Blue-green economy is not just an economic evaluation of our resources, rather it is philosophy of our being amidst bountiful resources that nature has bestowed us. In the ancient India, the resources were managed well and the Vedic way of living emphasized contentment and avoid violence and stay in harmony with nature. This research article delves into the value of our nature and strives to make the readers aware and get sensitive toward our nature. Nature is our source and the source is viewed as resource and when it is overused and abused, it effects our blue-green economy. Nature, resource, and sustainability have an additive effect toward our society. It is how we measure and value the measure that can define our behavior toward our natural sources.

Keywords Blue-green economy · Consumption · Industry · Planet · Sustainability

1 Blue and Green as Nature's Canvas

One argument is blue-green economic philosophy, which is gaining momentum, rests on the primary tenet that nature has an intrinsic color and the color seems to get faded. Human influence has been maximized during the past century and it shows no signs of abatement. Our interaction with the planet earth is innate and we are in

S. Roy (✉)
Balaji Institute of Marketing Management, Sri Balaji University, Pune, India
e-mail: Sumitroyg212@gmail.com

M. N. Alike
Department of Philosophy, Nnamdi Azikiwe University Awka, Awka, Nigeria
e-mail: mn.alike@unizik.edu.ng

R. Kumar and S. N. Mohanty (eds.), *Sustainable Economy Models in the Age of Industry 5.0*, https://doi.org/10.1007/978-981-96-4104-8_5

continuity to preserve this this innateness. Our values, beliefs, and behaviors drive us and also leave a human signature on the Earth. The anthropomorphic activity and a single species domination raise an ethical question: As a species, how long we can take the Earth for granted? The human species is just one among the several lakh species found on this Earth, and we as a species are found to be the most intelligent and most disturbing to the Earth. Civilization progressed because of taming nature and making nature to adapt the rising population. But this is something that cannot continue unabated as global climate change is showing enough symptoms of alert. The lush green forests are being targeted for agricultural produce and excessive emissions from the industries are greying our blue sky. The blue economy comprises our oceans and its reserves like fish, corals, weeds, minerals, renewable energy, maritime transport, and many more. Overfishing through trawling at deep seas level, waste dumping, using detonators for fishing are deteriorating our blue economy. Humans collectively at a planetary level function as a sloth species without consciousness of the environment we live in. The planet Earth is only one and we are not sure whether life exists on other planets or not, our planet Earth is sacred and holds divine power that embraces everything to sustain life. Our planet has a dharma of its own which reveres and generates life. This dharma in the present context is negated and taken for granted. Humans approach toward our planet has to go through a paradigm change and that would be a radical transformation in human morality. Morality is compassion or the control of senses, indriyanigraha. A society based on consumption and a hedonistic way of living is a deviation from rta, a principle that maintains order at a planetary level. Manu the law giver is given samanya dharma (general principles) and they are ahimsa, asteya, satya, saucha, and indriyaninigraha (Tiwari, 2017). These general principles alone have the potential to safeguard our blue-green economy if implemented. Nonviolence, non-stealing, truthfulness, cleanliness, and control of our sensory apparatus have become an urgency to abide by. Our opportunities in the blue-green economy are like a panorama of opportunities. Figure 1 shows The Panorama of Blue-Green Opportunities.

1.1 The Liquid Horizon and the Blueness of Our Earth

'Fluidity' is the quality of liquids and gases. What distinguishes both of them from solids is that fluids travel easily. Liquid economy is the term I coin here which explains that our economy is always in a flow state. Liquid has its property which is flow very opposite to a state of static. Our static resources have flows in an intrinsic domain that is not visible. All matters are building blocks of atoms which are further composed of subatomic particles and they are always in motion. The ocean currents, winds, storms, sunlight, photosynthesis, pollination, and biogeochemical cycles always flow to keep the ecosystem intact. Our water coverage in our planet is around 71%, it is our ocean that makes our planet appear blue when seen from space. The oceans are teaming with life and as of now only we have explored just 5% of the ocean and the rest is unknown. Our ocean facilitates the economy through the rich availability

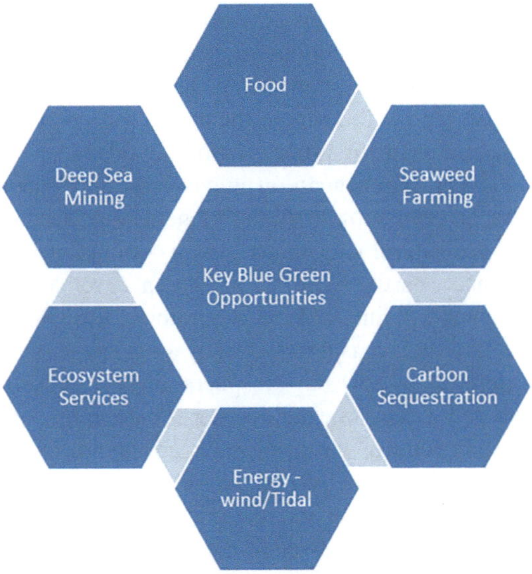

Fig. 1 The Panorama of Blue-Green Opportunities

of minerals, salt, fish, oil reserves, tidal waves for electricity generation, coastal tourism, and scenic and aesthetic beauty. The blueness of our planet has a majestic wonder and its economic and sustenance importance is unfathomable. The horizon of the sea beckons to the wonder of our mother Earth and it fills us with awe that nature is so vast and is an abundant supply of resources. The flip side of the wonder is taking the vastness for granted. Millions of tons of plastic end up in the ocean each year. Plastics are enduring and have limited degradation, and their processing through weathering and fragmentation results in the formation of micro plastics, which can cause more adverse effects on the health of marine organisms (including inflammation, blockage of intestinal tracts, oxidative stress, reproductive impact, and hormone disruption). Our Earth is now well outside of the safe functioning space to dwell. Ocean acidification is close to being broken, while aerosol filling regionally exceeds the margin. Earth system modeling of different levels of the lapse of the climate and land system change boundaries exemplifies that these anthropogenic impacts on Earth system must be considered in a systemic concern. Our Earth has exceeded six out of nine planetary boundaries. (Katherine et al., 2023).

1.2 Living Sustainably in Our Biosphere

The concept of sustainable development has a history though it was used for the first time in the year 1987 in the Brundtland Report. It was Lester Brown who created an organization named World Watch Institute where he did research on the said theme and published a report in the year 1984 named State of the World Report. This report was a revelation of how the present economic model is impacting the biosphere. In the year 1987, the commission created a proposal where the concerns of the biosphere can be tackled under the name Our Common Future popularly known as 'Brundtland Report' and in this report, the manifestation of the word 'Sustainable Development' appears for the first time (Gadotti, 2008)

The present generation enjoys the fruits of the biosphere and this generation must have an obligation to the future generation in providing them with all the resources and the natural laws intact. So this obligation demands a framework of living. The biosphere is the realm of the green cover and the blue cover that forest and the ocean as a maintainer and generator of life which have its intrinsic intelligence and this intelligence is interrupted by anthropogenic activities. Getting to understand the difference in how our biosphere and our industry function can give us an insight into how environmental disaster is caused by a mechanized way of orientation. The mechanized way has given its value and now what is demanded is to change the way we live. This living is not just for here and now but for the other species and for our generation to come and that harks for living sustainably. Nature gives us value and these values are utilized at the maximum but nature has its way of replenishing it. But with the present speed of extraction, anthropogenic activities are deteriorating the very capability of nature's replenishment. Value is being created from nature and human's device ingenious ways to extract value resources. 'Value chain' is a word coined by Michael Porter in his book Competitive Advantage in the year 1985. Considering the 'Value chain' from resource extraction and its usage like the way ores are being processed to get minerals, metals, and finished products, a lot gets wasted. Insead Business School professor Robert Ayres once calculated that over 95% of all resources extracted from the environment become waste within six months from harvest (Gregory, 2010).

Now that is a colossal loss of resources that is not valued. 'Value chain' as a concept is to be understood from a biosphere perspective. 'Value chain' as a concept and practice is hammered to the industry is a narrow way of seeing and fueling the economy. What is required is a value cycle. Nature's function is cyclical. Creating a value cycle can be a challenge. Marketing in the industrial area is focused on way things are and what is seen lies in the near proximity. Humans have evolved to survive and look for stability. The resource that lies within the proximity gets tapped. Industrialization demands the generation of raw resources. As man has developed, it acquired tools and technology to harness nature. The biosphere is considered an endless resource repository. This is a mind-set that is dominating. Our industries are based on a linear model that is extract, make, process, use, and waste. This is a linear model where the efficiency of a firm is judged through financial output that is in terms of revenue generated. Extraction and exploitation and making big is the mantra of the industries and the same is grooved in the minds of the managers and the same is trained to the management graduates and they repeat the same what their predecessor has done. This has become a pattern and breaking the pattern is becoming difficult.

The biosphere as a whole has a repository of resources that are made available not only for humans but also for all kinds of beings that are present on this planet. Industrialization, though has given immense convenience to humanity, has come with huge side effects. Reflecting and striving to understand the state of affairs is the demand of the time. The Greek philosophers were trying to understand the way nature is and striving to make society mature and be reflective in our actions. Plato exhorted that nature had a universal power that produces maximal diversity and abundance in the natural world and regulates changes by the uniform law. Socrates developed a method of inquiry where one can get to know the why of things, Aristotle insisted on virtue. Delving into the state of affairs of the industries would reveal the linear model of industrialization production and revenue-oriented and it fails to see the diversity of the biosphere where many beings are dependent. This linear model has taken traction and its momentum is now unstoppable. It is observed now that many institutions and forums give importance to entrepreneurial ventures. More entrepreneurial ventures, more is industrialization and the dictum that is prevailing is to fight poverty, entrepreneurship is the solution. It is not so. More ventures can create more disparity and at the far end, it is harmful to the biosphere. Industries are manned by the engineers. Engineers design and produce products that ultimately come to the market for usage and they have a moral responsibility to reflect and not simply accept the orders and tread the path that is going on. Engineers man the industries and it is the collective duty of the engineers, the industrialist, and the common man to make a conducive livelihood for a sustainable world. Our biosphere is engineered intelligently and is the product of nature's evolution and engineers in the industry can be the stewards of the biosphere, can engineers align the production process and not just accept the conventions of take, make, use, and waste. The present linear model of the industries has to be reconsidered. Our source that is nature is the product of infinite ingenuity and evolution. Engineers and Managers can be more responsible in their doing until they know what they are ruining. What comprises sustainable living? Let's get the habitable area on our Earth. 10% of the world is roofed by glaciers, and an additional 19% is barren land—deserts, dry salt areas, beaches, dunes, and exposed rocks. Forests account for a little over one-third of the habitable land area. This is around one-quarter of the total (both habitable and uninhabitable) land area (World in Data). We have destroyed one-third of the world's forest for the expansion of agricultural land (World in Data). As the population increases, the burden of feeding the population falls on the planet. Living sustainably is a collective inner call. We developing an appreciating eye to revere the blue and green shades of our nature is an imperative. Our actions and attitudes have a bearing on how long we can endure the greenness and blueness of our planet.

1.3 Green Belief

A civilization's well-being and its long-term survival depend on the beliefs that the civilization develops. It is the belief system that makes the living systems work. Our beliefs are driven by our interests and our interests at a primal level stem from our instinct which is for survival and comfort. Civilizations became a possibility because of agriculture and before agriculture it was hunting and fishing. Agriculture is a discovery and our biosphere made available every possible resource to make agriculture a success. Man's dependency and understanding the agricultural principles are the prime ingenuity that made humanity stable and propagate. This simple finding and applying it further for human feed is one of the greatest ingenious activities of humans. Green belief is the belief that the ecological cycles of nature make possible living through the continuous production of food, oxygen, water, and raw materials for our very existence.

As the population swells, feeding the nations becomes an imperative. Conventional and native methods of food production were questioned as to whether they were effective with the increase in the mass of people. The dichotomy between intrinsic and instrumental value comes into play as the situation commands us to take a stand. Here the intrinsic dimension of seeds and the crop cultivation method gets questioned. Science and technology urge to play almighty and take charge of which has evolved over a billion years. The primal find of the green belief got industrialized into the 'Green Revolution' where the name signifies creating abundance through science in the field of agriculture. The tenet of the Green Revolution is that technology is loftier to nature and nature can be manipulated to harness more and prevent the prevalent food crisis. Viewing nature as unable to supply the increasing demands of the rising population and giving a high hand to technology is paving the path toward ecological demolition. Nature is abundant and regenerative, but getting a tool to increase the yield of a seed has some unknown consequences.

A short-term high yield can play havoc with the agricultural ecology. Sustainability is embedded in the intrinsic value of the entities of nature that are invisible. By being invisible, the visible dimensions are taken care of. It is like the Tao principle where nothing needs to be done but nothing remains undone. Sustainable agriculture is based on the soil which has an inherent capacity to recycle itself through the nutrients that it harbors through humus, which is dead leaves, manure from the animals, the soil's ecosystem, earthworms, and millions of bacteria that add life to the soil. The Green Revolution model replaced the natural nutrient cycle with a reductionist and a linear view to increase production and make the agricultural discipline work with the industrial discipline which is entirely two different entities. The Green Revolution had an unintended consequence. The agricultural ecology is based on its own evolved intelligence. The Green Revolution is an external inducement that disregards the intelligence of nature and aims to use chemicals and technological methods to make the discipline of agriculture a commodity that can be marketed. Marketing of agricultural products follows mass production through unnatural means. Is this moral? Working against the intelligence of nature has a consequence. What that consequence can be that time would reveal in due time. What can be imbibed is the proper belief that can sustain the population through sustainable practices. Green belief can be a tool to walk the path of sustainability. Sustainable practices require accountability and the genesis of the sense of accountability is the respect for the Earth.

1.4 When Belief Becomes Economic Value

What can be the value of our blue-green economy? The very belief of our blue-green planet is that our productivity and survival depend upon the produce of our planet. The ecological cycles and the invisible work of our planet through biogeochemical cycles are invisible but the relentless work of nature is always on. The blue and the green color signifies the productive aspect of our planet and if the color fades, the possibility of us fading out can be stark. Humans understand the language of what is there for them. One way of getting the value of our natural resources is to take a holistic approach. The parameters are:

For a tropical forest, the Direct value includes Timber, non-timber, medicine, recreation, Botanical study, and habitat and the Indirect value includes nutrient cycling, watershed protection, air pollution mitigation, and microclimate. Option value is the combination of direct and indirect and Existence value is the very survival of several species that we are not even aware of and also the cultural and heritage value. Bequest value is the future value that is about the generation that is going to dwell and the way they are going to use natural resources.

Total Economic Value = Use Values + Non-Use Values
Use Values = Direct Use Values + Indirect Use Values + Option Values
Non-Use Values = Bequest Value + Existence Value

1.5 How Much Is Nature Worth Today?

A study by Costanza and colleagues estimated the global loss of ecosystem services due to land use changes between 1997 and 2011 to be an astounding US$4.3 trillion to US$20.2 trillion annually (Costanza, 2014). In the EU Biodiversity Strategy for 2030, it is indicated that more than half of the global gross domestic product—around 40 trillion euros—depends on nature. However, this value underestimates both the source value of nature and values as cultural values that are not immediately monetizable (Statzu, 2022). The World Bank estimates that the global economy could lose $US 2.7 trillion by 2030 (compared to business as usual) if certain ecosystem services collapse (pollination, carbon sequestration and storage, fisheries and timber provision) (World Bank Report, 2022). System of Environmental and Economic Accounts (SEEA) is an international statistical standard like the System of National Accounts (SNA). The SEEA framework has been endorsed by the UN Statistical Commission, which is a body of heads of statistical offices from all countries and international organizations like Eurostat, the IMF, OECD, UN, and the World Bank. Over time, countries are expected to implement statistical standards using SEEA, as they have done with the SNA. SEEA Framework—

Asset Accounts—Natural resources like the stock of water, land, forest, fish, and minerals in monetary terms.

Physical flow accounts—use of energy, water, material, and emissions by economic sector.
Monetary accounts—Environmental protection expenditure, Environment taxes, Natural resource management.

1.6 Sustainable Development vs. Sustainable Environment

Environment Marketing as a business activity is concerned with the endurance of the organization and it finds its ways for growth and development of the organization. Entrepreneurs and marketers strive to make their organizations sustainable. As seen, sustainable development is a broader context and it refers to the resources that can be harnessed for future generations and at the same time keep consuming and reaping the fruits of scientific development. Organizational goal is limited in a spatio-temporal context and the agents of the organizations can think in a given period which is marked by goals, targets, and budgets. Too much involvement in the pursuit of revenue and profits makes the managers devise strategies to achieve more and more as if to sustain is to chase profits and generate revenues. Organizations' existence is necessary as they are the vehicle for economic growth and facilitating the generation of goods and services for mankind and what gets negated is the holistic domain where the business operates. It is the part in the whole that exists, makes the part meaningful, and if the whole is collapsing because of the behavior of the part, then the concern for the behavior is to be looked upon. One of Hegel's basic conceptions is that of the relation of the whole to its parts. Any part of a whole is what it is because of its relation to the system as a whole and the other parts (Kelly, 1963).

An organism can function effectively because the parts of the organism are in tune with the whole organism. Parts can be sustained only when the whole is healthy. Every truth or fact is dependent upon and helps in its turn to determine every other truth or fact. This is known as organic theory of truth and reality. Sustainable development is much discussed across the organization's leaders and this is visible through various conferences and workshops that are held as an event but the focus remains on the sustainability of the business rather than the environment and if it is discussed, it gets a scant attention and very little implementation. One of the fundamental error is that sustainable resource is that the planet Earth is viewed as a reserve that is endless and this is in abundance. It seems that industries and humanity can take as much from the available abundance and the paradigm that is held is that the abundance is going to stay. Sustainable development is embedded in the environment, if the environment is viewed as a steady supply of resources and every tool and technology is used to harness it to the maximum, then the whole that is the environment will be depleted. Matter and energy have their pattern and it circulates dissipates and regenerates in a given condition. Regulation of matter and energy is an intelligent process carried out by nature. This very intelligent process is disrupted by humans. It is a paradox that as a species, humans are the most intelligent among millions of species that exist in the planet Earth and this intelligence is failing to grasp the intelligence of nature and if it is understood, it is among the scientific community and environmental activist. So the intriguing question is what prevents the majority of humanity to come to terms with the reality of the workings of the environment? How many environmental awareness forums and educational programs are failing to grasp some of the fundamental issues of the climate crisis and resource diminution? It may be that man by very nature is an unsustainable species and underdeveloped in the context of overarching intelligence that governs life. Sustainable development can be a possibility only in a sustainable environment though this seems obvious and radical unfortunately, this concept gets negated and man loves to dwell in the realm of being unconscious. Seeing the present state of affairs it seems Homo sapiens as a species may be a work in progress. Progress may not descend on man abruptly but it may gradually evolve as a dawn of truth through man facilitating the progress of its kind by finding an extension of him who is far superior. The same is evident through digitization and technology that is ubiquitous in its application.

1.7 Indian Philosophic Tradition and the Blue God

Lord Krishna is depicted in blue color and we revere our most beloved God. Our nature is mostly blue. The very development of humans is based on the use of ocean content, Can we revere our oceans as our God? How life originated on our Earth is still a mystery, life mostly originated from water and the photosynthesis activities made our Earth abundant with oxygen which facilitated the propagation of many species on our Earth. The creation philosophy versus evolution science is the most debated topic. Our blue-green resources are the genesis of other life principles and that is a creation tenet in this perspective our nature is like the almighty and it demands reverence and not just exploitation. Garrett Hardin coined the word 'Tragedy of commons' which goes with what we are doing to the blue economy. The five oceans are our 'common' resource. We are in an extracting mode from the oceans. The Aids accumulated by people, businesses, or governments from the oceans are magnanimous. This leads to overconsumption—the shared cost of which could reach $428 billion annually by 2050 (World Economic Forum, 2020).

One of the prominent religions of the East with a concern for the biosphere is Jainism. Jainism focuses on conquering self and seeing life everywhere, so why to be violent on our nature that is life based. Our action on nature accumulates karma which leads to the sufferings of mankind. Again, this is a matter of self-manifestation. Any harm imposed on nature will have a boomerang consequence on humanity. Education has to get back to the concept of karma through the experiential mode of learning. Once we learn by being in nature and through nature, then we are ontologically grounded which can raise our consciousness for the planet. Man relying on the concept of karma and treating nature as sacred is a moral and this morality can ascend man and prevent him from harming nature. Our present education can draw ample metaphors, symbols, and ancient practices to inculcate in the present curriculum and these can develop intuitive learning which modern day pedagogy and curriculum lack. The dangers of ocean content are alarming disappearance of coral reefs in certain ocean regions. The experiential component in our education like seeing the mangroves and coral reefs can create a visceral effect. A seamless connection at our attitudinal level with the planet, biosphere, values, and behavior can alter the ways to make our home planet livable at a robust level.

2 Sustainable Biosphere

The ethic of the Earth is a compelling demand and one of the demands is remaining sustainable. The definition of sustainability includes a responsibility toward the generations to come. The present generation enjoys the fruits of the biosphere and this generation must have an obligation to the future generation in providing them with all the resources and the natural laws intact. So this obligation demands a framework of living. The question may arise that it is one life and I should live the way I want and why should I bother. The attitude behind this kind of living may come from that nature is bountiful and corrects itself so let me be a *bhogi* as *charvaka* philosophy endorses life is for pleasure and everything is sensory based. But a deeper peep makes the concrete reality get manifested in the form of some evidence like the Earth's temperature is rising. Sea levels are rising and it is estimated a rise of 3–6 feet in another seventy years. It is dreadful even to imagine the displacement of such a large number of people. The above picture only project what can happen in the future and also in the present times, we are experiencing some of the onslaught from nature. Nature is resilient and also fragile. It is resilient as it can take or absorb the domination and beyond a point, nature snaps and shows its *tandava*. Humans look for order and peace. The species Homo sapiens wants to progress and grow and at the same time be in harmony. It is a puzzle that some of the behavioral dynamics create a biosphere turbulence in the future. So what can be done? Can we think of a value-driven living? And what can that be? Man left free can do and would do whatever he feels and thinks like. This feeling and the thinking as seen at times flawed, as human behavior gets generated from the foundation of his thinking and feeling. So at this context, a spontaneous question that comes to me is what about having a framework or a blueprint for sustainability? The hazards of the biosphere are being experienced and the after-effects on humans can be very perilous and the sufferings involved would be unbearable. So a Sustainability framework based upon restrained behavior can be beneficial where education can play a key role. It is being found that the Different sustainability pathways, such as 'green and blue economy', 'de-growth', 'Earth stewardship', 'nature protection', and alternatives arising from diverse world-views and knowledge systems, including 'Living well' and other philosophies of good living have varying prioritizations of specific values (instrumental, relational, and intrinsic) (Pascual, 2023). The use values in many cases are getting quantified and the non-use values are intangible and difficult to quantify. The process of appraising the worth of nature is complex. Formal methodologies to valuation in policy and result making involve understanding the relationship between costs and benefits of a projected change. Blue-green economy has monetary and non-monetary dimensions.

The Limitation of the study: Nature's work is very intricate and its ecosystem is still under study. Many ecosystem functions like pollination by bees, photosynthesis, carbon sinks, and many more have a tremendous value and it cannot be comprehended by any economic model nevertheless its value remains and humans require to intuit and revere the value.

3 Conclusions

The terrestrial arena is already invaded by urbanization and the race toward industrialization has depleted our forest areas; now economy and long-term sustainability are badly affected though measures are being taken to safeguard our natural resources. The blue economy which comprises the oceans is not simply a place for the withdrawal of resources as if the resources are infinite and an infinite space where dumping of waste products can be done time immemorial. The ocean values its services in economic modeling and decision-making processes and demands further research and its value to be disseminated. The blue and green economy paradigm constitutes a sustainable development framework for all nations that can help to address issues such as equity, rational usage, habitat protection, and development. The overall measure of our natural economy is unfathomable but what can be counted has to be made aware to people as data speaks and people can develop seriousness on the output and the sustenance of its own being.

Funding Statement No funding is being received from anywhere for this Research work.

Conflict of Interest There is no conflict of interest with any parties.

References

Costanza, R., de Groot, R., Sutton, P., van der Ploeg, S., Anderson, S. J., Kubiszewski, I., Farber, S., & Turner, R. K. (2014). Changes in the global value of ecosystem services. *Global Environmental Change (Human and Policy Dimensions), 26*, 152–158. https://doi.org/10.1016/j.gloenvcha.2014.04.002

Gadotti, M. (2008). What we need to learn to save the planet. *Journal of Education for Sustainable Development, 2*(1), 21–30.

Gregory, U. (2010). *Using nature's rules to build sustainable profits*. Harvard Business Press.

Kelly, W. (1963). *A history of modern philosophy*. The Macmillan Company.

Pascual, U., Balvanera, P., Anderson, C. B., et al. (2023). Diverse values of nature for sustainability. *Nature, 620*, 813–823. https://doi.org/10.1038/s41586-023-06406-9

Richardson, K., et al. (2023). Earth beyond six of nine planetary boundaries. *Science Advances, 9*(37), eadh2458. https://doi.org/10.1126/sciadv.adh2458

Tiwari, N. K. (2017). *Classical Indian ethical thought: A philosophical study of Hindu, Jaina and Buddha morals* (2nd ed.). Motilal Banarsidass Publishers.

Toward a Blue Economy: A Promise for Sustainable Growth in the Caribbean, World Bank Group. (2016). *United Nations 2012.* https://www.undp.org/sdg-accelerator/background-goals
World Bank Report. (2022). Finance for an equitable recovery, International Bank for reconstruction and Development. The World Bank, 2022.https://doi.org/10.1596/978-1-4648-1730-4

Embracing Sustainability in Supply Chain Management in the Era of Digital Industrial Revolution 5.0

Ram Singh, Vinay Pal Singh, Rohit Bansal, Vandana Madaan, Kamal Ahmad, and Fazla Rabby

Abstract With the introduction of the Digital Industrial Revolution, we are witnessing a significant transformation in the field of supply chain management. This transformation has significantly transformed supply chain management, utilizing advanced technologies like artificial intelligence, data analytics, IoT, big data, business intelligence, robotics, and blockchain. This has led to increased efficiency and agility in supply chain operations. Embracing sustainability in SCM ensures environmental concerns are addressed and offers operational and economic benefits. This study identifies strategies to integrate sustainability into supply chain processes, enhancing efficiency, resilience, and environmental stewardship. In this era of Industry 5.0, new technology can offer ways to integrate sustainability into SCM. By leveraging these technologies, organizations can strengthen efficiency, reduce emissions, and can also reduce waste while promoting sustainable sourcing which would not only help in embracing sustainability but will also provide a competitive edge to the organizations. Industry 5.0 aims to adopt environmentally responsible and human-centric practices within ethical boundaries, aligning technological advances with sustainability. Challenges faced by organizations in integrating sustainability in supply chain processes are highlighted. Case studies and real-world applications demonstrate how organizations can encourage sustainability in their supply chain processes, maximizing resource utilization, reducing carbon

R. Singh (✉)
Symbiosis University of Applied Sciences, Indore, India
e-mail: ramsinghcommerce@gmail.com

V. P. Singh
Quantum University, Roorkee, Uttarakhand, India

R. Bansal
Vaish College of Engineering, Rohtak, India

V. Madaan
Maharishi Markandeshwar (Deemed University), Mullana-Ambala, India

K. Ahmad
HNB Garhwal University, Srinagar, Garhwal, Uttarakhand, India

F. Rabby
Stanford Institute of Management & Technology, Sydney, NSW, Australia

R. Kumar and S. N. Mohanty (eds.), *Sustainable Economy Models in the Age of Industry 5.0*, https://doi.org/10.1007/978-981-96-4104-8_6

footprints, and minimizing waste. The study also underlines the evolution of SCM and provides recommendations for businesses to navigate this era while adhering to their environmental and social responsibilities.

Keywords Sustainability · Supply Chain Management · Industry 5.0 · Digital Industrial Revolution · Integration

1 Introduction

AI-driven technologies have been introduced as a result of Industry 5.0 that help in the optimization of supply chain processes and foster sustainability, for instance, IoT helps in real-time tracking of the goods while ensuring that transparency is maintained in the processes and the work is carried out efficiently with minimum wastage. Industry 5.0 emphasizes the adoption of agility for making supply chains flexible which is possible via decentralized production models. Decentralized production improves resilience and supports sustainability by reducing environmental impact.

Artificial intelligence can also help in better decision-making and would improve efficiency, thereby ensuring supply chain sustainability (Dubey et al., 2019). AI-driven technologies can also help in the optimization of routes for transportation and also help to navigate the environmental conditions with the use of IoT devices to ensure environmental stewardship and maintenance of sustainability standards. While Industry 4.0 focused on technological advancements, Industry 5.0 adds environmental and social considerations by adopting a human-centric approach through the adoption of COBOTS, i.e. collaborative robots that work along with humans and support them instead of replacing them.

Organizations today are implementing digital supply chain management systems by using AI-driven technologies, analytics, and IoT sensors to carry out their operations and to monitor the processes. In this era, businesses are shifting their focus toward sustainability and are recognizing their responsibility toward the environment by integrating sustainable practices in the supply chain that minimize emissions and optimize resource utilization. The integration of blockchain technology can also lead to better transparency. Blockchain can help the companies in sustainable sourcing of materials and comply with sustainability standards.

According to Ivanov et al. (2015), the introduction of Industry 5.0 can enhance sustainability in the supply chain processes by ensuring better utilization of resources and big data analytics can help in forecasting the demand effectively. AI-driven technologies can help the companies in optimization of inventory processes and provide accuracy along with transparency. These technologies also provide greater visibility, thereby ensuring that the standards of sustainability are adequately met. The integration of technology within the supply chain can also help in reducing the cost that is associated with operations at the same time it can also improve adaptability and efficiency.

However, several challenges are related to integrating sustainability in supply chain processes in the context of the Digital Industrial Revolution; for example, as discussed by Pagell and Shevchenko (2014), the supply chain involves various complexities that can bring obstacles to integrating sustainability in supply chain processes. There is a huge cost connected to integrating sustainable practices via technology, especially for small and medium-sized businesses.

2 Literature Review

Raz (2009) emphasizes the organization's recent attempts to enhance internal procedures and processes in response to these anticipated difficulties. To maintain their competitive advantage and increase their supply chain surplus, firms need to implement and embrace the concept of SCM. It has been highlighted that there are still opportunities available in the digital supply chain to explore. The opportunities can pave the way to move toward sustainability as in the era of digitalization using technology can enhance and help in sustainable development. The study highlighted the challenges that are involved in integrating sustainability in SCM to create a term called sustainable supply chain, he also highlighted the new technology and how it can help in achieving the standards of sustainability.

Zhong et al. (2016) concluded that new technology particularly AI-driven technology and big data helps in ensuring faster and more accurate data collection, and they are the major sources of moving toward sustainability. Integration of this term in supply chain is being increasing at a fast pace as people are becoming more conscious about sustainable development. It emphasized the crucial role that blockchain and AI-driven technologies play in reducing negative environmental effects and achieving the UN's sustainable development objectives. This research stressed how these technologies may assist in digitizing the supply chain and how they can help us achieve sustainability and efficiency. The concept of simulation in operations research has been increasingly used in supply chain and the idea to integrate it with the supply chain has been taken from the study of Okada et al. (2015) because it offers the advantage of outlining potential future scenarios and their impact on both the strategic and operational levels (production, warehousing, etc.) of an organization (such as supply chain). Valverde and Saadé (2015) also, draw attention to the technological advancements that have occurred in the field of digitization and the ways that trends in digital technologies have affected supply chain management. The authors address the trend toward product dematerialization and its implications for supply chain organizations in their general introduction. The change that will take place in the supply chain due to integration with technology will be fueled by the rise in digitalization over the coming decades. The growing demand for digitization and supply chain visibility is compelling organizations to integrate ethical supply chain management into their company's operating model and strategy. It has been emphasized how the new technology in the food chain, particularly in the context of agriculture, can bring sustainable development and how these technologies can be used in the supply chain.

The need for future research has also been highlighted in their study. Kumar et al. (2016) emphasized the support for interconnectivity by describing how integrative technologies like big data and the Internet have led to significant changes in supply chain operations. This study explained how the introduction of these technologies in the supply chain will blend normal production with customer and environmental centricity.

3 Sustainability in Supply Chain Management

The concept of sustainability has become important in supply chain management for businesses today as a result of the growing environmental concerns and a step of moving ahead toward environmental stewardship. Since global concerns and environmental issues like environmental degradation, pollution, and climate change are rising, the need to integrate sustainability in supply chain processes is imperative in today's era. In addition to legal obligations to adhere to sustainability standards, consumer knowledge and desire for sustainable goods and services also drive the trend toward sustainability.

Industry 5.0 has a focus on a human-centric approach and it prioritizes working within an ethical framework and ensuring workplace safety, fair labor practices, and upskilling the workers. This is because a skilled workforce works toward a more resilient supply chain and can easily grab the opportunities for environmental protection. Sustainable supply chain management requires considering the environment, economic, and social impacts of supply chain processes and seeking strategies to minimize adverse impacts while achieving positive outcomes. One of the key components of embracing sustainability in SCM is green procurement, which involves sourcing materials that have minimal negative impact on the environment. This includes promoting the use of recycled materials, selecting suppliers who follow environmentally friendly practices, and choosing eco-friendly products. Green procurement is a way that can reduce the environmental footprint of the supply chain and encourage suppliers to adopt sustainable practices (Walker & Brammer, 2009).

Other ways also include sustainable manufacturing, and eco-friendly logistics companies are now on their way to explore transportation options like rail and sea transport which have lower emissions when we compare them to air and road transport. It has been highlighted that efficient logistics help to minimize the adverse effects on the environment and can also benefit the company in terms of cost savings.

3.1 *The Supply Chain Incorporated in Industry 5.0, SCM May Be Digitally Replicated Using 5.0*

Large-scale supply chain management modifications are made possible by Industry 5.0 technologies like Virtual twins, 5G, Machine Learning, Internet of Things, and Edge Computing, etc. which can help enterprises match demand and supplies more quickly. These technologies can produce a digital version of the supply chain, featuring assets, storage areas, inventory locations, and logistics. Collaboration, interpersonal interaction, automation, data analytics analysis, and identification, etc. are all emphasized in Dmitry Ivanov's Industry 5.0 theme. The promise of digital change in SCM is discussed by Varriale et al. (2023), who also stress the need for additional conceptual and empirical research.

3.2 *Early Stages of Supply Chain 5.0*

To create more intelligent, robust, and environmentally friendly operations, Supply Chain 5.0 is the latest stage in SCM that integrates cutting-edge technologies like robots, augmented reality (AR), blockchain, artificial intelligence (AI), and IoT. Early on, Supply Chain 5.0 seeks to improve automation, connection, and transparency to revolutionize conventional paradigms. IoT devices offer real-time visibility and monitoring of items, forming interconnected supply chains, while AI-driven analytics and robotics improve inventory management, shipping, and warehousing. Although widespread adoption is still in its early stages, blockchain offers increased transparency by creating unchangeable archives regarding product origins, quality, and ethical norms. Additionally, augmented reality is starting to help with maintenance and training by giving workers remote support and real-time data.

4 Technological Advancements in SCM that Drive Sustainability

The introduction of electronic technologies had a profound effect on the supply chain processes in various ways. Figure 1 shows Embracing Supply Chain Sustainability with Digital Technologies.

4.1 *Internet of Things (IoT)*

Studies have examined the positive connection between sustainability and IoT (Lopez-Vargas et al., 2020). The growth of the IoTs, which have many uses across

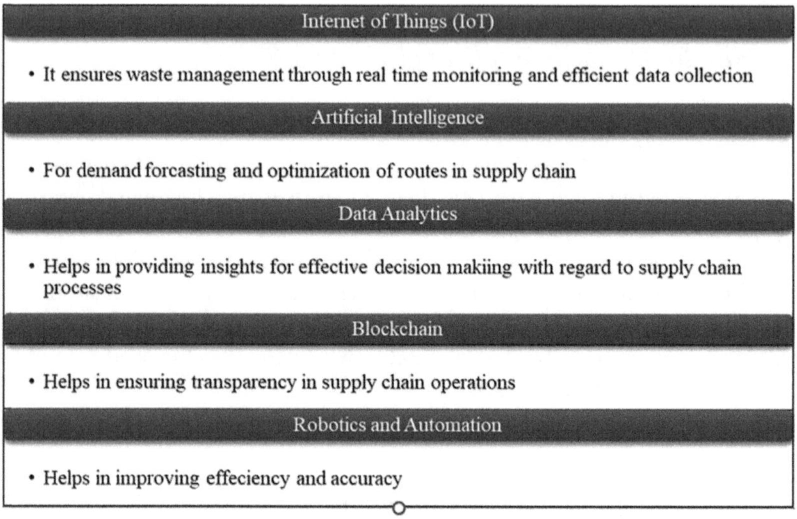

Fig. 1 Embracing Supply Chain Sustainability with Digital Technologies. *Source* Authors' own compilation

several industries, has shown to be quite beneficial (Abu Ghazaleh & Zabadi, 2020). "From minimizing water-related disasters as well as economic damages to conserving energy to better and bigger-scale connectivity, and more productive water management decision making," the Internet of Things can help with any environmental issue (Salam, 2020). High-tech methods for putting strategies and solutions required for sustainable development into practice were also mentioned, including digitization, ICT, IoT, the Internet of Business, the Internet of Energy, and the Internet of Manufacturing (Doost Mohammadian & Rezaie, 2020). The swift development of the Internet of Things is creating "a requirement for an updated definition of sustainability." In the fields of industrial ecology (Zhu & Zhao, 2018), ecologic engineering (Schneider, 2019), green engineering (Maksimovic, 2017), earth systems technology (Lee et al., 2014), and environmental sustainability, researchers have increasingly focused on developing innovative technologies and integrated approaches to address complex environmental challenges and promote sustainable development. IoT can assist in achieving the UN Sustainable Development Goals and the Paris Agreement's objectives (Salam, 2020). According to Laine (2014), networked gadgets and the Internet of Things were thought to be the main forces behind transformation. The corporate world is being forced by recent developments to add the word "globality" to its vocabulary. The below Table 1 shows various aspects of IoT in SCM.

Table 1 IoT implementation in SCM with a focus on sustainability

Sustainability	Description
Environment	Recovering and reusing materials, optimized waste management
Social	Better work, health, and safety condition for workers
Economic	Tracking products, automation of delivery and inventory process

Source Authors own work

4.2 Artificial Intelligence (AI)

AI is pivotal in improving the operations that are related to management and control of supply chain processes like the optimization of the supply chain operations that paves the way to embrace sustainability; for example, technological innovation, particularly the advancement of machine learning, helps in the forecasting of demand that helps the businesses in ensuring that overproduction and underproduction are avoided. Data analytics that are done via AI-driven technologies can also aid in the optimization of the logistics routes which would further help to cut fuel use and emissions. Technologies that are driven by artificial intelligence can also help in identification of the sustainable suppliers and sustainable sourcing techniques by analyzing data on their environmental and social practices.

By anticipating demand and arranging for the freshest products available, artificial intelligence (AI) can help with shelf restocking for perishable commodities and support environmentally friendly supply chains. To maximize the quality of the finished product, AI can also track important variables affecting the quality of the product at every stage of the supply chain. By finding the optimal supplier, AI applications can balance energy use and completion of projects time, leading to quicker and more environmentally friendly production. The sustainable characteristics of integrating AI into SCM are summed up in Table 2.

Table 2 Sustainable features of AI integration in Supply chain Management

Sustainability	Description
Environment	Optimization of resource consumption and reducing waste for perishable product
Social	Improved collaboration with supply chain partner and better planning and control of demand forecasting
Economic	Better planning and increased safety for the worker

Source Authors own work

Table 3 Sustainability of blockchain implementation in supply chain management

Sustainability	Description
Environment	Recycling programs, reducing perishable product waste
Social	Reduced time and cost for specific activities, automation of operation management
Economic	Tracking safety and fair practices

Source Authors own work

4.3 Blockchain

Blockchain technology helps to ensure that the supply chain processes are transparent and comply with the ethical and sustainability standards which are very important for verification of the sustainability of the products. By the use of blockchain technology, we can easily record every transaction on a nonchargeable ledger blockchain technology may ensure that products are sourced and transported responsibly at the same time in an ethical manner.

Blockchain technology prevents fraud and opportunistic conduct by establishing a cooperative marketplace for green producers. With tokenizer benefits for collectors, it may be applied to reverse logistics tasks such as recycling plastics. Order management is automated, delays are decreased, and contractual responsibilities are enforced using smart contracts. They lower the danger of bankruptcy, enhance credit opportunities, and boost cash flow. There are long-term advantages to using blockchain technology in supply chain management. The sustainable characteristics of integrating blockchain in SCM are summed up in Table 3.

4.4 Data Analytics

Big data analytics enable companies to make better and more accurate choices based on a comprehensive data analysis using several technologies. By examining data from various sources, businesses can find out the areas where they are inefficient and need to improve. They can also predict current and future trends, which would help in optimizing supply chain operations.

Big data can also help reduce waste by analyzing consumption patterns and by adjusting inventory levels. At the same time, it can also help provide information about customers' preferences for buying sustainable products and their willingness to adopt them. Big data analytics can also provide relevant guidance in product development and marketing strategies. Table 4 integrates sustainable aspects of data analytics in supply chain management.

Table 4 Use of data analytics in SCM in a sustainable manner

Sustainability	Description
Environment	Energy Efficiency and waste reduction
Social	Improved decision-making and enhanced innovation
Economic	Transparency and employee well-being

Source Authors own work

4.5 Advanced Robotics and Automation

Advanced robotics and automation can significantly help organizations in manufacturing efficiency and maintaining accuracy along with ensuring a substantial reduction in waste generation while optimizing resource usage. Automation helps in performing repetitive tasks in an accurate manner that in turn helps in minimizing errors and defects. Automation can also improve the recycling process by ensuring that materials are sourced and processed effectively Moreover, robots can operate in hazardous environments, protecting human workers and ensuring safety compliance. It can also be used for water conservation, sustainable cities, and protection of marine water (IFR International Federation of Robotics, 2024). These examples highlight the diverse ways that robots might help achieve the SDGs and demonstrate their revolutionary potential in several industries. Robots' involvement in sustainable growth will continue to develop as technology progresses, providing fresh approaches to pressing global issues. Table 5 integrates sustainable aspects of robotics in SCM.

5 Challenges of Sustainability in SCM Within the Framework of the Digital Revolution

Several challenges are to be faced while embracing sustainability in supply chain management, especially regarding the Digital Industrial Revolution. This is because as now new technology is emerging, we are integrating it at every stage of business, which also includes supply chain management. Five primary types of SCM difficulties were identified through synthesis, as illustrated in Fig. 2, even though several particular and detailed problems were also raised. Embracing sustainability in supply chain management during the Digital Industrial Revolution presents several

Table 5 Use of robotics in supply chain management in a sustainable manner

Sustainability	Description
Environmental	CO_2 emission reduction
Economic	Reducing labor and transport cost
Social	Enhancement of consumer satisfaction

Source Authors own work

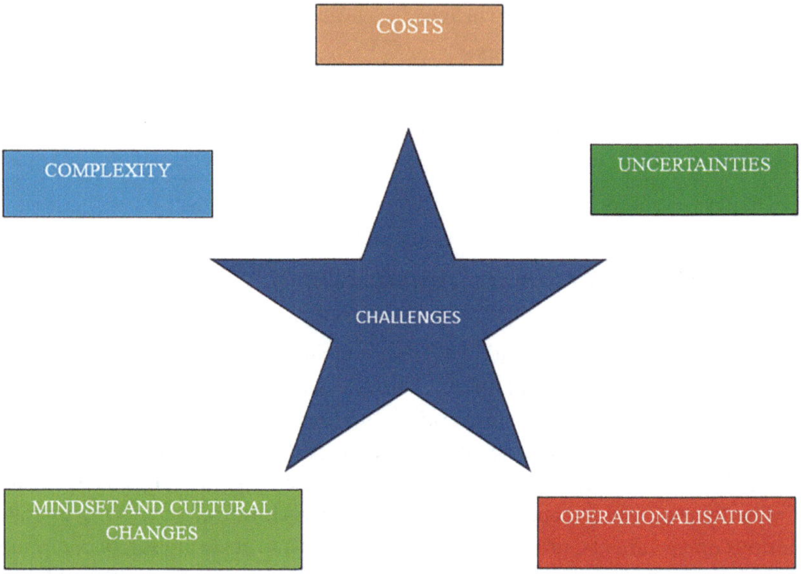

Fig. 2 Challenges of Sustainability in SCM. *Source* Author's own work

challenges. One of them is the complexity of the supply chain process. This is because the new technology may be complex to integrate and understand by the existing employees. Ensuring that each supplier adheres to sustainable practices requires rigorous monitoring and verification, which can be resource-intensive. The initial investment in sustainable technologies, such as renewable energy sources or waste reduction systems, can also be substantial. Smaller companies, like local garment manufacturers, may struggle to afford these investments, putting them at a competitive disadvantage.

Another challenge is the need for cultural and organizational change. Implementing sustainable practices requires a shift in mindset and behavior among all stakeholders. For example, Unilever's commitment to sustainability through its Sustainable Living Plan required significant internal changes, from sourcing raw materials to production processes. This transformation can be met with resistance due to entrenched practices or skepticism about the benefits of sustainability in Industry 5.0. Additionally, regulatory and compliance hurdles add another layer of complexity. For instance, the European Union's stringent environmental regulations can pose challenges for companies operating in multiple regions, as they must navigate and comply with varying standards and requirements for sustainability.

5.1 Costs

Cost is the issue with sustainable growth in supply chains that is brought up the most. Reducing the environmental impact of corporate operations while maintaining financial costs presents a challenge, as noted by McIntyre et al. (1998). The idea that being green has to be financially rewarding is presented in most of the publications. In most SCM domains, the topic of cost is brought up. For instance, in procurement, the challenges of including environmental considerations in purchase criteria and the problem of purchasing from low-cost vendors are discussed (Simpson & Power, 2006; Walker & Brammer, 2009). Welford et al. (2003) highlight that there is a connection between economic growth and well-being which appears to be an essential principle of global supply networks; therefore, we should consider economic concerns while ensuring that they meet sustainability standards. As a result, even while environmental issues and corporate social responsibility are seen as critical to the operations of supply chain and logistics, however, the primary factor influencing supply chain development remains, and would remain the cost. This is problematic since new ideas and models that are more in line with sustainability may be more expensive initially due to their complexity and unpredictability, necessitating the use of additional metrics besides cost.

5.2 Complexity

Many organizations face the dilemma of complexity in SCM that has taken place due to the prevailing concerns regarding sustainable development. This issue is emphasized in previous research as well wherein we have seen that sustainable development has brought it with complexity like what products to choose and how to procure them in a sustainable manner (green procurement). Not only this, it also involves green logistics, routes that would reduce fuels, reducing emissions, etc. Analyzing the effect of SC (supply chain) processes on the environment is also equally important. Ensuring how supply chain processes are carried out at every stage plays a part in varying degrees in the sustainability of the economy, environment, and society. Other difficult examples that bear mentioning are trade-offs between levels of service (Yang et al., 2005) and impact on the environment and delivery delays (Holweg & Miemczyk, 2002).

In addition, several challenges make it more difficult to quantify and evaluate the impacts brought about by various activities and processes inside supply chains. Although life cycle assessments offer reliable data regarding the environmental impacts of various products, their application is restricted to current product flows and certain regions (Matthews et al., 2008; Vieira & Horvath, 2008). The conflict between both environmental and social sustainability is examined (Jonsson & Johansson, 2006). Mobility and resource consumption may rise as a result of infrastructure and accessibility improvements, which could worsen the state of the environment. With the advent of new technology also, the employees may feel that the processes are becoming more complex than easy in case they do not adapt themselves with the technology and find out ways to not willingly adopt it. The awareness of sustainability and green products is increasingly becoming popular among people as they are becoming conscious of the environment. This puts pressure on organizations to integrate sustainable practices into their operations. Although this is necessary, overcoming the challenge of complexity cannot be fully avoided, however, it can be reduced to some extent by creating such ways that can reduce the complexities and problems that may arise in the supply chain operations and technology that is used in supply chain processes.

5.3 Operationalization

It was highlighted that many organizations struggle with the compatibility of new technologies with legacy systems which leads to operational inefficiencies. The adoption of new technology necessitates employee adaptation; if employees are unwilling to learn, accept, and understand the new technology, its operationalization becomes challenging. Finding a balance between the benefits of digital integration and the need for stringent cybersecurity measures is an operational challenge. As we advance toward greater technological integration, the risk of cybersecurity attacks increases. Ensuring a balance to avoid data privacy issues is crucial for successful operationalization; otherwise, it will pose significant challenges. Digital supply chains require technological integration, and ensuring compatibility with new technologies can be a substantial challenge. Operationalization also depends on the workforce's skills, specifically how they convert strategies into operational tasks. Addressing sustainability in operations is a major issue today. Although sustainability concerns are highlighted and strategies are developed during the strategy implementation stage, if these strategies are not followed in operations and not brought into action, the efforts and time invested will be wasted.

5.4 Mindset and Cultural Changes

Adapting to changing global, national, and organizational mindsets and cultures presents additional difficulties for ecologically friendly logistics. One aspect of this is the senior management's lack of involvement in environmental issues according to Preuss (2009), but the problem is more extensive than that. Another aspect is the resistance by the employees. In the context of the digital revolution, employees might show resistance toward change in technology as a result of gaps in their skills and knowledge concerning new technology. Another significant component in this regard is the manager's vision and leadership toward their workers, which can assist in changing people's mindsets. If individuals refuse to alter their thinking, it will be difficult to accept technology and incorporate it into the supply chain to ensure sustainability. Therefore, fostering a collaborative culture and creating value is essential to positively influence people's mindsets.

5.5 *Managing Uncertainties*

One of the most critical concerns in SCM, particularly in the context of the digital revolution, is managing uncertainty at various levels of the supply chain. Together with operational possibilities, it also includes supply and demand uncertainty. In addition, there are several uncertainties in the environment at the macro level that might result from modifications to the natural environment. Demand variability may lead to uncertainties. To handle these uncertainties and variations, the organization should make sure that inventory levels are maintained appropriately and that they have buffer stock on hand in case of an emergency.

Technology in the supply chain can provide innovative approaches to deal with uncertainty; however, some external uncertainties, such as economic fluctuations, are beyond a company's control. According to Ivanov et al. (2019), businesses should create a variety of risk management frameworks to increase supply chain processes' flexibility in response to these fluctuations.

6 Findings

It has been observed that the integration of digital technologies in supply chain processes can bring efficiency, resource optimization, and waste reduction. However, organizations face several challenges when adopting these technologies, such as cost and complexity, which are discussed in this study.

For this study, we conducted an extensive literature review and found numerous studies focused on integrating sustainability into supply chain processes within the context of the digital revolution, using data analytics, IoT, automation, and AI-driven technologies. However, there is a need for implementation to fully enjoy these benefits. The future of sustainable SCM lies in advancements in digital technologies, which help foster a collaborative culture of innovation while considering environmental, social, and economic concerns.

We are yet to integrate these technologies at middle-sized enterprises which can be done by spreading awareness. Despite the challenges that have been highlighted in this study regarding the integration of sustainability in supply chain processes, if these technologies are successfully integrated with the supply chain processes, it will pave the way to achieve environmental stewardship and would also benefit the organizations in the long run.

7 Conclusion

The integration of sustainability into supply chain management is not just necessary for providing a strategic advantage to businesses but is also an essential business practice. This study highlights the transformative impact that Industry 5.0 has had in ensuring sustainability in supply chain processes. The findings show the advantages of adopting digital technologies despite their challenges. Technologies such as advanced data analytics, IoT, and AI-driven solutions enable real-time tracking, and optimization in logistics, and contribute to environmental protection by reducing carbon footprints. Additionally, technologies like blockchain and other digital solutions have enhanced transparency and compliance with sustainability standards, emerging as crucial components of modern supply chain management. Industry 5.0 aims at integrating the circular economy into chain of supply processes to promote the minimization of waste and also integrate a people-centric approach.

Cloud-based platforms have transformed information sharing and coordination, creating an adaptive environment that helps respond to environmental challenges and advances environmental stewardship. However, the journey toward sustainable development faces challenges and obstacles such as high costs, data security issues, complexities, and the need for skilled individuals open to adopting new technologies. Organizations must address these challenges through proper investment in technology and talent, along with strategic planning to leverage technology in the supply chain and provide a way for sustainability.

Embracing sustainability in SCM during the digital revolution is promising, offering numerous benefits to organizations while addressing environmental concerns. Integrating sustainability into supply chain processes using technology will bring long-term success to organizations in this era of digital revolution, where environmental consciousness is increasingly prevalent. The digital transformation within Industry 5.0 is a catalyst for ensuring a more sustainable and resilient supply chain within an ethical framework that satisfies business requirements as well as the needs of our planet.

Funding Statement No funding is being received from anywhere for this Research work.

Conflict of Interest There is no conflict of interest with any parties.

References

Abu Ghazaleh, M., & Zabadi, A. M. (2020). Promoting a revamped CRM through Internet of Things and Big Data: An AHP-based evaluation. *International Journal of Organizational Analysis, 28*(1), 66–91. https://doi.org/10.1108/ijoa-12-2018-1602

Doost Mohammadian, H., & Rezaie, F. (2020). I-sustainability plus theory as an innovative path towards sustainable world founded on Blue-Green ubiquitous cities (Case Studies: Denmark and South Korea). *Inventions, 5*(2), 14. https://doi.org/10.3390/inventions5020014

Dubey, R., Gunasekaran, A., Childe, S. J., Bryde, D. J., Giannakis, M., Foropon, C., Roubaud, D., & Hazen, B. T. (2019). Big data analytics and artificial intelligence pathway to operational performance under the effects of entrepreneurial orientation and environmental dynamism: A study of manufacturing organisations. *International Journal of Production Economics, 226*(1), 107599. https://doi.org/10.1016/j.ijpe.2019.107599

Holweg, M., & Miemczyk, J. (2002). Logistics in the "three-day car" age. *International Journal of Physical Distribution & Logistics Management, 32*(10), 829–850. https://doi.org/10.1108/09600030210455438

IFR International Federation of Robotics. (2024). *The transformative role of robotics in achieving sustainable development goals*. IFR International Federation of Robotics. https://ifr.org/post/the-transformative-role-of-robotics-in-achieving-sustainable-development-goals

Ivanov, D., Dolgui, A., & Sokolov, B. (2019). The impact of digital technology and Industry 4.0 on the ripple effect and supply chain risk analytics. *International Journal of Production Research, 57*(3), 829–846. https://doi.org/10.1080/00207543.2018.1488086

Ivanov, D., Dolgui, A., Sokolov, B., Werner, F., & Ivanova, M. (2015). A dynamic model and an algorithm for short-term supply chain scheduling in the smart factory industry 4.0. *International Journal of Production Research, 54*(2), 386–402. https://doi.org/10.1080/00207543.2014.999958

Jonsson, D. K., & Johansson, J. (2006). Indirect effects to include in strategic environmental assessments of transport infrastructure investments. *Transport Reviews, 26*(2), 151–166. https://doi.org/10.1080/01441640500332406

Kumar, M., Graham, G., Hennelly, P., & Srai, J. (2016). How will smart city production systems transform supply chain design: A product-level investigation. *International Journal of Production Research, 54*(23), 7181–7192. https://doi.org/10.1080/00207543.2016.1198057

Laine, M. (2014). Defining and measuring corporate sustainability: Are we there yet? *Social and Environmental Accountability Journal, 34*(3), 187–188. https://doi.org/10.1080/0969160x.2014.967956

Lee, Y., Kim, S., Kim, M., & Choi, J. (2014). Antecedents and interrelationships of three types of pro-environmental behavior. *Journal of Business Research, 67*(10), 2097–2105. https://doi.org/10.1016/j.jbusres.2014.04.018

Lopez-Vargas, A., Fuentes, M., & Vivar, M. (2020). Challenges and opportunities of the internet of things for global development to achieve the United Nations Sustainable Development Goals. *IEEE Access, 8*, 37202–37213. https://doi.org/10.1109/access.2020.2975472

Maksimovic, M. (2017). Greening the future: Green internet of things (G-IoT) as a key technological enabler of sustainable development. *Studies in Big Data*, 283–313. https://doi.org/10.1007/978-3-319-60435-0_12

Matthews, H. S., Hendrickson, C. T., & Weber, C. L. (2008). The importance of carbon footprint estimation boundaries. *Environmental Science & Technology, 42*(16), 5839–5842. https://doi.org/10.1021/es703112w

McIntyre, K., Smith, H. A., Henham, A., & Pretlove, J. (1998). Logistics performance measurement and greening supply chains: Diverging mindsets. *The International Journal of Logistics Management, 9*(1), 57–68. https://doi.org/10.1108/09574099810805744

Okada, T., Namatame, A., & Sato, H. (2015). An agent-based model of smart supply chain networks. *Proceedings in Adaptation, Learning and Optimization*, 373–384. https://doi.org/10.1007/978-3-319-27000-5_30

Pagell, M., & Shevchenko, A. (2014). Why research in sustainable supply chain management should have no future. *Journal of Supply Chain Management, 50*(1), 44–55. https://doi.org/10.1111/jscm.12037

Preuss, L. (2009). Addressing sustainable development through public procurement: The case of local government. *Supply Chain Management: An International Journal, 14*(3), 213–223. https://doi.org/10.1108/13598540910954557

Raz, G. (2009). *Introduction to supply chain management*. https://store.hbr.org/product/introduction-to-supply-chain-management/UV5138

Salam, A. (2020). *Internet of Things for sustainable community development wireless communications, sensing, and systems*. Springer.
Schneider, S. (2019). The impacts of digital technologies on innovating for sustainability. *Palgrave Studies in Sustainable Business in Association with Future Earth*, 415–433. https://doi.org/10.1007/978-3-319-97385-2_22
Simpson, D. F., & Power, D. J. (2006). Use the supply relationship to develop lean and green suppliers. *Supply Chain Management: An International Journal, 10*(1), 60–68. https://doi.org/10.1108/13598540510578388
Valverde, R., & Saadé, R. G. (2015). The effect of e-supply chain management systems in the North American electronic manufacturing services industry. *Journal of Theoretical and Applied Electronic Commerce Research, 10*(1), 79–98. https://doi.org/10.4067/S0718-18762015000100007
Varriale, V., Cammarano, A., Michelino, F., & Caputo, M. (2023). Industry 5.0 and triple bottom line approach in supply chain management: The state-of-the-art. *Sustainability, 15*(7), 5712. https://doi.org/10.3390/su15075712
Vieira, P. S., & Horvath, A. (2008). Assessing the end-of-life impacts of buildings. *Environmental Science & Technology, 42*(13), 4663–4669. https://doi.org/10.1021/es0713451
Walker, H., & Brammer, S. (2009). Sustainable procurement in the United Kingdom public sector. *Supply Chain Management: An International Journal, 14*(2), 128–137. https://doi.org/10.1108/13598540910941993
Welford, R., Chan, C., & Man, M. (2003). Improving environmental performance through effective communications and management. *Journal of Environmental Management, 70*(3), 283–289. https://doi.org/10.1016/S0301-4797(03)00105-8
Yang, B., Yang, Y., & Wijngaard, J. (2005). Impact of postponement on transport: An environmental perspective. *The International Journal of Logistics Management, 16*(2), 192–204. https://doi.org/10.1108/09574090510634502
Zhong, R. Y., Newman, S. T., Huang, G. Q., & Lan, S. (2016). Big Data for supply chain management in the service and manufacturing sectors: Challenges, opportunities, and future perspectives. *Computers & Industrial Engineering, 101*, 572–591. https://doi.org/10.1016/j.cie.2016.07.013
Zhu, N., & Zhao, H. (2018). IoT applications in the ecological industry chain from information security and smart city perspectives. *Computers & Electrical Engineering, 65*, 34–43. https://doi.org/10.1016/j.compeleceng.2017.05.036

Circular Economy Concept for Small Medium Enterprises Through Industry 5.0

K. Devender, Kafila, Gurunadham Goli, and M. Shravan

Abstract Industry evolution has moved toward Industry 5.0, embodying a synergy between human perception and emerging technologies, while strongly emphasizing sustainable growth and innovation. This gives a chance for the SMEs to adapt to the practices of circular economy which in turn enhances its efficiency by reducing wastages in terms of promoting eco-friendly environment thus maximizing the utility of resources. The incorporation of Artificial Intelligence (AI), Internet of Things (IoT), and Robotics with circular economy supports the Small and Medium Scale Enterprises (SMEs) to enhance the ability of operation and supports them to sustain the competition and set their strong foot in the market for a long time. Industry 5.0 has given a platform to develop Human–Machine interface models to upgrade the operations in small and medium-sized enterprises (SMEs). This chapter explores interdependence of Industry 5.0 and circular economy. It also focuses on the opportunities that provide a dominance to the small and medium enterprises (SMEs) by this integration. SMEs can have a paradigm shift from the traditional processes to modern manufacturing systems thus attaining sustainable growth. The concepts of Industry 5.0 and Circular Economy make small and medium-sized enterprises (SMEs) socially and economically viable. The study provides inputs to succeed in this competitive Industry 5.0 environment.

Keywords Circular economy · Industry 5.0 · Resource efficiency · Sustainable growth · Small and medium enterprises

K. Devender (✉) · Kafila · Gurunadham Goli · M. Shravan
School of Business SR University, Warangal 506371, India
e-mail: devenderapple143@gmail.com

Kafila
e-mail: kafila@sru.edu.in

Gurunadham Goli
e-mail: gurunadhan.g@sru.edu.in

M. Shravan
e-mail: modemshravan20@gmail.com

R. Kumar and S. N. Mohanty (eds.), *Sustainable Economy Models in the Age of Industry 5.0*, https://doi.org/10.1007/978-981-96-4104-8_7

1 Introduction

Transition of industries from traditional units to empathetic manufacturing units to promote sustainability creates ample of scope for SMEs during this paradigm shift from Industry 4.0 to 5.0. The world is looking out for industries that can contribute toward minimizing global warming thus promoting sustainable economic growth through application of new principles of circular economy leading to improvement in overall efficiency in utilization of resources. Implementing environmentally conscious practices, supported by innovations like robotics, artificial intelligence (AI),and the Internet of Things (IoT), enables businesses to promote environmental conservation while staying competitive and resilient in dynamic market conditions. This book chapter focuses on simplification of the above technologies for easy adoption by the SMEs thus creating new windows for enhanced growth value creation and flexibility in operations.

Small and medium-sized business can improve their sustainability by utilizing Industry 4.0 technologies (Santos, 2024). The implementation of circular economy (CE) practices is reshaping production systems in order to close production cycles (Sell, 2023). A review of techno-economic analyses of whey recycling processes helps identify efficient process configurations and optimal plant sizes, highlighting the potential involvement of SMEs in utilizing whey for various food applications. (Lavelli, 2022). Lately, there has been a notable surge in the number of (CE) startups. An examination of the sector's evolving circular economy model over time reveals its potential to enhance resource efficiency, reduce waste, and support sustainable development (Ma, 2014). A supply chain-wide network of interconnected devices and systems enables intelligent manufacturing according to customer-specific requirements in the Industry 5.0 (I5.0) revolution (Dwivedi, 2023). They have a chance to embrace circular economy thanks to Industry 5.0 which helps to combine the perceptions of humans with the actions of machines. Robotics, artificial intelligence (AI), and the internet of things (IoT) allow SMEs to cut down on waste, increase resources efficiency, and propel innovation. How small and medium-sized enterprises (SMEs) can use Industry 5.0 to achieve circular growth is the focus of this chapter.

2 Literature Review

The effects and consequences of circular principles and IoT technologies on contemporary society have contributed to the ever-increasing fascination with digital circular economy models (Voulgaridis et al., 2022). Industry 5.0 opens up a world of possibilities for African manufacturing companies, where the widespread adoption of new technologies can accelerate sustainable manufacturing and contribute significantly to the achievement of the Sustainable Development Goals (SDGs) (Matenga, 2023). These are the primary economic areas and industries that are doing well (De Pascale et al., 2023). Cleaner production (CP) has emerged as a key tactic for energy-intensive manufacturing sectors to stay competitive in the wake of the sustainable development goals (Ma et al., 2024). The significance of incorporating sustainability principle into the Industry 5.0 is growing in a time of rising global environment concerns and climate change (Rame et al., 2024). The goals of the circular economy (CE) include reducing underutilization of processing byproducts, increasing the rate of reuse, and alleviating pressure on systems and natural resources (Cooney et al., 2023). Business operations within the framework of Industry 5.0. Industry 5.0 is obviously promoted sustainability alongside protecting the environment, (Mesjasz-Lech, 2023). With the advent of human-centric innovation brought about by Industry 5.0, small and medium enterprises (SMEs) are able to embrace circular economy practices, which help to alleviate waste and maximize resource utilizing. Small and medium-sized enterprise can improve their sustainability and efficiency by utilizing state of art technologies learn how small and medium-sized enterprises (SMEs) can benefit from Industry 5.0, circular economy.

2.1 Importance of Circular Economy for SMEs

Circular economy throws ample of scope for the Small and medium-sized enterprises (SMEs) to become self-reliant and break from traditional methods of making and dispatching the products as per received orders. The prime idea behind circular economy is to recycle and make the materials reusable. Disposal of used material only creates havoc to the environment, whereas the circular economy tries to eliminate this by reusing the disposed materials thus reducing the cost of new materials. This may help in effective resource utilization and making SMEs to reduce the dependence on suppliers during the times of emergency in turn resulting in reduction of operational expenses. Circular economy helps in creating a credibility in the market by enhancing our Goodwill through sustainable models addressing societal problems causing a damage to the environment. This also helps SMEs to attract grants and funding for their enterprises from the government. International funding can also be availed as the multi-national firms are more aligned to economic growth of the enterprises aiming toward sustainable growth thus encouraging FDIs and partnerships. Thus, we can say that circular economy benefits an SME through government funding and subsidies that promote green finance.

The idea is that small businesses form an integral part of developing countries. It explores how SMEs could implement circular economy principles, along with the opportunities they may leverage and the challenges they might encounter in the process (Ormazabal, 2018). Aside from the obvious monetary and ecological advantage, circular economy models have the potential to spur innovation among SMEs, which in turn can help build supply chains that are less susceptible to disruptions caused by shortage of key resources or fluctuations in their price. Table 1 indicates objectives of the circular economy. Analyzing the actions, obstacles, and enablers, that influence circular economy (CE), company strategy, and success, small and medium-sized businesses (SMEs) are preparing for the possibilities and threats

Table 1 Objectives of the circular economy

Objectives of the circular economy concept for small medium enterprises through Industry 5.0
• To investigate the influence of Industry 5.0 technologies (AI, IoT, Robotics) with circular economy principle in SMEs • To identify the results and scope of integrating circular economy SMEs in the Industry 5.0 era • To examine the challenges SMEs face in transitioning to circular economy models through Industry 5.0 technologies • To provide practical strategies and insights for SMEs to succeed in the circular economy within the Industry 5.0 framework

posed by the CE (Mura et al., 2020). In the end, the circular economy aids SMEs in striking balance between making a profit and doing the right thing, setting them up for success in the future of sustainability-focused business.

3 Key Elements of Circular Economy SMEs Industry 5.0

Figure showing Key elements of circular economy SMEs Industry 5.0 (author own creation by using infographics).

3.1 Human–Machine Collaboration

Among the hallmarks of Industry 5.0 is the interplay between men and the equipments, with technologies like AI, robotics, and the IoT complementing human ingenuity and decision-making. Industry 5.0 is all about human and machine co-creation, wherein both can use their strengths to make better things than what was possible with Industry 4.0's emphasis on digital era. The possibility for increased efficiency, sustainability, and productivity makes this partnership ideal for small and medium-sized businesses (SMEs). To better utilize resource, cut down on waste, and adopt more sustainable practices, SMEs can benefit from human–machine collaboration with the circular economy. Through the use of artificial intelligence and the internet of things, machines can gather and inspect massive volumes of inputs, which can shed light on production processes. When faced with problems like worker safety and lack of available workers, the dynamic construction industry is turning to human–robot collaboration (HRC) as a game changer (Chauhan et al., 2024). Understanding energy consumption and resource flows is essential for optimizing production processes, minimizing waste, and promoting sustainable industrial practices (Ben-Assuli et al., 2023) (Fig. 1).

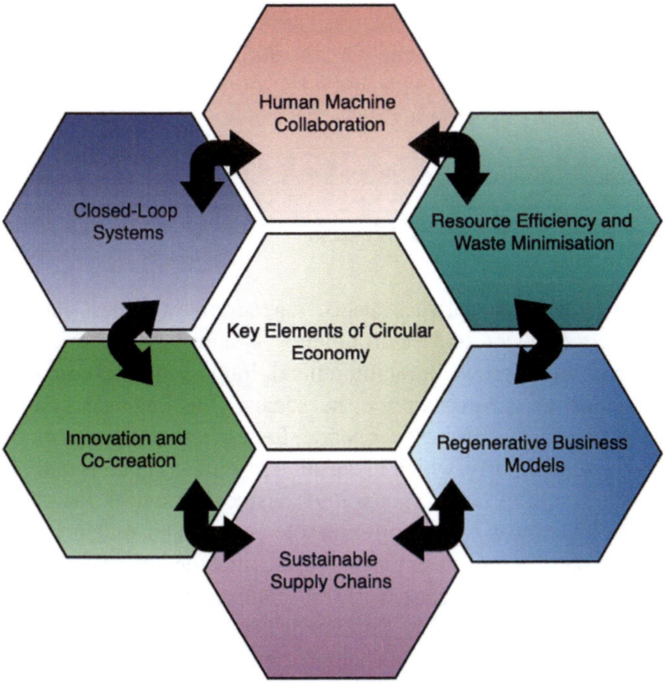

Fig. 1 Key elements of circular economy SMEs Industry 5.0 (author own creation by using infographics)

3.2 *Resource Efficiency and Waste Minimization*

Sustainable development and environmental conservation rely on minimizing waste and making efficient use of resources. In order to minimize waste, these methods center on making the most efficient use of energy, materials, and natural resources. The rapid speed of growth in industries, rising need for food, and concerns about hygiene all contribute to food wastage, which poses a serious threat to global sustainability (Sarkar, 2023). The goal of resource efficiency is to maximize the value obtained from resources by utilizing them more effectively and minimizing waste throughout their lifecycle. Not only does this help save money, but it also helps preserve limited resources. By reducing trash during production, consumption, and disposal, waste minimization helps boost resource efficiency. At its core, waste reduction strategies include product redesign, process improvement, and the use of circular economy principle, which call for reuse, recycling, or repurposing of materials. These practices can help business, particularly SMEs, cut operational costs and lessen their impact on the environment. One method of waste management is known as "waste minimization" and its primary motto is to reduce both the percentage of waste and the toxicity of hazardous waste (Saxena, 2024). More and more, rules and policies that encourage resource efficiency and trash reduction are being adopted by governments and business around the globe. Organizations can enhance their profitability and competitiveness in a world with limited resources while also contributing to a more sustainable future by adopting these strategies.

3.3 Regenerative Business Models

In addition to reducing negative impacts, regenerative business models work to improve and restore social and environmental systems. By restoring ecosystems, reviving communities, and encouraging ethical, long-term economic growth, these models seek constructive growth for the society, the financial system, and the surroundings in which the businesses operate. Regenerative business models revolve around the idea of circularity, which means that resources are constantly being recycled and waste is minimized. Regenerative business practices include increasing biodiversity, decreasing use of non-renewable inputs, and making greater use of renewable resources. Some examples of these include reforestation and carbon sequestration, which are solutions based on nature, and sustainable sourcing and restorative agriculture. Addressing global nutrition and food security issues is critical for ensuring the health and well-being of growing populations, particularly in the face of climate change and resource constraints (de Lauwere, 2024). It is essential for business to prioritize fair labor practices, equitable resource distribution, and community empowerment in order to promote social regeneration. Companies can strengthen their relationships with stakeholders and promote innovation that benefits society by aligning with regeneration principles. In light of the many pressing sustainability issues of our day, the circular economy is presently enjoying widespread support (De Angelis, 2024). It is possible for business, particularly SMEs, to gain an edge in today's dynamic market by embracing regenerative business models. These models ensure that companies can secure long-term access to resources while also contributing to the development of a more sustainable future.

3.4 Sustainable Supply Chains

From the sourcing of materials to their final disposal, sustainable supply chains take into account economic, social, and environmental factors during the lifecycle. The growing need for long living products in international markets is driving the development of supply chains that prioritize ethical labor practices, minimize environmental impact, and guarantee financial sustainability. In the fight against environmental degradation and its detrimental social and economic effects, sustainability and the circular economy have emerged as critical components of supply chains across all sectors and areas of endeavor (Lozano-Oviedo, 2024) Reduced waste and maximized efficiency are the cornerstones of environmentally friendly supply chains. Some ways to accomplish this goal include using energy-efficient production methods, reducing carbon emission during transportation, and obtaining raw materials from certified sustainable sources. While reducing environmental impacts and conserving valuable natural resources, long-lasting method for perishable food supply chains (FSCs) aims to provide sustenance to all parties involved (Shetty et al., 2024). Reducing material consumption and increasing material circulation are key components of the circular economy that contribute significantly to more sustainable supply chains. Sustainability in society is just as critical. All along their supply chains, business should be making sure that workers are paid a fair wage and that they are safe from harm. If businesses want to earn customers' trust and keep their good name, they need to use ethical sourcing practices. This is particularly true in the food and clothing industries.

3.5 *Innovation and Co-Creation*

Companies that want to be relevant and competitive in today's fast-paced global market must prioritize innovation and co-creation if they want to survive. The grail of innovation is to increase net worth through the creation of new goods services and processes. Co-creation takes innovation to the next level by involving all relevant parties (customers, employees, partners, and communities) in the process, which in turn encourages teamwork and collective responsibility for the product (Zhang, 2024). Business can benefit from a wider range of ideas and viewpoints through co-creation, which in turn leads to better solutions. Companies can gain a better understanding of user needs, increase customer satisfaction, and decrease the risk of market failure by involving customer directly in product development or service improvement. As stakeholders take an active role in the companies, their relationships and loyalty are strengthened through this collaborative approach. To thrive in modern agile business world, companies should be nimble and adaptable. One way to achieve this is through innovation, particularly through co-creation (Kumar et al., 2024). Inspiring people to try new things and take calculated risks frequently leads to innovative solutions that set business apart from the competition. The impacts of Digital Green Innovation Performance (DGIP) on digital green network embedding are explored in the study by (Yin 2024), highlighting the key role of innovation in enhancing sustainability through digital networks. Another benefit of co-creation is the long-term sustainability it drives by encouraging a culture of shared responsibility and constant improvement. Embracing innovation and co-creation allows business to grow faster, adapt to changing consumer demands, and benefit society and their own organization (Mohanty et al., 2024).

3.6 Closed-Loop Systems:

Circular economy is closed-loop systems, which aim to maximize efficiency by minimizing waste through the reuse, recycling, or regeneration of resources. Materials can be reincorporated into the production cycle through processes such as upcycling or remanufacturing,contributing to sustainability and reducing waste. In contrast, conventional liner models deplete resources and damage the environment because they adhere to the take, make, dispose paradigm, under the influence of competitiveness in the market among manufacturing SMEs in developing nations (Rehman, 2023). Designing products for durability, modularity, and easy disassembly is key to closed-loop systems, which focus on extending materials lifecycle. This makes sure that when a product is no longer needed, its parts can be reused or recycled without creating a lot of trash (Pogue, 2023). To lessen their negative environmental impact, textiles and packaging industries are increasingly embracing closed-loop systems to promote recycling and sustainability. Implementing closed-loop systems can help businesses save money on material, cut waste management costs, and improve their sustainability credentials. Technological advancements that aim to establish a relationship between men and technology to develop more resilient production models are highly congruent with these systems. Companies can help ensure that our planet's natural resources are around for the next generation by implementing closed-loop systems. Table 2 explores circular economy concept for small and medium enterprises (SMEs) through Industry 5.0. In this table, we can see how SMEs can implement key principles of the circular economy using Industry 5.0 technologies.

4 Methodology

A multi-disciplinary approach is reflected in the methodology, which includes both theoretical exploration and practical recommendations. Used Scopus data and science direct as sources for secondary data used literature. Creating visual representations by means of the authors, originals work, presented in a clear and concise manner.

5 Results

Opportunities for SMEs in Industry 5.0 and Circular Economy

Opportunities for small and medium-sized enterprises (SMEs) to drive growth, innovation, and sustainability a bound in Industry 5.0 and the circular economy. To start, online tools like the internet of things (IoT) and artificial intelligence (AI) can help small and medium-sized enterprises (SMEs) save money by improving resource efficiency. Furthermore, small and medium-sized enterprises (SMEs) have met the need for eco-friendly and adaptable products because human–machine collaboration

Table 2 Synergies between Industry 5.0 Technologies and circular economy practices for SMEs

Industry 5.0 technology	Application in SMEs	Circular economy benefit	Example use case
Internet of Things (IoT)	Real-time monitoring of resource use, supply chain optimization	Improved resource efficiency, waste reduction	Smart sensors in factories to track energy and material use
Artificial Intelligence (AI)	Predictive maintenance, demand forecasting	Minimization of waste and overproduction	AI-powered algorithms to optimize inventory and reduce waste
Robotics	Precision manufacturing and automation	Enhancing recycling, reducing errors and waste	Automated recycling systems to separate and process materials
Big Data & Analytics	Analyzing production processes for optimization	Identifying areas for reuse, refurbishing, or recycling	Using data to design products with longer life cycles
3D Printing	Custom, on-demand manufacturing	Reducing material waste, enabling repair and reuse	SMEs producing spare parts for their products to extend life
Collaborative Robots (Cobots)	Human–machine collaboration in production	Increased efficiency in resource handling and manufacturing	Cobots assisting workers in disassembling products for recycling
Cloud Computing	Shared platforms for circular economy initiatives	Resource sharing, efficient coordination among SMEs	Cloud-based marketplaces for sharing surplus materials

makes personalization and sustainability possible. Third, by participating in regenerative models like product-as-a-service or remanufacturing, small and medium-sized enterprises (SMEs) can open up new avenues of income generation through the reuse of resources. Fourth, small and medium-sized enterprises (SMEs) can innovate and stand out from the competition by incorporating advanced technologies. A look at Employment 5.0 through the lenses of political economy, skill bias, and socio-technical analysis (Kolade,2022) reveals key insights into the evolving nature of work. In contrast to the value-driven approach of Industry 5.0,the technology-driven Industry 4.0 focuses primarily on technological advancements (Xu et al., 2021). *This allows them to reduce their reliance on raw materials and build adaptive, sustainable operations; SMEs that embrace circular economy principles strengthen their resilience against supply chain disruptions. This coming together of sustainability and industry gives small and medium-sized enterprise a chance to succeed in a dynamic global market.* Figure 2 shows opportunities for SMEs in Industry 5.0 and circular economy (author own creation by using lucid chart).

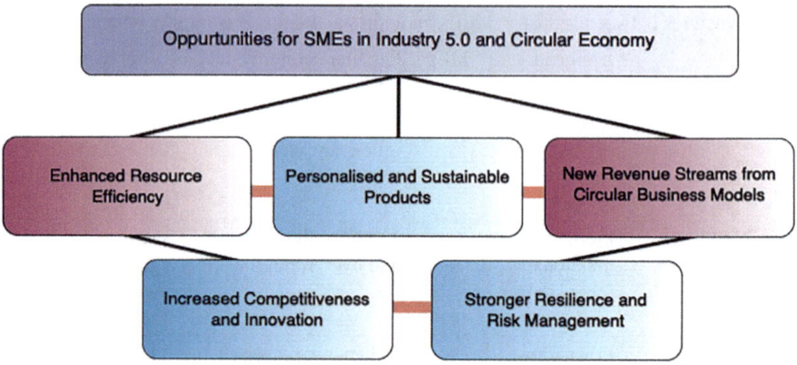

Fig. 2 Opportunities for SMEs in Industry 5.0 and circular economy (author own creation by using lucid chart)

5.1 Enhanced Resource Efficiency

Using real-time material and energy consumption tracking, SMEs can optimize resource use with the help of Industry 5.0 technologies such as the internet of things (IoT), artificial intelligence (AI), and robotics. Exporters and industries can take advantage of new avenues for increased global competitiveness brought about by the digital economy's revolution of international trade (Yuan, 2024). Because of this, companies can lessen their financial and ecological footprints by cutting down on waste, increasing material reuse, and improving operational efficiency. When resources are used strategically, they can be more efficient in a way that reduces waste, energy consumption, and environmental impact, all while maximizing productivity. This method allows companies to increase their output while decreasing their inputs, which is essential for sustainable development. Reducing operational costs and increasing competitiveness are two outcomes of resource efficiency initiatives like recycling processes, energy-efficient technology, and waste minimization. By reducing waste and increasing product longevity, it helps pave the way for circular economies. Organizations can improve their resilience in a world with dwindling resources and reduce their impact on the environment by putting an emphasis on efficiency.

5.2 Personalized and Sustainable Products

When humans' machines work together, small and medium-sized enterprises (SMEs) can meet customer demands for greener, more personalized products. The procedure by which producers create customized goods for sale to consumers is known as mass customization, allowing for personalized products while maintaining production efficiency (Cheng, 2024). Sustainable competitiveness in manufacturing enterprises is greatly influenced by the performance of regular and consistent suppliers, as highlighted by Liao et al. (2023). With the increased manufacturing flexibility brought about by Industry 5.0, it is now much simpler to design products that adhere to the principle of circular economy, such as extending the life of products and getting them recyclable. Sustainable products that cater to individual taste while also taking into account environmental concerns are a win–win for consumers and the planet. Customer satisfaction is higher and experiences are more memorable when products are tailored to individual preferences. Concurrently, these goods are manufactured with a minimum influence on the environment thanks to the incorporation of sustainable practices. Some ways to achieve this goal include recycling and reusing items, adopting circular economy principles, and making use of environmentally friendly materials. Business can appeal to consumers increasing desire for ethical consumption, strengthen customer loyalty, and advance global sustainability objectives by providing sustainable, customized products.

5.3 New Revenue Streams from Circular Business Models

By replacing linear thinking with systems that put an emphasis on reusing, recycling, and regenerating resources, circular business models create new avenues of income. Instead of following the age-old procure, produce, and dispose model, circular business models aim to increase product longevity, decrease waste, and find new uses for used materials. Business can tap into untapped markets by providing services like product as-a-service, where clients pay to use an item on an as-needed basis and then return it for a refurbishment. Other ways to make money include recycling, remanufacturing, and selling used or upcycled goods. In addition, companies can benefit from new material recovery technologies and long-term collaborations that transform trash into something useful. Membership subscriptions, donor funding, and conference fees for network events are the main models that were observed for funding (Muzamwese, 2024). In addition to increasing profits, these models improve brand image, appeal to environmentally conscious customers, and help achieve sustainability objectives in the long run. Companies can diversify their revenue streams and promote environmental stewardship through circular business models.

5.4 Increased Competitiveness and Innovation

Innovation is enhanced by combining circular economy practices with new technologies in Industry 5.0. To stand in a crowded global market, small and medium-sized enterprises (SMEs) can provide sustainable solutions and products that consumers are increasingly demanding. Environmental regulations impact exports in two ways: first, by fostering innovation, and second, by emphasizing the importance of reducing pollution intensity to mitigate the negative effects of stringent regulations on innovation and global competitiveness (Fabrizi et al., 2024). To thrive in today's ever-changing market, companies must be more innovative and competitive than ever before. Business can gain an advantage over rivals through motivating the workforce to think beyond the box and create novel goods, services, and procedures. When companies innovate, they are better able to meet the needs of their customers, take advantage of new technology, and come up with solutions that add value.

5.5 Stronger Resilience and Risk Management

Adopting circular economy principle can help SMEs become more resilient in the face of supply chain disruptions and resource shortages. Businesses can improve their strategies to manage and minimize risk with the help of technologies, which enhance risk prediction, resource management, and the implementation of sustainable practices to guarantee business continuity. Management of resilience offers guidance on how businesses can integrate into their operations for improved performance (Liu et al., 2024). In the face of uncertainty and disruption, companies must strengthen their resilience and improve their risk management. Companies need to minimize the risk as well as increase the flexibility in operations. The firms need to be ready to tackle various risks from the environment in which they operate like socio-economic, political, demographical, and geographical. The SMEs need to acquaint themselves to deal with natural and economic shocks in the prevailing markets which are always volatile and cause disruptions in the supply of materials. Sustainable growth being a long-term goal of the enterprise can be achieved only through proper anticipation of such risks and diversifying it by becoming adaptive to the uncertainties in implementation of robust plans. Thus, SMEs can withstand the dynamic changes by becoming adaptive to the conducive environment and continue the operations to meet the competition and survive during such uncertainties.

6 Insights on the Success of SMEs in the Global Green Scenario

Global green scenario is all about focusing on the sustainable growth of green industries globally. Small and medium-sized enterprises (SMEs) can be successful by employing circular economy practices by clearing the footprints caused into the environment due to their robust operations thus conquering a special place in the global market. Small and medium-sized enterprises (SMEs) can be in the forefront of the development process by producing products and services which are eco-friendly and

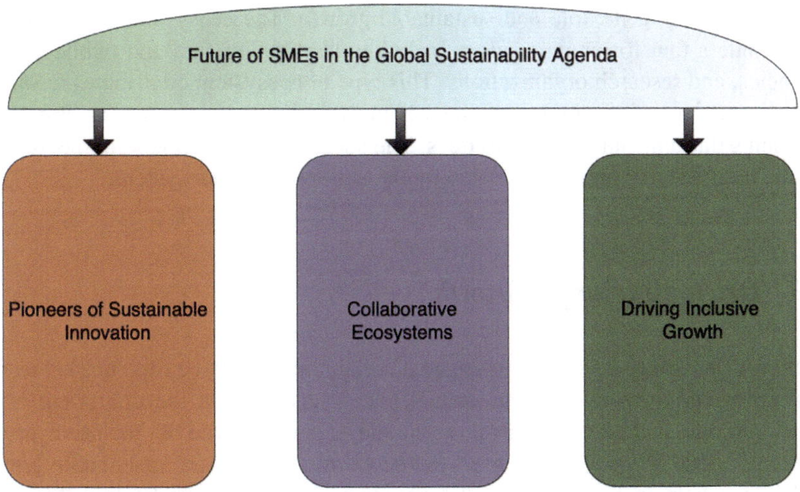

Fig. 3 Future of SMEs in the global sustainable agenda (author own creation by using lucid chart)

non-hazardous to the environment and in turn grab the market share of the competitors and lure the government through sustainable growth initiatives. Figure 3 shows future of SMEs in the global sustainable agenda.

6.1 Pioneers of Sustainable Innovation

SMEs can always rely on innovation due to its small size manufacturing which is easily adaptable to novel technologies to produce innovative products. The **"Small is Beautiful"** concept can give a chance to SMEs to come up with many new products to match the customer tastes and preferences to meet the requirements of the market. Integration of circular economy with Industry 5.0 helps in promoting sustainability pertaining to environment and economy thus reducing the obstacles caused by the bureaucrats. Such SMEs become pioneers in renewable energy products, challenging large corporations through their ethical practices and commitment to promoting a humanitarian approach in business.

6.2 Collaborative Ecosystems

Foster a collaborative ecosystem involving all the stakeholders from different locations to have a mutual exchange of the expertise practices for attainment of a common goal. This can be achieved through effective communication for exchange of ideas to understand diversified business practices for successful collaboration that can help in promoting flexible and sustainable growth. The ecosystem should involve stakeholders that foster innovation, including startups, private and public funding agencies, and research organizations. This type of ecosystem contributes to sharing of best practices of several businesses which can help in overcoming major business problems through sharing of resources. The success of such collaborations is based on the transparency and trust created by the entities in the ecosystem.

6.3 Driving Inclusive Growth

Inclusive growth is a strategy to create a collaboration between the market leaders and followers so as to promote mutual benefit and help in the upliftment of the firms running in losses to ensure equitable economic growth. Inclusive growth removes social inequalities to meet the economic policies for sustainable growth. SMEs can help in fostering social innovations through diversification thus stimulating digital transformation for enhancing the standard of living of developing economies resulting in the economic prosperity of the Nation.

7 Implications

The integration of the concepts of circular economy in SMEs with Industry 5.0 has the following implications:

Optimization of resources for SMEs can be possible through the concept of Industry 5.0 by practicing recycle and reuse thus reducing the cost and wastage.

The collaboration of circular economy and Industry 5.0 will help in designing and creating user-friendly and personalized products to satisfy the customers.

Smart Technologies like IoT, AI, etc. help in tracking the recyclable and reusable material thus providing data to improve the durability and shelf life of the materials. This will make the products sustainable and create trust among the stakeholders.

The collaboration also helps in job creation in green industries thus paving a way for human-centric approach.

Strategy implications include exchange of knowledge and resources in the ecosystem creating a mutual benefit.

Government policy frameworks should include recognition of such SMEs through awards, rewards, and incentives thus increasing scalability of SMEs.

8 Conclusion

An ocean of opportunities is available for SMEs to contribute to the objective of sustainability both locally and globally through enhancing its competitive spirit by overcoming the challenges by inculcating the concepts of circular economy. It can drive to innovation which can mitigate the risks of getting obsolete. The regenerative business models help in the effective allocation of resources and reduction in wastages through recycling and reuse of disposals. Industry 5.0's overall objective is met by increasing the operational efficacy through the integration of human knowledge with developing technologies such as Artificial Intelligence (AI) and Internet of Things (IoT). The human–machine interface provides a platform for innovation in SMEs. This safeguards the SMEs from the pressures created by the regulatory bodies to cope with scarcity of resources.

All the technological advancements through circular economy pave a way for the transformation of SMEs. Industry 5.0 aims at green economy through reducing the carbon footprints left over by the industries in their manufacturing process. This chapter contributes toward the various aspects of digital transformation using circular economy to increase operational efficiencies of SMEs through reduced wastages, mitigating risks and attaining global economic and environmental sustainability in the context of Industry 5.0.

Declarations: Ethics approval and consent to participate
Not applicable.

Funding Statement: There is No Specific Funding Received in this Chapter.
Declaration of competing interest:
The authors have no conflict of interest to declare.

Abbreviations

SMEs	Small and Medium Enterprises
AI	Artificial Intelligence
IoT	Internet of Things

References

Ben-Assuli, O., Heart, T., Klempfner, R., & Padman, R. (2023). Human-machine collaboration for feature selection and integration to improve congestive Heart failure risk prediction. *Decision Support Systems, 172*(113982), 113982. https://doi.org/10.1016/j.dss.2023.113982

Chauhan, H., Jang, Y., & Jeong, I. (2024). Predicting human trust in human-robot collaborations using machine learning and psychophysiological responses. *Advanced Engineering Informatics, 62*(102720), 102720. https://doi.org/10.1016/j.aei.2024.102720

Cheng, B.-Y., Duan, J., Shi, X.-Y., & Zhou, M. (2024). Integrated optimization of design and production process with personalization level of products. *Optimization Letters*. https://doi.org/10.1007/s11590-024-02099-9

Cooney, R., de Sousa, D. B., Fernández-Ríos, A., Mellett, S., Rowan, N., Morse, A. P., Hayes, M., Laso, J., Regueiro, L., Wan, A. H. L., & Clifford, E. (2023). A circular economy framework for seafood waste valorisation to meet challenges and opportunities for intensive production and sustainability. *Journal of Cleaner Production, 392*(136283), 136283. https://doi.org/10.1016/j.jclepro.2023.136283

De Angelis, R., & Vesci, M. (2024). Circular economy business models, value creation and humane entrepreneurship: A micro-sized and social enterprise perspective. *Circular Economy and Sustainability*. https://doi.org/10.1007/s43615-024-00419-w

de Lauwere, C., Smits, M.-J., Dijkshoorn-Dekker, M., Brummelhuis, A.K.-T., & Polman, N. (2024). Understanding circular and nature-inclusive agricultural business models. *Circular Economy and Sustainability*. https://doi.org/10.1007/s43615-024-00433-y

De Pascale, A., Di Vita, G., Giannetto, C., Ioppolo, G., Lanfranchi, M., Limosani, M., & Szopik-Depczyńska, K. (2023). The circular economy implementation at the European Union level. Past, present and future. *Journal of Cleaner Production, 423*(138658), 138658. https://doi.org/10.1016/j.jclepro.2023.138658

Dwivedi, A., Agrawal, D., Jha, A., & Mathiyazhagan, K. (2023). Studying the interactions among Industry 5.0 and circular supply chain: Towards attaining sustainable development. *Computers & Industrial Engineering, 176*(108927), 108927. https://doi.org/10.1016/j.cie.2022.108927

Fabrizi, A., Gentile, M., Guarini, G., & Meliciani, V. (2024). The impact of environmental regulation on innovation and international competitiveness. *Journal of Evolutionary Economics, 34*(1), 169–204. https://doi.org/10.1007/s00191-024-00852-y

Kolade, O., & Owoseni, A. (2022). Employment 5.0: The work of the future and the future of work. *Technology in Society*, *71*(102086), 102086. https://doi.org/10.1016/j.techsoc.2022.102086

Lavelli, V., & Beccalli, M. P. (2022). Cheese whey recycling in the perspective of the circular economy: Modeling processes and the supply chain to design the involvement of the small and medium enterprises. *Trends in Food Science & Technology, 126*, 86–98. https://doi.org/10.1016/j.tifs.2022.06.013

Liao, H., & Wen, Z. (2023). Capturing attitudinal characteristics of decision makers in multi-criterion sorting problems for performance evaluation of sustainable and circular suppliers. *Annals of Operations Research*. https://doi.org/10.1007/s10479-023-05441-z

Liu, Y., Du, J., Kang, T., & Kang, M. (2024). Establishing supply chain transparency and its impact on supply chain risk management and resilience. *Operations Management Research*. https://doi.org/10.1007/s12063-024-00499-9

Lozano-Oviedo, J., Cortés, C. E., & Rey, P. A. (2024). Sustainable closed-loop supply chains and their optimization models: a review of the literature. *Clean Technologies and Environmental Policy, 26*(4), 999–1023. https://doi.org/10.1007/s10098-023-02730-w

Ma, S., Ding, W., Liu, Y., Zhang, Y., Ren, S., Kong, X., & Leng, J. (2024). Industry 4.0 and cleaner production: A comprehensive review of sustainable and intelligent manufacturing for energy-intensive manufacturing industries. *Journal of Cleaner Production, 467*(142879), 142879. https://doi.org/10.1016/j.jclepro.2024.142879

Ma, S.-H., Wen, Z.-G., Chen, J.-N., & Wen, Z.-C. (2014). Mode of circular economy in China's iron and steel industry: a case study in Wu'an city. *Journal of Cleaner Production, 64*, 505–512. https://doi.org/10.1016/j.jclepro.2013.10.008

Matenga, A. E., & Mpofu, K. (2023). Africa and industry 5.0: Challenges and opportunities in the future of manufacturing. *Procedia CIRP, 120*, 1256–1261. https://doi.org/10.1016/j.procir.2023.09.158

Mesjasz-Lech, A. (2023). Can Industry 5.0 be seen as a remedy for the problem of waste in industrial companies? *Procedia Computer Science, 225*, 1816–1825. https://doi.org/10.1016/j.procs.2023.10.171

Misso, A. (2024). Enhanced resource allocation strategies to improve the spectral efficiency in massive MIMO systems. *Journal of Electrical Systems and Information Technology, 11*(1). https://doi.org/10.1186/s43067-023-00132-y

Mura, M., Longo, M., & Zanni, S. (2020). Circular economy in Italian SMEs: A multi-method study. *Journal of Cleaner Production, 245*(118821), 118821. https://doi.org/10.1016/j.jclepro.2019.118821

Muzamwese, T. C., Heldeweg, M. A., & Franco-Garcia, L. (2024). Financing and business models for scaling up sustainable business networks—Building a circular economy. *Circular Economy and Sustainability, 4*(3), 1655–1667. https://doi.org/10.1007/s43615-024-00348-8

Ormazabal, M., Prieto-Sandoval, V., Puga-Leal, R., & Jaca, C. (2018). Circular economy in Spanish SMEs: Challenges and opportunities. *Journal of Cleaner Production, 185*, 157–167. https://doi.org/10.1016/j.jclepro.2018.03.031

Pogue, E. A., Alexander New, McElroy, K., Le, N. Q., Pekala, M. J., McCue, I., Gienger, E., Domenico, J., Hedrick, E., McQueen, T. M., Wilfong, B., Piatko, C. D., Ratto, C. R., Lennon, A., Chung, C., Montalbano, T., Bassen, G., & Stiles, C. D. (2023). Closed-loop superconducting materials discovery. *Npj Computational Materials, 9*(1). https://doi.org/10.1038/s41524-023-01131-3

Rame, R., Purwanto, P., & Sudarno, S. (2024). Industry 5.0 and sustainability: An overview of emerging trends and challenges for a green future. *Innovation and Green Development, 3*(4), 100173. https://doi.org/10.1016/j.igd.2024.100173

Rehman, F. U., Gyamfi, S., Rasool, S. F., Akbar, F., Hussain, K., & Prokop, V. (2023). The nexus between circular economy innovation, market competitiveness, and triple bottom lines efficiencies among SMEs: evidence from emerging economies. *Environmental Science and Pollution Research International, 30*(58), 122274–122292. https://doi.org/10.1007/s11356-023-30956-0

Santos, A. de M., & Sant'Anna, Â. M. O. (2024). Industry 4.0 technologies for sustainability within small and medium enterprises: A systematic literature review and future directions. *Journal of Cleaner Production, 467*(143023), 143023. https://doi.org/10.1016/j.jclepro.2024.143023

Sarkar, B. D., Sharma, I., & Gupta, S. (2023). Waste minimization in agri-food supply chain: perspective for sustainable development. *Environment Development and Sustainability*. https://doi.org/10.1007/s10668-023-04130-y

Saxena, Y. K. (2024). Waste minimization at source. In *From Waste to Wealth* (pp. 71–83). Springer Nature Singapore.

Sell, L. G., Tedesco, M., Simioni, F. J., Sehnem, S., Soares, J. F., & Coelho Junior, L. M. (2023). Circular economy practices using the ReSOLVE framework: An assessment by sector and scale in the Brazilian planted tree industry. *Journal of Cleaner Production, 433*(139780), 139780. https://doi.org/10.1016/j.jclepro.2023.139780

Shetty, L., Srivastava, S., Dwivedi, A., Pamucar, D., & Patil, A. (2024). Shaping sustainable paths for perishable food supply chains—Contemporary insights and future prospects. *Environment Development and Sustainability*. https://doi.org/10.1007/s10668-024-05067-6

Mohanty, R., Samal, A., & Kumar, R. (2024). Green finance supports sustainable environment: Empirical approach towards applicability in tourism sector. *Theoretical and Practical Research in Economic Fields, 15*(1), 114–126.

Kumar, R., Goel, R., Singh, T., Mohanty, S. M., Gupta, D., Alkhayyat, A., & Khanna, R. (2024). Sustainable finance factors in indian economy: Analysis on policy of climate change and energy sector. *Fluctuation and Noise Letters, 23*(2), 2440004–339.

Tsironis, G., Daglis, T., & Tsagarakis, K. P. (2024). The 21 most practiced RE-s of circular economy from LinkedIn company profiles on a global scale. *Resources, Conservation & Recycling Advances, 21*(200202), 200202. https://doi.org/10.1016/j.rcradv.2024.200202

Voulgaridis, K., Lagkas, T., Angelopoulos, C. M., & Nikoletseas, S. E. (2022). IoT and digital circular economy: Principles, applications, and challenges. *Computer Networks, 219*(109456), 109456. https://doi.org/10.1016/j.comnet.2022.109456

Xu, X., Lu, Y., Vogel-Heuser, B., & Wang, L. (2021). Industry 4.0 and Industry 5.0—Inception, conception and perception. *Journal of Manufacturing Systems, 61*, 530–535. https://doi.org/10.1016/j.jmsy.2021.10.006

Yin, S., & Zhao, Y. (2024). Digital green value co-creation behavior, digital green network embedding and digital green innovation performance: Moderating effects of digital green network fragmentation. *Humanities & Social Sciences Communications, 11*(1). https://doi.org/10.1057/s41599-024-02691-5

Yuan, H., Peng, G., Song, C., Wang, L., & Lu, S. (2024). Enhancing digital economy: Optimizing export enterprise markup and resource allocation efficiency. *Journal of the Knowledge Economy*. https://doi.org/10.1007/s13132-024-02142-1

Zhang, X., Yang, L., Gao, T., & Zhou, W. (2024). The coordination mechanism of value co-creatison between developers and users in digital innovation ecosystems. *Electronic Markets, 34*(1). https://doi.org/10.1007/s12525-023-00685-w

From Linear to Circular: A Holistic Strategy for Sustainable Development in Emerging Economies

Shanu Singh, Aditi Anand, Kalyani, and Yashmita Awasthi

Abstract The immediate upheaval for transitioning and slipping from mundane linear system to deliberately choosing sustainable and equitable circular economy system has been in airs, suggesting its shared benefits and changed outlook. The current research paper proposes a framework "RE-GENO" which supports the design of responsible circular economy practices and argues that this framework can act as a preliminary tool for improving and expanding the decision-making context for a more equitable transition through use of artificial intelligence. The research papers have discussed the idea of transitioning from linear economy that results in generation of large amount of waste to a circular economy which leads to sustainable form of consumption and have further highlighted the challenges followed by it and application of artificial intelligence in its transition. They have given different frameworks which could form base of transitioning into it but at the same time, there has not been enough insight on implementation of it in Indian based economy. This research paper brings up a modified version for a framework whose main aim is to provide a new base for moving to circular economy and focusing on strategies on its efficient implementation in developing nation like India. It is found that application of this framework in India requires multifaceted approach that integrates all aspects such as technology, education, government initiatives and monitoring through risk analysis to foster sustainable growth, production, and consumption. Hence, the paper suggests that successful implementation might open the way for more sustainable and efficient resource management strategies across a variety of businesses.

S. Singh (✉) · A. Anand · Kalyani · Y. Awasthi
Christ University, Bengaluru, India
e-mail: shanu.singh@christuniversity.in

A. Anand
e-mail: aditi.anand@bcomp.christuniversity.in

Kalyani
e-mail: kalyani@bcomp.christuniversity.in

Y. Awasthi
e-mail: yashmita.awasthi@christuniversity.in

R. Kumar and S. N. Mohanty (eds.), *Sustainable Economy Models in the Age of Industry 5.0*, https://doi.org/10.1007/978-981-96-4104-8_8

Keywords Artificial intelligence · Circular economy · Framework · Sustainable practices · Framework · SDG

1 Introduction

In recent years, concern on topics associated with climate change and environmental degradation has been in airs which has further highlighted the immediate call for sustainable practices across industries. The immediate call for this need has been intensified by government initiatives to reduce global emissions, pointing to various instances like Green Deals of the European Union (Greif et al., 2024). Further technology is advancing, which can be witnessed with its speeding impact on socio-economic transition from linear economic model which has dominated industrial and economic activities for centuries, leading to massive environmental degradation, resource depletion, and unsustainable waste. In the recent decade, the Circular Economy (CE) has been in news within business, policy, and academic spheres with its aim to suggest set of practices targeting decreased dependence on unstainable models. It can be better understood as process of making a transition to a sustainable and equitable economic system which takes into essentials the needs and rights of all stakeholders (Purvis et al., 2023). The concept of CE is contrary to that of the prevailing linear economy model, based on consideration that that the available resources are abundant for consumption and production of goods, and has better ways of disposal (Nowicki, 2020; Prieto-Sandoval et al., 2018; Roberts et al., 2022). One of the most important ways to create sustainable growth and integrate corporate social responsibility is to design and implement Circular Economy (CE) plans (Pacheco et al., 2024).

Many physical and intangible services, including increased opportunities for cooperation, market access, product/service customization, waste reduction, and long-term value retention, can result from the use of CE. Furthermore, the CE aims to reduce the risks associated with resource shortages and improve environmental performance. According to Raman et al. (2024a), Raman et al. (2024b), the adoption of CE adds value to materials and products and revitalizes them till they reach the end of their useful lives.

2 Literature Review

In recent years, Artificial intelligence (AI) and the circular economy have been hot topics. Numerous studies have examined how AI could enhance resource efficiency, reduce waste, and advance a circular economy. The review demonstrates the ability of artificial intelligence to transform classic traditional economic frameworks by enabling more effective utilization of resources and increasing product life cycles. This transformation not only helps the environment, but also improves the economy by reducing costs. Furthermore, research identifies a few models that actively promote the circular economy. These models use AI technology to improve recycling processes, forecast product lifecycle stages, and minimize waste. This paper reviews the existing paper on how AI can contribute towards a circular economy and identify key findings and points out various needs for the future.

2.1 Circular Economy (CE) Transition and Sustainability

Niwalkar et al. (2022) has brought out the fact how shifting towards greener fuels and biofuels and limiting the impacts of rising energy demand through use and implementation of circular economy strategies by focusing on various sectors like industrial, domestic, and transportation of the National Capital Territory (NCT) of India. Rejeb et al. (2022) has thrown light on shift from linear economy, which is characterized by excessive waste, to circular economy which brings out opportunities for collaboration, market access, product and service innovation, waste minimisation, and long-term value retention. This paper explains that circular economy can positively contribute to sustainable development goals, close the resource loops, and achieve a harmonious integration of economic performance. Rejeb et al. (2023) has stressed about application of circular economy principles in the COVID-19 pandemic ranging from recent recycling and upcycling activities to the production of sanitizers from residual materials and the manufacturing of face masks from textile remnants for healthcare facilities. This paper highlights the relevance of circular economy model during a pandemic, as supply chains have been disrupted and the demand for certain goods has shifted, leading to a need for more adaptive and flexible economic models. Pacheco et al. (2024) investigates the role played by British Standard (BS) 8001:2017 in interpreting the transition from linear economy to circular economy in reference to MSMEs from economies still in development. Despoudi et al. (2023) points the obstacles to adoption of Industry 4.0 for CE policies implementation in fruit and vegetable SMEs in India from a resource-based perspective as linear-based model has caused resource scarcity.

2.2 Artificial Intelligence (AI) and CE Integration

Roberts et al. (2022) has explained that how a transitioning to a circular economy (CE) model of development can be adopted as a sustainable alternative to reduce high levels of resource extraction, consumption, and waste that are creating adverse impact on positive environmental outcomes. Fang et al. (2023) highlights the application of artificial intelligence in waste-to-energy, smart bins, waste-sorting robots, waste generation models, waste monitoring, etc. which results in reduction in transportation distance, cost savings, and time savings as well. Pathan et al. (2023) assists in knowing intersection of AI and CE and policies around implementing CE principles using AI with discussion taken from initiatives taken by the Irish government to adopt circularity and explore the role AI plays in these. Pathan et al. (2023) had explained the application of AI in Ireland to enhance recycling, minimize waste, and maximize the usage of resources. Chi et al. (2023) through their paper has explained the various ways in which AI might support the circular economy. It looks on improving resource efficiency and reducing waste through the use of blockchain, IoT, and artificial intelligence. Tripathi et al. (2024a), Tripathi et al. (2024b) has discussed challenges pertaining to sustainability ranging from local to global level, stressing on use of artificial intelligence and big data.

2.3 Challenges and Ethical Considerations in CE and AI

Purvis et al. (2023) depicts that circular economy as a process of socio-economic transformation grounded in principles of social and environmental justice that faces the danger of creating or exacerbating existing injustices, and certain groups or individuals may be disproportionately affected by its implementation which can be solved through adopting Responsible Research and Innovation (RRI). Grief et al. (2024) has explained how artificial intelligence helps to tackle sustainability challenges. The study has conducted a SWOT analysis where it discusses drawbacks like energy usage and data biases, strengths like improved monitoring and resource efficiency, and opportunities like climate modelling and smart infrastructure. Martini et al. (2024) draws attention to the difficulties in establishing a balance between the growth of technology, human well-being, and a sustainable environment. The paper talks about how artificial intelligence (AI) might reduce the environmental impact and increase the productivity. Onyeaka et al. (2023) had explained how food wastage can be addressed and the circular economy can be supported by artificial intelligence (AI). This paper tells at how artificial intelligence can be used to improve resource efficiency through optimizing supply chains and food spoiling prediction.

2.4 Technological Innovations Supporting CE Goals

Keulemans et al. (2024) puts across changes required in techniques, technologies, labour, and practice within the design industry that greatly impact circular economy through emerging and established digital technologies of Building Information Management (BIM), 3D scanning, and artificial intelligence (AI) for their capacity to ease and improve transformative, design-led reuse practices for various purpose like interiors, furnishings, architecture, and building. Roberts et al. (2022), in his paper, has talked about the earlier research on AI's uses for waste reduction, resource optimization, and recycling. It highlights the effects of AI on employment. It points out areas for both industry and research gaps that require attention. Zota et al. (2024) explores the use of chatbots in smart and green cities to enhance waste disposal. It focuses at how citizens may sort and recycle waste more efficiently by using chatbots. The paper discusses the advantages of incorporating AI-driven to improve productivity. Adewale et al. (2024) has explained the various applications of AI in sustainable buildings. It talks about how AI can be used to design structures that use less resources and enhance the operations.

2.5 Digital Transformation and Emerging Technologies in Sustainability

Rusch et al. (2021) provides a more comprehensive side to the fact of current and potential instances of DT applications in SPM for example product design/assessment, business models, etc. which further allows for new insights in moving towards a circular economy. Marquez et al. (2024) has explained methods for advancing sustainability and managing natural resources in the context of Society 5.0 that talks about integrating modern technologies in order to minimize impact on the environment and manage resources effectively. Lamas et al. (2021) has explained how transition can be made digitally with the use of Edge AI and Green IoT (Internet of Things) effectively that focuses on their contributions to the creation of a waste-free circular economy by improving data processing, networking, and decision-making. Nleya et al. (2024) has explained about Industrial Metaverse. It explains how such technology integrates the physical and virtual worlds to improve industrial processes and operations. The paper raises issues about consumption of resource while adopting metaverse technologies in industries.

3 Research Methodology

The study adopts both descriptive and case study approaches, divided into two parts. Through this two-part methodology, the present study investigates both the thematic dimensions of circular economy and hence framework synthesis has been adopted where previous frameworks have been combined to create a refined and comprehensive framework that includes integrative framework studies, conceptual framework development papers, synthesized and build own conceptual theories. The second part of the study throws light on the case study where framework which is being created has been applied theoretically in FMCG industry to showcase how the synthesis of the framework may happen once it is created.

3.1 *Research Design*

The current study employs a mixed-method approach combining qualitative analysis of existing frameworks and case study application. The research is divided into two parts: The first part includes analysis and modification of existing frameworks to create a new, adapted framework, and the second part includes application of the modified framework in a case study.

3.2 *Framework Analysis and Creation of Modified Framework*

The study employs a qualitative research approach, utilizing a descriptive research design to explore the integration of artificial intelligence (AI) within the circular economy, specifically within the Indian context. The primary objective is to develop a comprehensive framework that will lead to the successful implementation of AI in the circular economy. To achieve this, extensive review of 50 academic research papers is conducted on similar topics, carefully selected, based on their relevance to the Indian economy and their contribution to understand the interplay between AI technologies and circular economic principles. These research papers are accessed from Scopus database and Web of Science to ensure the quality of data being selected and accessed. The location for this research paper is confined to developing nations like India for which yet suitable framework is not in existence, considering its social and economic key factors whereas duration of research paper collected from the Scopus database and the Web of Science dates back from 2015 to 2024. Based on which the current paper has conducted Framework Analysis which is the systematic approach for analyzing qualitative data. This method allows researchers to systematically break down complex information into simpler forms, in order to realize important findings.

The first part of study aimed at identifying the key themes, methodologies, strategies, and limitations of the existing frameworks which later have been used to create the modified version of the framework. The approach establishes a theoretical foundation for understanding circular economy flow for emerging countries like India where sustainable future and economic growth are interlinked. The first research objective of the study is:

> "To create comprehensive and all inclusive framework for the flow of circular economy on the basis of *key themes, methodology and limitations of existing frameworks*"

To substantiate the above research objective, the existing frameworks have been identified in terms of their themes, methodology used, limitations, and further scope, respectively. It has been done in structured form by identifying the scope and area of further development.

3.2.1 Frameworks Comparison and Adaptation

Responsible Research and Innovation (RRI) principles can act as a leading system for making a dramatic shift to a CE by making sure that the transition is socially responsible, inclusive, anticipatory, and reflexive. The AIRR framework comprises the matter pertaining to combination of alternative thinking and reflection, and perspectives from series of stakeholders. As democratic decision-making processes prevail in various countries like India, steps will have to be initiated to confront societal norms, theories of equality, winner/loser settings, and other social factors (Purvis et al., 2023). Whereas, among the different sustainable frameworks recognized to transform the linear economy into circular economy, the ReSOLVE framework hits differently because of its effective methods required for such transitions. The Ellen MacArthur Foundation (2015) states that "the ReSOLVE framework provides a detailed plan for transforming into a circular economy, making sure each step towards managing resources and sustainability is covered. The paper focuses on the need of an hour that requires immediate attention towards sustainability which calls the various organizations to implement different strategies and actions which will help them to grow sustainably. According to Bocken et al. (2016), the ReSOLVE framework's emphasis on particular steps such as sharing, repetition, and optimizing makes it extremely useful for the organizations that want to incorporate techniques into their business which will lead to the circular economy. However one must be innovative and ready to adopt the technological changes, when it comes to survival in such competitive era. Technology is an important component of the circular economy, and the ReSOLVE framework works to encourage the growth and implementation of innovative solutions. ReSOLVE framework is interconnected with various Sustainable Development Goals (SDG) promoting fuller utilization of resources and minimized waste generation which help the organizations to grow sustainably. As economic growth treads along the footpath of linear economic model, India is almost on the path of industrialization that warns to worsen existing negative externalities and various environmental challenges like pollution, greenhouse gas emissions, and water and food scarcity. In India, 150,000 tonnes of municipal solid waste (MSW) is produced per day where only 83% of the waste created is collected and only 30% is treated (Kaalari, 2021). The current economic growth impact in India is negative on basis of disparity in income and resource allocation. However, implementing a circular economy model would create a better provision for resource consumption but also can intensify the equitable growth (Kaalari, 2022). After considering various frameworks AIRR framework and ReSOLVE framework, this paper proposes better and modified framework ReGen framework for implementation of circular economy which brings its key components starting from regeneration of natural resource and ending at monitoring the entire circular economy through vivid risk analysis and efficient system for its mitigation. This framework brings in positive impact on various social factors which scale down to all income groups of society, therefore, creating the component of accessibility and inclusivity as shown in Fig. 1.

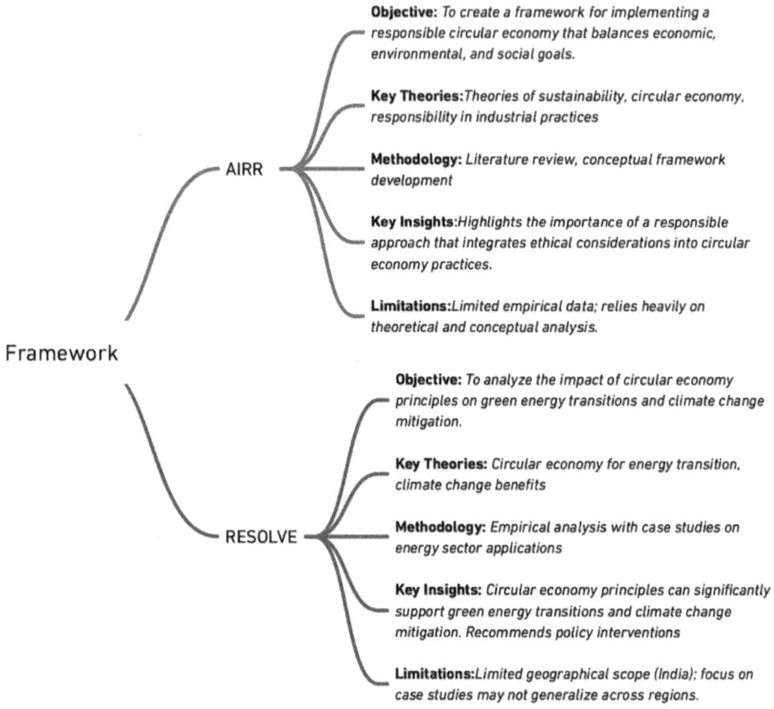

Fig. 1 Existing frameworks. Author's own

Figure 2 discusses Accountability, Inclusivity, Responsibility, and Reflexivity, or AIRR. It is a concept designed to help organizations and policymakers innovate in a way that is morally and socially responsible, especially when it comes to sustainability and emerging technology. Every element has a distinct function: Accountability guarantees that companies are held responsible for their deeds and the effects they have on the environment and society. Including a range of perspectives, particularly those influenced by decisions, is crucial for ensuring a well-rounded perspective, as inclusion emphasizes (Mohanty et al., 2024). By following the AIRR framework, which emphasizes social responsibility and openness, organizations can more successfully negotiate the complex ethical landscape of sustainable practices and innovation. This is shown in Fig. 2

Figure 3 illustrates the RESOLVE framework, which was developed specifically for putting circular economy efforts into practice, to help companies transition from linear production patterns to more sustainable, circular processes. The acronym RESOLVE stands for Regenerate, Share, Optimise, Loop, Virtualise, and Exchange—all crucial processes in promoting circularity. Regenerate seeks to revitalize and repair ecosystems through the use of renewable resources. By encouraging resource sharing, sharing reduces the demand for new products. Optimise prioritizes waste minimization, as well as efficient production and consumption.

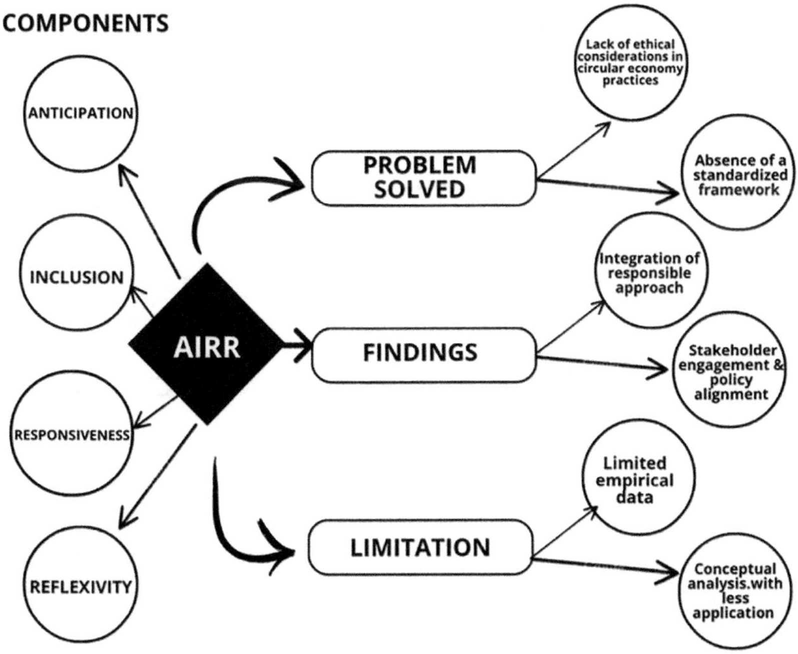

Fig. 2 AIRR framework. Author's own

To keep materials in continuous circulation, loop entails recycling and remanufacturing them. Exchange encourages the replacement of less sustainable materials and energy sources with more ecologically friendly alternatives, whereas Virtualise seeks to replace tangible goods with digital services whenever feasible.

3.2.2 Comprehensive Framework Development for Circular Economy

The REGENO Framework for risk analysis further provides assistance in collecting potent information regarding potential challenges and existing as well as upcoming vulnerabilities popping in circular systems, such as disruptions in supply chain, fluctuations prevailing in market, technological limitations, climate change, pollution, and overexploitation that could create an obstacle in regeneration. Through assessment of these risks, stakeholders can devise mitigation strategies to bring into factors of resilience and adaptability. This combo of monitoring and risk analysis allows to surface a feedback loop that aids the dynamic and ongoing enhancement in circular economy practices, assisting in achievement of long-term sustainability and economic viability as shown in Fig. 4.

Revitalize is the first component in the process which deals with the core notion of renewal and refreshment of ecosystems that get depleted in traditional economies. Revitalize reflects how AI can identify opportunities for renewal and restoration of

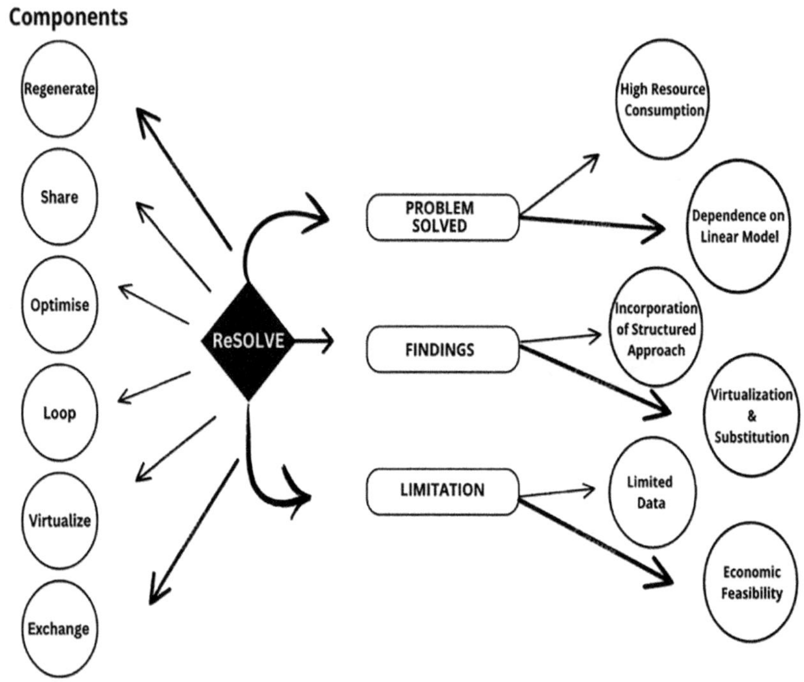

Fig. 3 ReSOLVE framework. Author's own

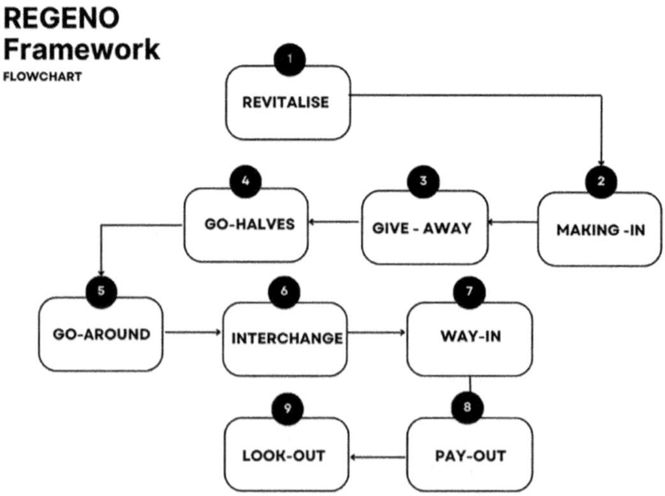

Fig. 4 REGENO framework. Author's own

ecosystems. This component of the framework can enable the application of technology in ways that do less harm while efficiently seeking the reversal of loss within a country like India—where depletion and degradation of resources and the environment are pressing concerns (Khan et al., 2023). AI-powered tools—predictive analytics and remote sensing, for example—allow real-time monitoring of environmental health (Paliwal et al., 2024), allowing the early diagnosis of ecosystem degradation to guide focused intervention tactics. By integrating AI into such regeneration efforts, the framework advances this idea.

It is followed by second component, making in which discusses efficient production is the major component in the shift from a traditional economy towards a more circular economy (Kumar et al., 2024). This inlays strong emphasis on making production procedures clean and reducing waste, alongside the maximization of resources extracted. Unlike the traditional models of linear production, often associated with high resource depletion and bad environmental impacts, efficient production seeks to improve resource use, ease the procedures, and reduce general environmental impacts (Badjeena et al., 2024). This may involve making resource-efficient products that result in less waste besides increasing production efficiency (Chou, 2024). For example, artificial intelligence (AI) systems may analyze real-time data from production lines to identify inefficiencies, predict equipment failures, and recommend adjustments to lower resource consumption (Kerkeni et al., 2024). Predictive maintenance driven by AI extends the lifespan of equipment and avoids unplanned downtime (Kerkeni et al., 2024). Give Away, the third element in the flow, emphasizes the deliberate allocation of resources to maximize their impact and worth. This concept's primary objective is to allocate extra resources to areas where they are most needed in order to avoid wasting them (Debrah et al., 2022). AI could significantly enhance the concept of more precise and efficient resource distribution (Singh et al., 2024). In order to optimize the distribution process, artificial intelligence can use machine learning algorithms and sophisticated data analytics to spot patterns in resource availability and demand.

Go-Halves, the fourth element in the flow, encourages the sharing economy, in which multiple people can pool resources, goods, and services rather than owning them individually. By making greater use of what is already available, such a plan helps to decrease waste and the turnover of new items. It aims to increase resource durability and make sure that resources are used in ways that benefit the community as a whole through shared utilization. Next comes, Give Away which is closely related to this philosophy. Following their distribution through the Give Away concept, the Go-Halves component ensures that these resources are not only allocated but also used jointly. A loop that increases the value of reallocated resources and reduces waste is created by actively using them (Muller et al., 2024). Sharing has financial advantages since it lowers the cost of ownership and promotes more equitable access to resources (Ding et al., 2024). Shared-use models could also give rise to collaboration and community involvement in trying to create a setting socially more cohesive and sustainable. Through the integration of this element with the Give Away, the framework puts in place a holistic mode of resource management that amplifies efficiency and sustainability. Fifth core component is Go Around which articulates that resources and their associated goods are cycling on a continuous basis. This way of thinking encourages recycle, reuse, and repair to gain maximum period for which the resources remain in service.

Next component is Interchange that plays an integral part for the flow of circular economy as it assists in the optimum use of resources and reduction in waste. Interchanging explains the practice or process of replacing or substituting one resource or product for another to optimize, use, and increase its lifecycle (Hernández-Chover et al., 2024). Followed by seventh component, Way—in that discusses key aspect of accessibility while implementing a circular economy comes into picture, making sure that sustainable practices are equally inclusive and widespread. This encompasses making affordable and easy-to-use programmes targeting recycling, giving widespread access to services like repair and refurbishment, and providing that information about sustainable practices is readily available and the public can easily comprehend (Negrutiu, 2024). Eighth component is Government Payout where Government initiatives are instruments to bring a circular economy into existence through devising, proceeding, and implementing procedures and policies that encourage sustainable practices and resource efficiency. Financial Assistance like subsidies, grants tax incentives and reforms for all sectors and business enterprises who are onto the path of adopting circular economy principles can significantly decrease the financial burden and boost innovation in various aspects and dimensions pertaining to recycling, remanufacturing, and eco-design. Ninth and final component in the process is identified as Look out for which monitors the rhythm of circular economy and analysis of risk associated with its adoption and implementation are significant for its successful functioning, ensuring that initiatives are efficient and sustainable. Figure 4 summarizes all the components of framework created in order to include untouched aspects of circular economy, especially for emerging countries like India where the potential is high (Table 1).

3.3 Application of the Modified Framework

The second part of study is to identify the potential areas where the framework may be used. The approach establishes a theoretical foundation for understanding circular economy flow for fast-moving consumer goods industry (FMCG) in India. The second research objective of the study is:

> "To identify potential areas within the FMCG sector in India where the theoretical framework of circular economy flow can be effectively applied, enhancing sustainability practices and resource efficiency".

To substantiate the above research objective, the existing frameworks have been identified in terms of their themes, methodology used, limitations, and further scope, respectively. It has been done in structured form by identifying the scope and area of further development. This research adopts a qualitative research approach with an emphasis on secondary data analysis. The focus is on identifying potential areas

Table 1 Development of modified framework (Author's own)

S. no	REGENO dimension	Issues regarding transition	Key objectives	Potential strategies	Techniques & approach
1	Revitalize	· Lack of Consumer	· Revitalize natural resources	· Policies promoting Triple Bottom Line (TBL)	· Eco-Design and Preemptive Environmental Planning
		· Awareness and Acceptance	· Foster Integrated	· sustainability approaches	· Net Zero Strategies
		· High Upfront Costs	· Ecosystem Management	· Focus on achieving	· Permaculture
		–	· Increasing resilience	· carbon neutrality	· Closed Loop System
		–	–	· Net zero strategies and use of local materials	–
		–	–	· Implementing community-driven strategies to ensure	–
		–	–	· resilient and flexible infrastructure	–
2	Making In	· Technology Adoption	· Efficient utilization of resources	· Gain real insights from the market to line the production in response to demand	· Lifecycle Assessment
		· Quality Control and Consistency	· Minimize byproducts	· Implement advanced	· Cradle-to-Cradle Design
		· Scalability Issues	· Technology upgradation	· quality assurance systems to maintain product	· Circular Product
		–			
		–	–	· consistency and quality.	· Development
		–			
		–	–	· Improved Machinery for less waste generation	–
		–	–		
3	Give away	· Fluctuating inventories are challenging to manage	· Align supply with market demand	· Create an online platform for all stakeholders to collaborate, share, resource distribution efforts	· Reverse logistics systems

(continued)

Table 1 (continued)

S. no	REGENO dimension	Issues regarding transition	Key objectives	Potential strategies	Techniques & approach
		· Transporting resources over long distances is complex	· Optimize logistics & distribution	· Implement predictive analytics for demand & supply forecasting	–
		· Poor coordination causes inefficiencies in logistics	· Ensure resource quality and suitability	· Adopt Just-In-Time (JIT) inventory systems	–
4	Go-Halves	· Lack of trust hinders collaboration	· Sharing of assets	· Policies for reuse through resale	· Peer-to-Peer sharing platforms
		· Compatibility issues lead to inefficiencies	· Facilitate assets reuse	· Integrated design and operational teams throughout the life of the asset	· Freecycle Network
		· Ambiguity in maintenance responsibilities	· Foster integrated	· Encourage material reuse through resale policies	· Product-as-a-service (PaaS)
		–	· Asset management	· Foster lifecycle collaboration for shared ownership	· Leasing models for consumer goods
		–	· Maximum utilization of resources	–	–
5	Go around	· Lack of infrastructure	· Minimize waste generation through upcycling techniques	· Design for durability and repairability minimizing resource depletion	· Product Life Cycle Assessment (LCA)
		· Initial high costs	· Enhance resource efficiency through closed-loop systems	· Creation of reverse logistics systems for efficient material recovery	· Extended Producer Responsibility (EPR)
		· Regulatory hurdles	· Foster innovation technologies for materials recovery	· Extension of product life through 5Rs	· Material flow

(continued)

Table 1 (continued)

S. no	REGENO dimension	Issues regarding transition	Key objectives	Potential strategies	Techniques & approach
		· Difficulty in standardizing	· Bring sustainability in production and consumption patterns	· Initiate for take-back plans for retrieving used products for recycling	· Analysis (MFA)
		–	· Cultivate partnership among businesses government and consumers	· Devise eco-friendly, biodegradable packaging	· Urban mining
		–	–	–	· Industrial Symbiosis
6	Interchange	· Supply chain complexity	· Facilitate material flow	· Digital marketplaces for businesses	· Blockchain for transparency
		· Logistical challenges	· Reduce waste generation	· Cross-industry collaboration	· Reverse logistics
		· Standardization issues for materials and products	· Improved collaboration through efficient exchange systems	· Incentive programmes like credits, or tax benefit	· Product passports
		· Data transparency regarding material origins, qualities	· Ensure traceability for materials and products through supply chain	–	· Closed-loop supply chains
		· Financial & non-financial incentivization gaps	–	–	· Resource pooling
7	Way in	· Restricted access to sustainable resources	· Inclusive resource accessibility, affordability for all income consumers	· Subsidized circular economy programmes	· Resource sharing platforms

(continued)

Table 1 (continued)

S. no	REGENO dimension	Issues regarding transition	Key objectives	Potential strategies	Techniques & approach
		· Gaps in technology usage	· Policy development to foster inclusivity	· Local resource hubs for resource sharing, refurbishing, recycling	· Product-as-a-service models
		· Improper unregulated policy support	–	· Creation of micro-enterprises to boost small circular business	· Community-led recycling initiatives
		· Gaps in infrastructure for resource redistribution	–	· Public–private partnerships for access and affordability	· Inclusive modular design
8	Pay-out	· Budget constraints creating difficulty in fund allocation	· Financial support provision for circular economy practices	· Conditional funding programmes	· Impact assessments
		· Bureaucratic delays can decrease the flow government pay-outs distribution	· Efficiency and transparency for pay-outs distribution	· Performance-based incentives	· Digital payment systems for traceable distribution
		· Misallocation of funds to projects	· Supporting SMEs and local initiatives	· Public–Private Partnerships (PPPs)	· Targeted Financial Instruments like green bonds or circular economy-specific financial instruments
		· Poor monitoring and evaluation mechanisms	–	· Pilot Programmes to test the effectiveness before scaling up	· Monitoring & Evaluation (M&E) Frameworks
		–	–	· Direct subsidies and grants	· Capacity building programmes

(continued)

Table 1 (continued)

S. no	REGENO dimension	Issues regarding transition	Key objectives	Potential strategies	Techniques & approach
9	Look out for	· Data Inconsistencies impacting ability to monitor circular processes effectively	· Standardized Metrics to measure circular economy performance	· Integrate technology for efficient monitoring	· Circular Economy Dashboards
		· Lack of integrated systems across different sectors	· Real-time feedback monitoring	· Encourage cost-effective solutions for small businesses	· Lifecycle monitoring tools
			· Data accuracy and transparency	· Regulatory harmonization across different sectors	· Blockchain for transparency
			· Compliance with circular economy regulations and standards	· Open data platforms	· Automated reporting
			–	–	· Sustainability audits
			–	–	·Predictive analytics

where the circular economy (CE) framework can be applied effectively within the FMCG sector in India, by examining the sustainability practices of two leading FMCG companies, ITC and Unilever. The global economy's cornerstone, the fast-moving consumer goods (FMCG) sector, faces enormous environmental issues due to its traditional linear model of production and consumption, which generates a lot of waste and depletes resources. Industry contributes to about 50 per cent of the world's packaging wastes; thus, the demand for innovative approaches inducing some sustainability intervention is ever on the rise nowadays. The REGENO Framework integrates circular economy principles into the FMCG industry to offer an integrated solution to reduce waste, increase resource efficiency, and regenerate damaged ecosystems. This framework provides a means for FMCGs to reduce waste, maximize the use of resources, and spur environmental regeneration to tackle the problems surrounding excessive packaging waste and depletion of natural resources. Conversion of FMCG from a linear-based model into a circular one will not only reduce the environmental burden of the products but will also heighten the supply chain efficiency and the industry's resilience. The practices of the circular economy enabled by this framework can result in maintenance savings, boosted brand recognition, and the alignment of companies with consumer sentiment underpinning the growth of a demand for sustainability products. This case study seeks to investigate the REGENO Framework application in the FMCG sector, showcasing its potential to transform how products are designed, produced, consumed, and disposed of. The REGENO Framework, composed of elements that include everything from AI-driven production techniques to resource allocation and circularity of the product's design, is said to embody these reductions in environmental loading tremendously.

3.3.1 Implementing the REGENO Framework in the FMCG Sector in India

The FMCG sector is a bit tough in the context of developing nations like India; from resource constraints to sustainability pressure, this is considered so wide. This one unique approach to transform the sector into one that is more inclusive, resilient, and sustainable using this sustainable advanced model through a total of nine interconnected elements is called the REGENO Framework. The study will adopt a case study design involving secondary data from ITC and Unilever to analyze their adoption of circular economy principles in their operations. The research will discuss sustainability reports, environmental impact reports, and publicly accessible documentation on their resource management practices. The study also compares both companies' efforts and draws insights as to how the circular economy model could be improved in the Indian FMCG sector.

Application of Circular Economy Framework in ITC

Circular economy principles are integrated into ITC business practices through overall reduction of environmental impact, sustainable packaging, and waste management. ITC's Sustainability 2.0 programme aims to evolve systems of comprehensive and repeatable waste management that work to promote recycling in local communities and save landfills from being filled (Fig. 5).

The "Well Being Out of Waste" (WOW) programme is supported under the initiative of recycling and solid waste separation centred on homes and schools in India, extending to 4.6 million homes and schools in total. Among the innovations are biodegradable and recyclable packaging options, such as the "Filo" series and the "Bioseal" solutions, increasingly utilized in the food service and fast-moving consumer goods industries. A leader in this field, ITC has kicked off big projects like the Green Temple project that compost and convert waste from temple premises into biogas, collaborates with local governments to install best practices in waste management. This model tackles major issues of waste, resource consumption, and carbon footprint reduction demonstrating ITC's commitment to continue to build on a sustainable circular economy, additionally inching closer towards the national goal of Swachh Bharat (Clean India).

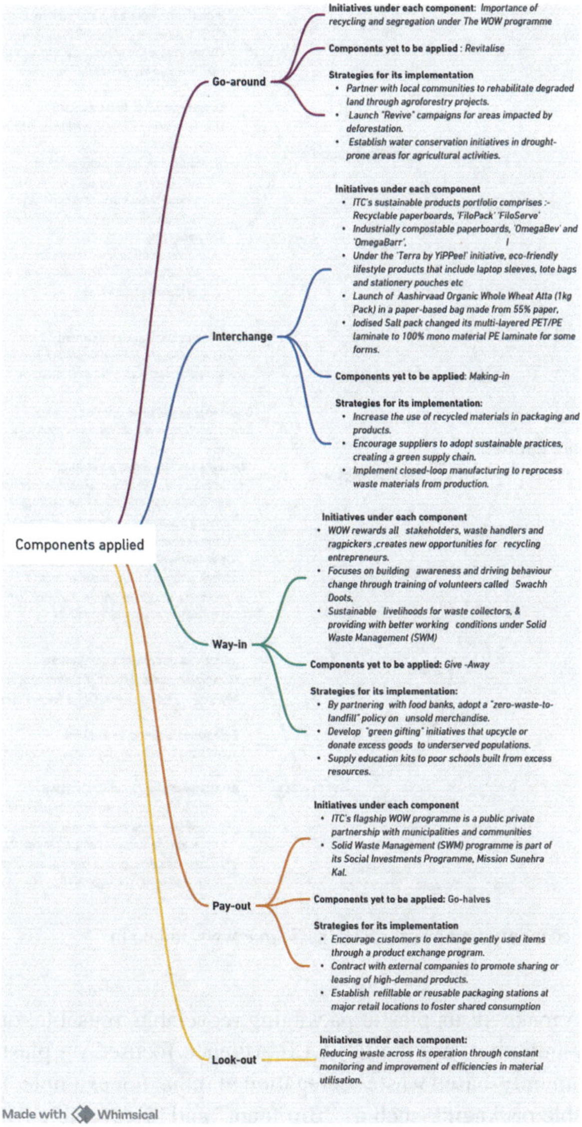

Fig. 5 Circular economy framework in ITC. *Source* www.itcportal.com

Application of Circular Economy Framework in HUL

The Hindustan Unilever (HUL) is inducting the principles of the circular economy into core business strategies in such a manner that they reduce the harmful impacts on the environment, provide sustainable packaging, and develop best practices of waste management. HUL has identified an ambitious commitment in its Sustainable

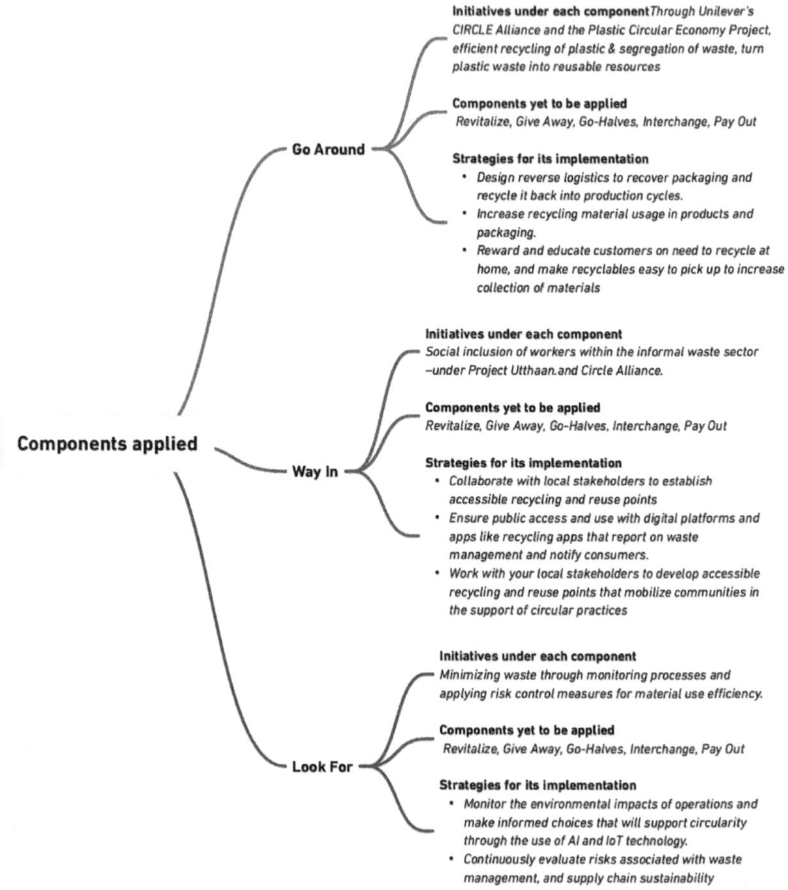

Fig. 6 Circular economy framework in HUL. *Source* www.hul.co.in

Living Plan to make all its plastic packaging recyclable, reusable, or compostable. Here, the key initiatives are the CIRCLE Alliance focused on plastic waste recycling and community-based waste segregation in India. For example, HUL innovates in biodegradable packaging such as "Bio foam" and "Recyclable Mono-materials" towards sustainable solutions for FMCG (Fig. 6).

4 Findings

The research states the implementation of the REGENO framework, a circular economy model, across developing nations, particularly India. This model presents a transformative approach to sustainable development with incorporation of nine interrelated components: revitalize, making-in (production), give away (distribute), go-halves (sharing), go around (recycling), interchange (substitute), way-in (accessibility), government payout, and look out for (monitor). The findings give an insight into opportunities and challenges related to the possibility of applying the framework to the fast-moving consumer goods sector in India—a very important industry with a lot of growth due to sustainability.

Circular economy concepts of the REGENO framework, such as supply chain management, production, and product design, can be applied to FMCG. Companies can work with sustainable raw materials (Interchange), make use of recycled packaging (Go Around), and also engage in product-sharing or refill systems (Go-Halves). Possible Collaboration: The cooperation between FMCG companies and government comprises accessing concessional financing or tax breaks (Government Payout) and tracking their environmental impact through data-driven solutions (Look Out For). With the adoption of a REGENO framework, an FMCG company lowers the associated waste, optimizes the source of the business model, and, thereby, advocates green operations (Table 2).

5 Managerial Implications

Implementing the REGENO Framework offers managers who want to lead sustainable initiatives a number of benefits. Using components like Revitalize and Making-in can give managers a competitive edge in resource-constrained markets by increasing productivity, reducing resource costs, and bolstering supply chain resilience. Additionally, managers can employ Give Away and Go Around to benefit from closed-loop distribution and recycling cost savings, which reduce garbage disposal expenses and foster an image of environmental responsibility, which in turn boosts customer confidence and loyalty. Furthermore, Go-Halves and Interchange components encourage creativity in product design and materials, enabling managers to differentiate their offerings and satisfy consumer demand for environmentally friendly alternatives. The Way-In component increases market reach and promotes growth by diversifying the consumer base across demographics. Through Government Payout, managers can obtain government incentives that will boost profitability by making up for the initial investment in sustainable practices. Last but not least, the Look Out For component assists managers in putting in place robust monitoring systems that reduce regulatory risks and improve the long-term sustainability of their companies by promoting accountability. Using REGENO, which offers a framework for strategy for attaining financial goals while attending to social and environmental responsibilities, managers may establish their organizations as leaders in the sustainable economy (Fig. 7).

Table 2 Generalized findings. Author's own

Component	Generalized findings
Revitalize	· emphasizes restoring natural resources and ecosystems
	· Challenges include high resource extraction rates, deforestation, and population pressures
	· Requires collaboration between public and private sectors for effective restoration
Making-in	· Prioritizes sustainable production methods to reduce waste
	· Focus on energy efficiency and the use of renewable resources
	· Green manufacturing practices gaining popularity globally
Give away	· Aims to shift from linear supply chains to closed-loop systems
	· Encourages reusable packaging and sustainable logistics
	· Scaling remains challenging in areas with developing infrastructure
Go-Halves	· Promotes shared consumption to reduce demand for new products
	· Aligns with traditional communal values but lacks widespread digital infrastructure
	· Rising urbanization offers potential for expansion
Go-around	· Integrates formal recycling with existing informal systems
	· Waste pickers play a key role in material recovery, especially plastics and metals
	· More innovation is needed for hard-to-recycle materials
Interchange	· Focuses on substituting harmful materials with sustainable alternatives
	· Biodegradable and compostable options gaining traction
	· Scale and cost barriers remain for mass adoption of alternatives
Way-In	· Ensures accessibility of sustainable solutions across all income levels
	· Emphasizes bridging urban–rural disparities
	Requires public awareness initiatives to drive demand for sustainable products
Government payout	· Highlights need for government incentives to support sustainable practices
	· Tax incentives, subsidies, and financial support are essential
	· Clearer regulatory frameworks are needed for effective adoption
Look out for	· Stresses accountability and monitoring of sustainability practices
	· Transparency in supply chains and compliance audits are necessary
	· Technology (e.g. blockchain) can enhance tracking and accountability

6 Conclusion and Scope for the Further Study

The REGENO Framework offers a comprehensive and practical roadmap for growing the circular economy in emerging nations like India. This strategy's nine interconnected elements—revive, make-in, give-away, go-halves, go around, interchange,

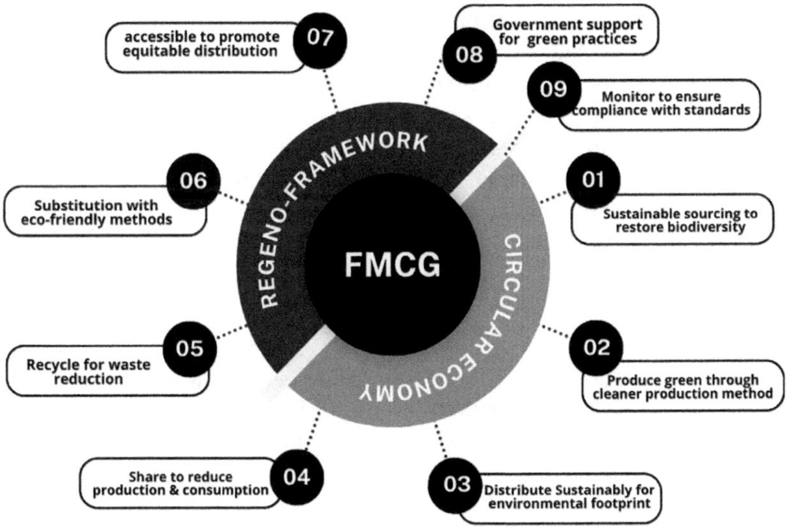

Fig. 7 Application of REGENO framework in FMCG sector. Author's own

way-in, government payout, and look out for—help the FMCG industry address the unique opportunities and challenges it encounters as it works towards sustainability. India's growing population, rapid urbanization, and increasing demand for consumer goods make it a vital and challenging environment for circular economy initiatives. The findings indicate that while India has made significant strides in putting circular principles into practice—for instance, by embracing sustainable production methods, promoting recycling, and supporting government programmes—there are still significant challenges to be addressed, including low consumer awareness, the high cost of sustainable alternatives, and inadequate infrastructure. However, the potential for public–private sector collaboration and rising consumer awareness offers a strong foundation for future growth.

As India continues to develop, the successful implementation of this strategy can act as a template for other developing nations seeking to reconcile economic expansion with environmental stewardship. The research revealed a number of limitations, despite the REGENO framework's encouraging promise for the implementation of the circular economy in India. First, infrastructural shortcomings, particularly in rural regions, hinder the broad adoption of sustainable practices like closed-loop distribution and efficient recycling. Many eco-friendly products and technologies, such as biodegradable packaging, are still too costly for widespread use, especially in the fast-moving consumer goods (FMCG) sector where profit margins are usually narrow. Additionally, the informal recycling sector—which is essential to material recovery—does not work in tandem with formal systems, leading to inefficiencies and missed opportunities for resource optimization. Another disadvantage is that the shift to sustainable production and consumption is sluggish since consumers and smaller businesses are not well-informed about the concepts of the circular economy.

Furthermore, bureaucratic inefficiencies that frequently impede their implementation have hampered the effectiveness of government incentives and programmes, even if they have been developed. Finally, without robust monitoring and accountability systems, it is challenging to assess progress and ensure adherence to the circular economy's tenets, particularly for smaller enterprises. These limitations highlight the need for more targeted interventions, increased collaboration, and stronger regulatory frameworks in order to effectively benefit from the REGENO framework in developing nations like India.

Acknowledgements We are grateful for the insightful criticism provided by our reviewers, which has greatly raised the calibre of this work.

Funding Statement The authors declare that they did not receive support from any organization for the submitted work. No funding was received to assist with the preparation of this manuscript. No funding was received for conducting this study. No funds, grants, or other support was received.

Conflict of Interest The authors declare that there are no conflicts of interest regarding the publication of this book chapter.

References

Bąk, J., & Żaba, T. (2024). Towards reducing the consumption of drinking water in buildings as part of the circular economy model: Strengths–weaknesses–opportunities–threats analysis and perspectives for implementation. *Energies, 17*(6), 1444. https://doi.org/10.3390/en17061444

Chauhan, A., Jakhar, S. K., & Chauhan, C. (2020). The interplay of circular economy with industry 4.0 enabled smart city drivers of healthcare waste disposal. *Journal of Cleaner Production, 279,* 123854. https://doi.org/10.1016/j.jclepro.2020.123854

Danish, M. S. S., & Senjyu, T. (2023). Shaping the future of sustainable energy through AI-enabled circular economy policies. *Circular Economy, 2*(2), 100040. https://doi.org/10.1016/j.cec.2023.100040

De Jesus Pacheco, D. A., Rampasso, I. S., Michels, G. S., Ali, S. M., & Hunt, J. D. (2024). From linear to circular economy: The role of BS 8001:2017 for green transition in small business in developing economies. *Journal of Cleaner Production, 439*, 140787. https://doi.org/10.1016/j.jclepro.2024.140787

Despoudi, S., Sivarajah, U., Spanaki, K., Charles, V., & Durai, V. K. (2023). Industry 4.0 and circular economy for emerging markets: evidence from small and medium-sized enterprises (SMEs) in the Indian food sector. *Annals of Operations Research.* https://doi.org/10.1007/s10479-023-05404-4

Filho, W. L., Salvia, A. L., Paço, A., Dinis, M. a. P., Vidal, D. G., Da Cunha, D. A., De Vasconcelos, C. R., Baumgartner, R. J., Rampasso, I., Anholon, R., Doni, F., Sonetti, G., Azeiteiro, U., Carvalho, S., & Ríos, F. J. M. (2022). The influences of the COVID-19 pandemic on sustainable consumption: an international study. *Environmental Sciences Europe, 34*(1). https://doi.org/10.1186/s12302-022-00626-y

Hernández-Chover, V., Bellver-Domingo, Á., Castellet-Viciano, L., & Hernández-Sancho, F. (2024). AI applied to the circular economy: An approach in the wastewater sector. *Sustainability, 16*(4), 1365. https://doi.org/10.3390/su16041365

Keulemans, G., Harle, J., Hashimoto, K., & Mugavin, L. (2020). Design considerations for the transformative reuse of a Japanese temple. *Proceedings of DRS, 1.* https://doi.org/10.21606/drs.2020.183

Pathan, M. S., Richardson, E., Galvan, E., & Mooney, P. (2023). The role of artificial intelligence within circular economy activities—A view from Ireland. *Sustainability, 15*(12), 9451. https://doi.org/10.3390/su15129451

Purvis, B., Celebi, D., & Pansera, M. (2023). A framework for a responsible circular economy. *Journal of Cleaner Production, 400*, 136679. https://doi.org/10.1016/j.jclepro.2023.136679

Raman, R., Pattnaik, D., Lathabai, H. H., Kumar, C., Govindan, K., & Nedungadi, P. (2024). Green and sustainable AI research: An integrated thematic and topic modeling analysis. *Journal of Big Data, 11*(1). https://doi.org/10.1186/s40537-024-00920-x

Roberts, H., Zhang, J., Bariach, B., Cowls, J., Gilburt, B., Juneja, P., Tsamados, A., Ziosi, M., Taddeo, M., & Floridi, L. (2022). Artificial intelligence in support of the circular economy: Ethical considerations and a path forward. *AI & Society, 39*(3), 1451–1464. https://doi.org/10.1007/s00146-022-01596-8

Sánchez-García, E., Martínez-Falcó, J., Marco-Lajara, B., & Manresa-Marhuenda, E. (2023). Revolutionizing the circular economy through new technologies: A new era of sustainable progress. *Environmental Technology & Innovation, 33*, 103509. https://doi.org/10.1016/j.eti.2023.103509

Tripathi, S., Bachmann, N., Brunner, M., Rizk, Z., & Jodlbauer, H. (2024). Assessing the current landscape of AI and sustainability literature: Identifying key trends, addressing gaps and challenges. *Journal of Big Data, 11*(1). https://doi.org/10.1186/s40537-024-00912-x

Falk, S et al. (2023). Challenging AI for sustainability: What ought it mean? *AI and Ethics,* 1–11.

Tripathi, S., et al. (2024). Assessing the current landscape of AI and sustainability literature: Identifying key trends, addressing gaps and challenges. *Journal of Big Data, 11*(65), 1–68.

Das, S et al. (2024). Clustering the research at the intersection of industry 4.0 technologies, environmental sustainability and circular economy: Evidence from literature and future research directions. *Circular Economy and Sustainability,* 1–32.

Grief, L., et al. (2024). Strategic view on the current role of AI in advancing environmental sustainability: A SWOT analysis. *Discover Artificial Intelligence, 4*(45), 1–19.

Raman, R., et al. (2024). Sustainable development goal 12 and its synergies with other SDGs: Identification of key research contributions and policy insights. *Discover Sustainability, 5*(150), 1–26.

Roberts, H et al. (2022). Artificial intelligence in support of the circular economy: Ethical considerations and a path forward. *AI and Society,* 1–14.

Pathan M et al. (2023). The role of artificial intelligence within circular economy activities—A view from Ireland. *Sustainability*, 1–18.

Rathore, A., et al. (2023). Deploying artificial intelligence for circular economy and its link with sustainable development goals. *International Journal of Policy Sciences and Law, 1*(4), 1–29.

Canedo, V., et al. (2024). A review of green artificial intelligence: Towards a more sustainable future. *Neurocomputing, 599*, 1–10.

Martini, B. et al. (2024). Human-centered and sustainable artificial intelligence in industry 5.0: Challenges and perspectives. *Sustainability, 16*(13), 1–17.

Zota, R., et al. (2024). Practical approach for smart and circular cities: Chatbots used in waste recycling. *Applied Sciences, 14*(7), 1–23.

Marquez, C. et el. (2024). Natural resource optimization and sustainability in society 5.0: A comprehensive review. *Resources, 13*(2), 1–20.

Chi, Z., et al. (2023). Driving circular economy through digital technologies: Current research status and future directions. *Sustainability, 15*(24), 1–28.

Onyeaka, H., et al. (2023). Using artificial intelligence to tackle food waste and enhance the circular economy: Maximising resource efficiency and minimising environmental impact: A review. *Sustainability, 15*(13), 1–20.

Rosario, A., et al. (2022). Sustainability and the digital transition: A literature review. *Sustainability, 14*(7), 1–18.

Lamas, P. et al. (2021). Green IoT and edge AI as key technological enablers for a sustainable digital transition towards a smart circular economy: An industry 5.0 use case. *Sensors, 21*(17), 1-36.

Chidepatil, A. et al. (2020). From trash to cash: How blockchain and multi-sensor-driven artificial intelligence can transform circular economy of plastic waste? *Administrative Sciences, 10*(2), 23, 1–36.

Adewale, B. et al. (2024). A systematic review of the applications of AI in a sustainable building's lifecycle. *Buildings, 14*(7), 2137, 1–28.

Penailillo, F. et al. (2024). Transformative technologies in digital agriculture: Leveraging internet of things, remote sensing, and artificial intelligence for smart crop management. *Journal of Sensor and Actuator Networks, 13*(4), 39, 1–26.

Mohanty, R., Samal, A., & Kumar, R. (2024). Green finance supports sustainable environment: Empirical approach towards applicability in tourism sector. *Theoretical and Practical Research in Economic Fields, 15*(1), 114–126.

Kumar, R., Goel, R., Singh, T., Mohanty, S. M., Gupta, D., Alkhayyat, A., & Khanna, R. (2024). Sustainable finance factors in Indian economy: Analysis on policy of climate change and energy sector. *Fluctuation and Noise Letters, 23*(2), 2440004–339.

Relation Between Green Economy and Sustainable Manufacturing Through Industry 5.0

Chunchu Nagaraju, Gurunadham Goli, and Rachoru Himani Srihita

Abstract Some of the necessary focuses for transition towards environmentally conscious industrial practices are the overlapping areas between the green economy and sustainable manufacturing. As Industry 5.0 develops while focusing on human-centric technologies that are environmentally sustainable, the synergy of these two paradigms all the more becomes important. Such principles seek resource efficiency, low emissions, and circularity—the aces for the goals of sustainable manufacturing to make an industrial ecosystem more resilient and eco-friendlier. Industry 5.0, which involves cooperation between intelligent machines and people, presents new solutions to conflicts with sustainability by optimizing processes related to production with low waste and the integration of renewable energy sources. By incorporating advanced technology like artificial intelligence, IoT, and robotics, Industry 5.0 hopes to build environmentally friendly and economically growth-enhancing smart factories. Some of the leading approaches, like cleaner production techniques or digital innovation that may encourage a rapid shift towards sustainable manufacturing in line with the imperatives of a green economy are outlined in this chapter. This chapter provides an overview of changes to such practices like industrial and technological, that could be the outcome of Industry 5.0, Synergy Between Sustainable Manufacturing and Industry 5.0, Technological Enablers for Circular Economy, as it unfolds in the cases of real life and new trends. Thus, Industry 5.0 can be finally positioned as a great enabler of the green economy by promoting sustainable development and by ensuring that economic prosperity and responsibility exist simultaneously for tomorrow's world.

Keywords Green economy · Sustainable manufacturing · Industry 5.0 · Human-Centric technologies · Circular economy

C. Nagaraju (✉) · G. Goli · R. H. Srihita
School of Business, SR University, Warangal, India
e-mail: 2302C70024@sru.edu.in

G. Goli
e-mail: gurunadhan.g@sru.edu.in

R. H. Srihita
e-mail: 2302C70018@sru.edu.in

R. Kumar and S. N. Mohanty (eds.), *Sustainable Economy Models in the Age of Industry 5.0*, https://doi.org/10.1007/978-981-96-4104-8_9

1 Introduction

This is Industry 4.0's shift to Industry 5.0, in which producing has become more focused on people, sustainable, and collaborative. Industry 5.0 aligns well with the goals of the green economy because it incorporates digital advancements, such as the Industrial Metaverse, with sustainable manufacturing practices. The transition is not only technology-driven but also has socio-technical elements that address environmental challenges while putting human welfare first. Yao et al. (2024) It could be debated, then, that Industry 5.0 puts much greater concentration on the formation of wisdom and includes a connected CPS that, through cooperation, generates collective intelligence as related to the creation of responsive, sustainable production conditions. Under such circumstances, an Industry 5.0 metaverse would lay out the conditions under which intelligent manufacturing systems adapt to the green economy objectives within fewer resources, using more environmentally favourable forms.

Industry 5.0 is further expanding the frontiers of digital manufacturing as its values are now aligned with those of stakeholder interests. Social and environmental issues are therefore integrated into production. For Sarkar et al. (2024) this is a human-centred framework based on key pillars developed through Industry 4.0 but enhancing the overall capacity of such integration at highly comprehensive digitization, more adaptability level, and strict compliance with sustainable development goals. Industry 5.0, therefore, not just revitalizes production methods, but also coincides them utilizing the ideas of a green economy by synchronizing technological innovations with environmental imperatives that promote resilience, sustainability, and human-centric values in the industrial environment.

2 Overview of the Green Economy

It refers to a green economy: this is an approach to the sustainable development model of economic development that pays attention to environmental protection and social welfare, addressing serious issues in the world today, including climate change, resource depletion, and ecological degradation. It advocates for an economy where growth cannot be paid for by harming the environment, focusing on resource efficiency, waste minimization, and sustainability of ecosystems Kannan et al. (2024). This means that the green economy aims at changing the industries towards low-carbon and sustainable methodologies, especially through renewable energy sources, circular economic models, and green technologies. Since manufacturing is one of the sectors with a high contribution to carbon generation, its benefits from the use of green economic principles will be highly substantial. Industry 5.0 facilitates this integration by harmonizing technological innovations with environmental objectives, thus fostering intelligent manufacturing systems that are oriented towards sustainability and human well-being. As stated by Leng et al. (2024) it is the congruence with principles of the green economy due to human-centred and environmentally efficient design methodologies that Industry 5.0 can rely on for its foundation of sustainable production practices. Blockchain-enabled circular economies make important contributions to the green economy through traceability, waste minimization, and product lifecycle management. The implementation of such systems faces technological complexity and regulatory constraints and requires strategic intervention for a full actualization of a green economy Kannan et al. (2024).

2.1 The Green Economy and Its Key Pillars

The concept of a green economy is fundamentally oriented towards the attainment of sustainable development through an emphasis on environmental integrity, social equity, and economic prosperity. The principal foundations of this paradigm include low-carbon advancement, optimized utilization of resources, and social inclusiveness, all of which facilitate sustainable production practices and promote responsible consumption behaviours. Rehman and Umar (2024). It aligns with the objectives of Industry 5.0 as it focuses on human-centric and sustainable practices that do not affect the ecological but rather contribute to resilience and innovation among manufacturing industries. Ghobakhloo et al. (2024), Sharma and Gupta (2024). These primary factors together form a concept that not only sustains ecological resources but also promotes equal growth at all levels.

2.2 Principles of the Green Economy

A green economy is characterized by basic principles of sustainable economic development, ecological responsibility, and equitable social outcomes. Under the premise of a green economy, economic development does not have to be at the expense of ecological damage but works towards sustainable stability by reducing dependency on non-renewable sources and carbon emissions. Shankar and Gupta (2024) analyze the degree to which the advances with Technologies of Industry 5.0 facilitate this new development by promoting resilience and sustainability in the methodologies of production. Environmental accountability in the green economy is, actually a key point that focuses on sustainable practice, preservation of resources, and climate achievement globally. Social equity as a main principle ensures inclusive growth along with equal opportunities and distributes the benefits towards various communities Agrawal et al. (2024). All these principles combined comprise a comprehensive development framework in which sustainable progress across economic, ecological, and social aspects is ensured.

2.3 Examples of Green Economy Initiatives Globally

Green economy-based international initiatives are proof that it is possible to carry out sustainable development based on modern and environmentally oriented techniques. So, for example, 'Evergreened' in Germany is an approach focused on the transfer of an economy from reliance on fuel sources to the exploitation of renewable energy. Carbon emission levels should be lowered along with sustaining an economy. Such policy was implemented as an exemplary framework for holistic environmental transition. Similarly, Industry 5.0 aligns with China's pursuit of smart manufacturing due to the fact that this shall integrate digital and autonomous technologies in various industrial settings from mining, where development of the autonomous driving is meant for sustainable operating models Chen et al. (2024), Horvat et al. (2024).

In Australia, employing strategic road mapping in a systemic manner for environmental sustainability has seen a variety of industries such as agriculture and manufacturing undergo changes. The recent initiative was related to technology based on Industry 5.0 Ghobakhloo et al. (2024). These efforts highlight resource optimisation, ecological well-being improvement, and the support of societal benefits as demonstrating the way technology-enabled activities are taking sustainable progress to new heights globally.

3 Introduction to Sustainable Manufacturing

Sustainable manufacturing involves the production of products through methods that do not have negative impacts on the environment, make maximum efficient use of energy and resources, and care for the welfare of workers and communities. It is aligned with the Industry 5.0 concept, which focuses on innovations aimed at human needs and sustainability. Integration of sustainable AI is the key component that the Industry 5.0 paradigm integrates to maximize resource usage and minimize wastes in the sector of manufacturing. It involves the integration of lifecycle evaluations, renewable resources of energy, and closed-loop processes to ensure further environmental responsibilities.

Another thing is that artificial intelligence-based innovations help develop adaptive manufacturing systems that can respond dynamically to the requirements of the environment and enhance sustainable production practices. Bécue et al. (2024) discuss how Industry 5.0 and artificial intelligence support the adaptability of manufacturing processes and develop sustainable development through efficient and flexible operating methodologies.

3.1 Defining Sustainable Manufacturing

It seeks to build environmentally sustainable and resource-efficient production frameworks. The guiding principles centre around the concept of resource efficiency, which is generally defined as lowering the amount of energy consumption and minimizing the use of materials to protect the environment. Waste reduction is another important focus through leanness in manufacturing, recycling, and other similar strategies focused on reducing unnecessary excess and pollutants. In addition, practices associated with the circular economy indicate material reuse and prolongation of product life cycles, thereby reducing reliance on resource supplies from newly sourced originations Ali and Johl (2024). This strategy is in line with Industry 5.0. model, which sustains socially responsible and human-centred manufacturing, whereby the principles of innovation-based technologies lead to sustainable production results Visvizi et al. (2024). Taken together, these tenets generate a strong and sustainable manufacturing industry that produces both social-structure and ecological-system benefits.

3.2 Sustainable Manufacturing: Goals and Challenges

It aims at designing products that cause minimal environmental impact, use resources with minimal waste, and ensure more social responsibility. Main goals include the minimization of waste, conservation of energy resources, and the encouragement of human-centred resilient production methods. However, challenges include the adoption of advanced technological systems and sustainability imperatives with the need for productivity. As per Dacre et al. (2024) it has been underlined that a human-centred and resilient focus of Industry 5.0 helps alleviate several of the challenges mentioned above, though sustainability in manufacturing remains an issue due to cost, technological differences, and more regulatory issues (Kumar et. al., 2024).

3.3 Challenges in Achieving Sustainability in Manufacturing

Factors and barriers of achieving sustainability in manufacturing sector achievement often encounter significant challenges, including high costs, technological inadequacy, and other market demands. Some sustainable strategies, for example, renewable energy utilization or waste minimization, demand considerable upfront financing, which deters companies, especially small and medium-sized enterprises, from achieving sustainability. Vacchi et al. (2024). Furthermore, diverse applications of advanced sustainable technologies often make it challenging for easy integration into environmentally friendly practices (Mohanty et al., 2024).

The human-centric and collaborative paradigms inherent to Industry 5.0 seek to address these failures in a way that there must be synergistic digital change within the manufacturing sectors as proposed by Olsson et al. (2024). However, these calls of market forces with which profits are needed are sought instantly, thus prompting disagreement towards long-term investment opportunities towards sustainability.

3.4 Importance of Digitalization in Sustainability

Digitalization contributes significantly to raising the sustainability of manufacturing sectors through optimization of production methods and optimal usage of resources. For instance, additive digital moulding methods enable greater flexibility in production while reducing waste simultaneously, aligning with the tenets of Industry 5.0, and support supply chains that are sustainable. Fernández-Miguel et al. (2024). Third, development in high technology is necessary because it supports optimal usage of energy and minimizes the emission through the entire production chain. Zhang et al. (2024). This means that high technologies will support firms to track the resources they use in real-time while increasing the efficiency in which they operate and adopt circular economy principles, thus supporting sustainability in production structures.

4 Integrating Green Economy & Eco-Friendly Production Through Industry 4.0

The integration of the green economy into sustainable manufacturing by Industry 5.0 strengthens environmental effectiveness and sustains the economy. This relationship provokes the adoption of the most sophisticated technologies and approaches that rely on resource efficiency and the tenets of the circular economy. By embracing the principles of Industry 5.0, such as human-centric design and co-manufacturing, manufacturers are significantly reducing waste and increasing the outcome of sustainability. Narkhede et al. (2024a, 2024b). Further, applying social network analysis would enhance the understanding of such integration, thus ensuring the right alignment of stakeholders for the sustainability initiatives. Slavic et al. (2024). Technological Enablers for Sustainable Manufacturing. **Figure 1 shows Technology Facilitators of Sustainable Production.**

Fig. 1 Technology facilitators of sustainable production (*Source* Author)

4.1 Technological Enablers for Sustainable Manufacturing

Technological enablers of sustainable manufacturing shown in Fig. 1 include Industry 4.0 solutions like IoT, AI, and large-scale data analytics that, while teamed with the spirit of Industry 5.0, can facilitate resource use efficiency as well as ecological sustainability. This technological development permitted the real-time monitoring and development of production systems for significant improvements in outcomes for sustainability Martín-Gómez et al. (2024).

4.1.1 AI and Machine Learning

The areas of sustainable manufacturing, including optimization of energy consumption, reduction in waste, and enabling predictive maintenance strategies, are very much integrated into the world of artificial intelligence and machine learning. Therefore, applying data analytics enables real-time oversight and regulation of operations, ensuring energy efficiency and reducing resource wastage. For example, predictive maintenance is done through sophisticated algorithms, which calculate the equipment performance and, therefore, will predict a breakdown even before it happens, hence lowering downtime and prolonging the lifespan of equipment. Such proactive measures increase operational efficiency and also are aligned to the principles of Industry 5.0 that considers sustainable and human-centricity in manufacturing. Awouda et al. (2024), Narkhede et al. (2024a, 2024b). This brings into the framework of manufacturing strong methodologies empowered by AI that can face modern challenges in sustainability.

4.1.2 IoT and Big Data

The amalgamation of big data analytics combined with the Internet of Things (IoT) within the manufacturing Monitoring in real-time of emissions, resource usage, and transparency in the supply chain enhances sustainability projects within this sector. IoT devices are capable of collecting massive data that could provide more information on the environmental impact as well as the efficiency of resources used. Analysis and processing of the data help improve efficiency in the running of the operations, decrease waste generation, and observe emissions towards an environment of green production. Furthermore, with big data analytics implementation, it provides complete supply chain visibility, and one may detect inefficiencies promptly. Organization is able to rectify those inefficiencies timely and takes corrective actions for any flaws. Moreover, being an Industry 5.0, more attention will be given towards human-centric approach, thus the development of these technologies would prove crucial in transparent and responsible manufacturing practices. Madzik et al. (2024), Movahed et al. (2024).

4.1.3 Advanced Robotics and Cobots

The advanced robotics and cobots domain play a very significant part in boosting operational effectiveness and reducing the ecological footprint in the manufacturing industry. Such technologies enable the streamlining of workflows by eliminating mundane tasks, thereby increasing productivity levels and reducing energy consumption. Cobots, designed with precision to work alongside human operators, enhance flexibility in production processes while simultaneously ensuring worker safety. These systems contribute significantly towards the achievement of sustainability goals in enhanced energy efficiency and reduction in waste. During this changeover of industries towards Industry 5.0, incorporation of cutting-edge robotics with human-centred methodologies forms a more environmentally friendly manufacturing paradigm and thereby reduces the ecological footprint for production systems Hashim et al. (2024), Baig & Yadegaridehkordi (2024).

5 Examples of Industry 5.0 in Sustainable Manufacturing

Technologies of Industry 5.0 are advancing remarkably and bring principles of green economy to the forefront in this domain of sustainable manufacturing. For example, Ghobakhloo et al. (2023a), evaluate how high-tech technologies, such as IoT and AI, facilitate companies in moving forward and increasing efficiency in resource usage through lesser waste generation. Empirical case studies illustrate the manner in which companies leveraged such technologies to move towards circular economy models with large cuts in their absolute carbon intensities. Furthermore, Gladysz et al. (2023) believe the transition to Operator 5.0 in which human operators are provided data analytics, allowing human operators to make informed decisions that help subsequently make the production methods used sustainable. **Figure 2 indicates outcomes of sustainable manufacturing and the green economy.**

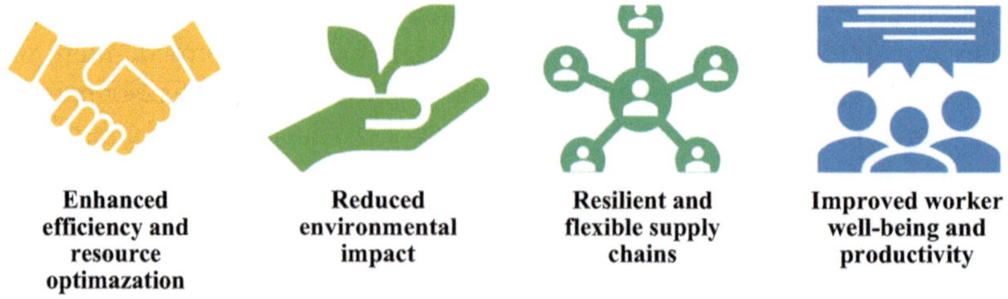

Fig. 2 Outcomes of Sustainable manufacturing and the green economy (*Source* Author)

6 Benefits of Industry 5.0 for Sustainable Manufacturing and the Green Economy

There are a tonne of chances in Industry 5.0 for the sustainable manufacture and the greening economy as it takes a strategy that is human-centred in the incorporation of technology as depicted in Fig. 2. According to Raja Santhi et al. (2023), this new paradigm encourages human actors and the advanced intelligent systems to act synergistically so that higher productivity will be realized concurrently with reducing the adverse impact on the natural world. The most notable benefits of the new technologies are resource productivity, significant waste reduction, and the utilization of the principles of the circular economy that push towards sustainable economic development. In addition, the latest Industry 5.0 technologies allow businesses to use real-time data analytics, which enables the implementation of strategic decisions in conformity with environmental sustainability goals.

6.1 Enhanced Efficiency and Resource Optimization

Industry 5.0 has, at its core the augmentation of efficiency in operations and the optimization of resources in the manufacturing sector with less resource consumption through advanced technologies and more precise regulation of manufacturing methodologies. Ghobakhloo et al. (2023a, 2023b). Big data and artificial intelligence with IoT help integrate in an improvement to the capacity of real-time surveillance and process analytics in the process, enabling manufacturers to optimize resource sharing and resource elimination. In doing so, it does not only add to productivity indices but also those of sustainability in terms of circulating economic models. Karmaker et al. (2023) further note that post-pandemic supply chain sustainability must overcome barriers, and the incorporation of the Industry 5.0 philosophy will greatly improve the resilience and adaptability to be more resilient and adaptable in managing resources for driving the advancement of even more sustainable manufacturing techniques.

6.2 Reduced Environmental Impact

The mitigation of ecological repercussions via Industry 5.0 underlines the need to reduce emissions, waste, and energy usage in manufacturing methodologies. According to De Giovanni (2023), the shift towards Industry 5.0 is to further sustainability into the metaverse by introducing resource efficiency and reducing carbon emissions. Besides, Masoomi et al. (2023) it would be relevant to mention here that the benefits in the form of advanced smart technologies and synergistic collaborative frameworks found in Industry 5.0 hold a tremendous potential for reducing the issues related to sustainability posed in the supply chain for renewable energy. Such technologies can, therefore, allow the minute control of production processes.

6.3 Resilient and Flexible Supply Chains

Adapting quickly to changing demands and disruptions with minimal environmental impact. Therefore, Industry 5.0 requires resilient and flexible supply chains in its adaptation to fast-changing demands and disruptions with minimal harm to the environment. Flexibility is promoted in human-centric smart manufacturing according to Zhang et al. (2023), which means allowing organizations to respond to fluctuating markets and supply chain breaks with efficiency. Ghobakhloo et al. (2023a, 2023b) emphasize that there is a need for a tactical road map to support the transformation in Industry 5.0, which involves sustainable practices that enhance supply chain resilience.

6.4 Improved Worker Well-Being and Productivity

Fifth industry will emphasize being a human-centric and there is an expected outcome to be increased benefits on worker welfare along with output in productivity. Mukherjee et al. (2023) the author clarifies that adoption challenges within the developing economics need to be surmounted about industry 5 and must tackle the strategy with methods enhancing the safety quotient along with satisfaction level within work premises. More to that, Dwivedi et al. (2023) this study focuses on the interaction between circular supply chains and Industry 5.0 in fostering sustainable growth, which eventually leads to a safer and more rewarding occupational environment. Focusing on the well-being of people can make operations efficient while concurrently having a favourable impact on employees.

7 Challenges

According to Barata and Kayser (2023), it is surrounded by technological deficiency, workforce preparedness, and the exigency for considerable financial commitment towards infrastructure development. Such elements cause an organization to not be thoroughly transformed into more sustainable, human-centred manufacturing paradigms, which negatively influences the efficiency and efficacy within the industrial sector. Figure 3 represents the technological challenges.

Fig. 3 Challenges (*Source* Author)

7.1 Technological and Infrastructural Challenges

There are huge technological and infrastructural hurdles that come in the way of the induction of Industry 5.0 technologies. Rajesh (2023) states the text explaining challenges pertaining to the inclusion of such sophisticated digital tools like the internet of things and artificial intelligence within pre-existing frameworks. These technologies need drastic improvement of both the infrastructural constituents and the human capabilities. Kumar et al. (2023) further emphasis the necessity of sustainability recycling methodologies and circular economy paradigms that require the establishment of new technological frameworks and supply chain structures.

7.2 Economic and Policy Barriers

Huge investment costs remain the biggest challenge for many companies, particularly SMEs as they do not have financial resources to carry out technological upgrading. Economic and policy barriers have considerably impeded the way towards Industry 5.0. Dhayal et al. (2023). Moreover, there is no policy or subsidy structure in place which enables these organizations to eliminate this financial barrier and, hence invest more in these sustainable practices. Further, the regulatory frameworks that might not align with Industry 5.0 would increase financial barriers and, therefore would decrease innovation and the implementation of green technologies.

7.3 Skills and Workforce Gaps

The development towards Industry 5.0 requires heavy up-skilling of the workforce so that it may respond well to the deficit in competencies created by the progress of technological advance. In this regard, Maddikunta et al. (2022) it is important to highlight training programmes to provide advanced technology like the internet of things and artificial intelligence, and robotics to equip the personnel for future professional roles. Adel (2022). Change towards human-centred solutions is needed, but not only on the technical skills level. The process of thinking changes in an employee's mentality. Continuous learning and professional development will be the emphasis for ensuring that the people working can communicate with these new systems and techniques easily.

7.4 Environmental Trade-Offs

As Industry 5.0 progresses, careful consideration must be taken into account regarding the environmental impacts of increased technological adoption. While these technologies will likely contribute to sustainability, they also introduce new environmental difficulties, including electronic waste as well as the energy needs connected to complex manufacturing systems. In how far do Huang et al. (2022) suggest that there is a growing need for a narrative around the need for integrated approaches that balance the 'silver linings' brought by technological change while being conscious of the corresponding ecological impacts, ensuring these technologies are deployed to bring about improvements in sustainability performances without losing ecological integrity.

8 Industry 5.0 Case Studies and Illustrations in Green Producing

Examples from case studies related to Industry 5.0 describe how the advanced technologies may ensure effective, environmentally friendly production and support sustainable development. For instance, Houkan et al. (2023) describe various projects in developing countries where IoT and AI improve energy consumption and reduce waste products. Such technological advancements allow for the monitoring in real-time and effective delivery of inputs, thereby reducing their ecological footprint. Moreover, Kolaro et al. (2023) this article looks at the competitive landscape of small economies that adhere to Industry 5.0 principles.

9 Future Directions

Future trends in Industry 5.0 are towards innovations in sustainability, including smart manufacturing facilities, zero-waste production methodologies, and circular economic frameworks. Gladysz et al. (2023) have argued that the transition towards Operator 5.0 will herald a shift towards human-centred frameworks that use advanced technologies to enhance operational efficiency, which is also an ecosystem of ecological sustainability. Note that Akundi et al. (2022) scholarly studies focus on including more sustainable approaches within productive systems. Specifically, waste minimization and effective usage of available resources have been indicated.

10 Conclusion

The integration of the green economy and environmentally friendly production through Industry 5.0 holds transformative potential in the achievement of environmental and economic goals. This chapter underscores the synergy between human-centric, cutting-edge technologies like AI, IoT, and collaborative robotics and sustainable production methods that put an emphasis on waste minimization, resource efficiency, and workforce well-being. Industry 5.0 is a critical enabler for aligning economic growth with environmental responsibility, enhancing resilience, and optimizing resources. However, this would only be possible if all these challenges are overcome—technological readiness, policy support, and workforce skills. Industry stakeholders can lead the transition towards a greener, more resilient manufacturing landscape by embracing Industry 5.0, focusing on sustainability, paving the way for a sustainable future aligned with green economy principles.

Funding Statement There is No Specific Funding Received in this Chapter

Declaration of competing interest The authors have no conflict of interest to declare.

References

Adel, A. (2022). Future of industry 5.0 in society: human-centric solutions, challenges and prospective research areas. *Journal of Cloud Computing, 11*(1), 40.

Agrawal, S., Agrawal, R., Kumar, A., Luthra, S., & Garza-Reyes, J. A. (2024). Can industry 5.0 technologies overcome supply chain disruptions?—a perspective study on pandemics, war, and climate change issues. *Operations Management Research, 17*(2), 453–468.

Akundi, A., Euresti, D., Luna, S., Ankobiah, W., Lopes, A., & Edinbarough, I. (2022). State of Industry 5.0—Analysis and identification of current research trends. *Applied System Innovation, 5*(1), 27.

Ali, K., & Johl, S. K. (2024). Driving sustainability in industry 5.0 through sociotechnical approach of quality management. *Total Quality Management & Business Excellence, 35*(13–14), 1567–1592.

Awouda, A., Traini, E., Bruno, G., & Chiabert, P. (2024). IoT-Based Framework for Digital Twins in the industry 5.0 era. *Sensors, 24*(2), 594.

Baig, M. I., & Yadegaridehkordi, E. (2024). Industry 5.0 applications for sustainability: A systematic review and future research directions. *Sustainable Development, 32*(1), 662–681.

Barata, J., & Kayser, I. (2023). Industry 5.0–past, present, and near future. *Procedia Computer Science, 219*, 778–788.

Bécue, A., Gama, J., & Brito, P. Q. (2024). AI's effect on innovation capacity in the context of industry 5.0: a scoping review. *Artificial Intelligence Review, 57*(8), 215.

Chen, L., Li, Y., Silamu, W., Li, Q., Ge, S., & Wang, F. Y. (2024). Smart mining with autonomous driving in industry 5.0: Architectures, platforms, operating systems, foundation models, and applications. *IEEE Transactions on Intelligent Vehicles.*

Dacre, N., Yan, J., Frei, R., Al-Mhdawi, M. K. S., & Dong, H. (2024). Advancing sustainable manufacturing: a systematic exploration of Industry 5.0 supply chains for sustainability, human-centricity, and resilience. *Production Planning & Control*, 1–30.

Dhayal, K. S., Giri, A. K., Kumar, A., Samadhiya, A., Agrawal, S., & Agrawal, R. (2023). Can green finance facilitate Industry 5.0 transition to achieve sustainability? A systematic review with future research directions. *Environmental Science and Pollution Research, 30*(46), 102158–102180.

Dwivedi, A., Agrawal, D., Jha, A., & Mathiyazhagan, K. (2023). Studying the interactions among Industry 5.0 and circular supply chain: Towards attaining sustainable development. *Computers & Industrial Engineering, 176*, 108927.

Fernández-Miguel, A., García-Muiña, F. E., Jiménez-Calzado, M., San Román, P. M., del Hoyo, A. P. F., & Settembre-Blundo, D. (2024). Boosting business agility with additive digital molding: An Industry 5.0 approach to sustainable supply chains. *Computers & Industrial Engineering, 192*, 110222.

Ghobakhloo, M., Iranmanesh, M., Morales, M. E., Nilashi, M., & Amran, A. (2023). Actions and approaches for enabling Industry 5.0-driven sustainable industrial transformation: A strategy roadmap. *Corporate social responsibility and environmental management, 30*(3), 1473–1494.

Ghobakhloo, M., Iranmanesh, M., Tseng, M. L., Grybauskas, A., Stefanini, A., & Amran, A. (2023). Behind the definition of Industry 5.0: A systematic review of technologies, principles, caomponents, and values. *Journal of Industrial and Production Engineering, 40*(6), 432–447.

Ghobakhloo, M., Iranmanesh, M., Foroughi, B., Tirkolaee, E. B., Asadi, S., & Amran, A. (2023). Industry 5.0 implications for inclusive sustainable manufacturing: An evidence-knowledge-based strategic roadmap. *Journal of Cleaner Production, 417*, 138023.

Ghobakhloo, M., Iranmanesh, M., Foroughi, B., Rejeb, A., Nikbin, D., & Tseng, M. L. (2024). A practical guide on strategic roadmapping for information and operations technology management: a case study on industry 5.0 transformation. *Journal of Industrial and Production Engineering, 41*(5), 397–421.

Ghobakhloo, M., Mahdiraji, H. A., Iranmanesh, M., & Jafari-Sadeghi, V. (2024). From Industry 4.0 digital manufacturing to Industry 5.0 digital society: a roadmap toward human-centric, sustainable, and resilient production. *Information Systems Frontiers*, 1–33.

De Giovanni, P. (2023). Sustainability of the Metaverse: A transition to Industry 5.0. *Sustainability, 15*(7), 6079.

Gladysz, B., Tran, T. A., Romero, D., van Erp, T., Abonyi, J., & Ruppert, T. (2023). Current development on the Operator 4.0 and transition towards the Operator 5.0: A systematic literature review in light of Industry 5.0. *Journal of Manufacturing Systems, 70*, 160–185.

Hashim, M. A. M., Tlemsani, I., Mason-Jones, R., Matthews, R., & Ndrecaj, V. (2024). Higher education via the lens of industry 5.0: Strategy and perspective. *Social Sciences & Humanities Open, 9*, 100828.

Horvat, D., Jäger, A., & Lerch, C. M. (2024). Fostering innovation by complementing human competences and emerging technologies: an industry 5.0 perspective. *International Journal of Production Research*, 1–24.

Houkan, A., Sahoo, A. K., & Nayak, S. (2023, August). Industry 5.0 and Sustainable Development in the Developing World. In *2023 IEEE 3rd International Conference on Sustainable Energy and Future Electric Transportation (SEFET)* (pp. 1–6). IEEE.

Huang, S., Wang, B., Li, X., Zheng, P., Mourtzis, D., & Wang, L. (2022). Industry 5.0 and Society 5.0—Comparison, complementation and co-evolution. *Journal of Manufacturing Systems, 64*, 424–428.

Kannan, D., Amiri, A. S., Shaayesteh, M. T., Nasr, A. K., & Mina, H. (2024). Unveiling barriers to the integration of blockchain-based circular economy and Industry 5.0 in manufacturing industries: A strategic prioritization approach. *Business Strategy and the Environment.*

Karmaker, C. L., Bari, A. M., Anam, M. Z., Ahmed, T., Ali, S. M., de Jesus Pacheco, D. A., & Moktadir, M. A. (2023). Industry 5.0 challenges for post-pandemic supply chain sustainability in an emerging economy. *International Journal of Production Economics, 258*, 108806.

Kolaro, K., Pitić, G., Vlačić, E., & Milosavljević, U. (2023). Competitiveness and sustainability in small and open economies in the age of industry 5.0. *Ekonomika preduzeća, 71*(1–2), 113–127.

Kumar, R., Kariminejad, A., Antonov, M., Goljandin, D., Klimczyk, P., & Hussainova, I. (2023). Progress in sustainable recycling and circular economy of tungsten carbide hard metal scraps for industry 5.0 and onwards. *Sustainability, 15*(16), 12249.

Kumar, R., Goel, R., Singh, T., Mohanty, S. M., Gupta, D., Alkhayyat, A., & Khanna, R. (2024). Sustainable finance factors in indian economy: Analysis on policy of climate change and energy sector. *Fluctuation and Noise Letters, 23*(2), 2440004–339.

Leng, J., Guo, J., Xie, J., Zhou, X., Liu, A., Gu, X., & Wang, L. (2024). Review of manufacturing system design in the interplay of Industry 4.0 and Industry 5.0 (Part I): Design thinking and modeling methods. *Journal of Manufacturing Systems, 76*, 158–187.

Maddikunta, P. K. R., Pham, Q. V., Prabadevi, B., Deepa, N., Dev, K., Gadekallu, T. R., ... & Liyanage, M. (2022). Industry 5.0: A survey on enabling technologies and potential applications. *Journal of industrial information integration, 26*, 100257.

Madzik, P., Falat, L., Jum'a, L., Vrábliková, M., & Zimon, D. (2024). Human-centricity in Industry 5.0–revealing of hidden research topics by unsupervised topic modeling using Latent Dirichlet Allocation. *European Journal of Innovation Management, 28*(1), pp. 113–138.

Martín-Gómez, A. M., Agote-Garrido, A., & Lama-Ruiz, J. R. (2024). A Framework for Sustainable Manufacturing: Integrating Industry 4.0 Technologies with Industry 5.0 Values. *Sustainability, 16*(4), 1364.

Martini, B., Bellisario, D., & Coletti, P. (2024). Human-centered and sustainable artificial intelligence in industry 5.0: Challenges and perspectives. *Sustainability, 16*(13), 5448.

Masoomi, B., Sahebi, I. G., Ghobakhloo, M., & Mosayebi, A. (2023). Do industry 5.0 advantages address the sustainable development challenges of the renewable energy supply chain? *Sustainable Production and Consumption, 43*, 94–112.

Mohanty, R., Samal, A., & Kumar, R. (2024). Green finance supports sustainable environment: Empirical approach towards applicability in tourism sector. *Theoretical and Practical Research in Economic Fields, 15*(1), 114–126.

Movahed, A. B., Movahed, A. B., & Nozari, H. (2024). Opportunities and challenges of smart supply chain in Industry 5.0. *Information Logistics for Organizational Empowerment and Effective Supply Chain Management*, 108–138.

Mukherjee, A. A., Raj, A., & Aggarwal, S. (2023). Identification of barriers and their mitigation strategies for industry 5.0 implementation in emerging economies. *International Journal of Production Economics, 257*, 108770.

Narkhede, G., Pasi, B., Rajhans, N., & Kulkarni, A. (2023). Industry 5.0 and the future of sustainable manufacturing: A systematic literature review. *Business Strategy & Development, 6*(4), 704–723.

Narkhede, G. B., Pasi, B. N., Rajhans, N., & Kulkarni, A. (2024). Industry 5.0 and sustainable manufacturing: a systematic literature review. *Benchmarking: An International Journal, 32*(2), pp. 608–635.

Narkhede, G., Chinchanikar, S., Narkhede, R., & Chaudhari, T. (2024). Role of Industry 5.0 for driving sustainability in the manufacturing sector: an emerging research agenda. *Journal of Strategy and Management*, (ahead-of-print).

Olsson, A. K., Eriksson, K. M., & Carlsson, L. (2024). Management toward Industry 5.0: A co-workership approach on digital transformation for future innovative manufacturing. *European Journal of Innovation Management, 28*(1), pp. 65–84.

Piccarozzi, M., Silvestri, L., Silvestri, C., & Ruggieri, A. (2024). Roadmap to Industry 5.0: Enabling technologies, challenges, and opportunities towards a holistic definition in management studies. *Technological Forecasting and Social Change*, *205*, 123467.

Raja Santhi, A., & Muthuswamy, P. (2023). Industry 5.0 or industry 4.0 S? Introduction to industry 4.0 and a peek into the prospective industry 5.0 technologies. *International Journal on Interactive Design and Manufacturing (IJIDeM), 17*(2), 947–979.

Rajesh, R. (2023). Industry 5.0: analyzing the challenges in implementation using grey influence analysis. *Journal of Enterprise Information Management*, *36*(5), 1349–1371.

Rehman, A., & Umar, T. (2024). Literature review: Industry 5.0. Leveraging technologies for environmental, social and governance advancement in corporate settings. *Corporate Governance: The International Journal of Business in Society*.

Sarkar, B. D., Shardeo, V., Dwivedi, A., & Pamucar, D. (2024). Digital transition from industry 4.0 to industry 5.0 in smart manufacturing: A framework for sustainable future. *Technology in Society*, *78*, 102649.

Shankar, R., & Gupta, L. (2024). Modelling risks in transition from Industry 4.0 to Industry 5.0. *Annals of Operations Research*, 1–46

Sharma, R., & Gupta, H. (2024). Harmonizing sustainability in industry 5.0 era: Transformative strategies for cleaner production and sustainable competitive advantage. *Journal of Cleaner Production*, *445*, 141118.

Slavic, D., Marjanovic, U., Medic, N., Simeunovic, N., & Rakic, S. (2024). The Evaluation of Industry 5.0 Concepts: Social Network Analysis Approach. *Applied Sciences*, *14*(3), 1291.

Vacchi, M., Siligardi, C., & Settembre-Blundo, D. (2024). Driving manufacturing companies toward industry 5.0: a strategic framework for process technological sustainability assessment (P-TSA). *Sustainability*, *16*(2), 695.

Visvizi, A., Malik, R., Guazzo, G. M., & Çekani, V. (2024). The Industry 5.0 (I50) paradigm, blockchain-based applications and the smart city. *European Journal of Innovation Management*, *28*(1), pp. 5–26.

Yao, X., Ma, N., Zhang, J., Wang, K., Yang, E., & Faccio, M. (2024). Enhancing wisdom manufacturing as industrial metaverse for industry and society 5.0. *Journal of Intelligent Manufacturing*, *35*(1), 235–255.

Zhang, C., Wang, Z., Zhou, G., Chang, F., Ma, D., Jing, Y., ... & Zhao, D. (2023). Towards new-generation human-centric smart manufacturing in Industry 5.0: A systematic review. *Advanced Engineering Informatics*, *57*, 102121.

Zhang, S., Wang, X., Xu, J., Chen, Q., Peng, M., & Hao, J. (2024). Green manufacturing for achieving carbon neutrality goal requires innovative technologies: A bibliometric analysis from 1991 to 2022. *Journal of Environmental Sciences, 140*, 255–269.

Significance of Sustainable Entrepreneurship for Inclusive Economic Growth

Mantena Siva Pavan Kumar Raju and Ramesh Babu Damarla

Abstract Sustainable entrepreneurship is gaining importance in the recent times and becoming increasingly crucial for inclusive economic growth in India. This chapter explores the relationship between sustainable entrepreneurship and inclusive economic growth by emphasizing the role of entrepreneurs in adopting or creating business models that promote sustainability, social equity, community development, and economic inclusivity. It delves into how sustainable practices like waste reduction, resource management, and social responsibility integrated into traditional entrepreneurship can contribute to inclusive economic growth that benefits all segments of society, reduces poverty, provides livelihoods, encourages broad participation, and sustainable development. The chapter explores policy frameworks, government initiatives, role of SMEs, innovation, and start-up ecosystems, which promote sustainable entrepreneurship and inclusive economic growth as well as the challenges and opportunities, especially in the developing economies like India. Besides this, it also highlights the recommendations of the role played by collaboration between the Public and private sectors in enabling Inclusive Economic Growth fostering Resilient, Equitable, and Environmental Protection.

Keywords Sustainable entrepreneurship · Inclusive growth · SMEs · Sustainability · Innovation · Public–private collaboration

1 Introduction

An emerging nation like India faces significant development challenges in economic inequality, worsening conditions of the environment, and social exclusion (Pradhan et al., 2021). It has extraordinary positive economic growth, but leaving a sizeable

M. S. P. K. Raju (✉)
Sasi Institute of Technology and Engineering, Tadepalligudem, AP, India
e-mail: mantenapavan@hotmail.com

M. S. P. K. Raju · R. B. Damarla
SR University, Warangal, Telangana, India
e-mail: rameshdamarla2009@gmail.com

R. Kumar and S. N. Mohanty (eds.), *Sustainable Economy Models in the Age of Industry 5.0*, https://doi.org/10.1007/978-981-96-4104-8_10

population deprived of the essential services needed for them in health and education sectors and clean energy sources being limited in marginalized parts of the country and rural regions (Parida et al., 2022). Economic disparities have surged to highly alarming levels, which reveals that there is an increasing gulf between the rich and the poor, especially in metropolitan cities (Malhotra & Kiran, 2023). Simultaneously, environmental degradation has turned into a matter of urgency; in the country, serious cases of air and water pollution, deforestation, and excess exploitation of natural resources prevail. These environmental challenges have been compounded by climate change impacts, seriously affecting vulnerable populations (Parida et al., 2022). Social exclusion remains an important issue in which marginalized communities, women, tribal groups, and lower castes are left behind in economic opportunities and initiatives for development (Ramadani et al., 2022). All these interlinked factors bring a strong call for innovative solutions that must not only bridge the economic disparities but also be sustainable and socially inclusive.

Sustainable entrepreneurship promises a way forward to solving such problems. Sustainable entrepreneurship offers a framework where the economic, social, and environmental goals are integrated so businesses can operate in ways that contribute to long-term societal welfare and environmental preservation (Vig, 2022). It focuses on creating value in ways that foster social equity, economic inclusivity, and environmental stewardship by making sure that people have access to opportunities for creating wealth and ensuring fair wages and quality of living for disadvantaged groups while also ensuring resource efficiency and environmentally friendly business practices (Spigel & Harrison, 2017). This enables sustainable entrepreneurship to develop solutions to the most pressing concerns such as resource scarcity, climatic change, and inequality in society through innovative business models (Pradhan et al., 2021). The entrepreneurs can create value that is both economic as well as positive for society. Focused on sustainability and inclusiveness, this will help entrepreneurs make more inclusive growth while preserving a better environment for future generations (Sharma et al., 2023).

This chapter aims to explore the role of sustainable entrepreneurship in fostering inclusive growth in India, and for that, it focuses on the intertwining challenges of economic inequality, environmental degradation, and social exclusion.

The objectives of this chapter are:

1. Defining sustainable entrepreneurship and discussing its potential to address India's social, economic, and environmental challenges.
2. To understand the mechanisms whereby sustainable entrepreneurship can be linked to inclusive economic growth through strategies for resource optimization, waste reduction, and community development.
3. Case studies of successful sustainable entrepreneurship in India supported by evidence of circular economy models and CSR programs with a positive impact.
4. Suggestions on how the policies of the government and institutional frameworks could be improved in support of sustainable entrepreneurship and scaling up SMEs and start-ups for innovation sustainably.

This chapter will demonstrate how sustainable entrepreneurship can be a transformative force for creating equitable, resilient, and sustainable economic growth in India.

2 Sustainable Entrepreneurship

Sustainable entrepreneurship is therefore a valid approach to entrepreneurship that entails the integration of both economic sustainability and social equity into environmental responsibilities (Schot & Steinmueller, 2018). Here, the entrepreneurship will focus not only on building value to shareholders but also, in all cases, benefitting different stakeholders, ranging from community, employees and the environment (Shane & Venkataraman, 2000). It surpasses the traditional profit-making goals and focuses on the long-term impact of its operations, ensuring that business contributes to a more equitable, inclusive, and sustainable society (Sharma et al., 2023). Economic sustainability is when a business remains financially viable for a long time while minimizing the negative effects on the environment and society. Social equity is the promotion of fair treatment, opportunities, and outcomes for all people, particularly marginalized or underserved groups (Saebi et al., 2018). Environmental responsibility involves reducing the ecological footprint of business activities by prioritizing resource efficiency, waste reduction, and sustainability.

The undermentioned three dimensions of integration play a very important role in making the sustainable entrepreneurship successful (Schot & Steinmueller, 2018). Just focusing on profit or sustainability alone won't be considered, true sustainable entrepreneurship would require business growth to deliver positive results to both society and the environment (Malhotra & Kiran, 2023). Figure 1 would represent the core area where economic growth, social equity, and environmental stewardship run into one.

Economic Dimension: This comprises financial viability, profitability, and resource use efficiency. It ensures that the businesses are economically viable and competitive while responding to the environmental and social impacts (Ramadani et al., 2022). The key elements for this dimension are financial viability, resource optimization, and long-term profitability.

Social Dimension: It entails fairness, social justice, and community well-being. In businesses, they need to think of what they do for the people in the communities, such as employees and society at large (Satar et al., 2023). They need to embrace equitable practice, which implies fair compensation, an all-inclusive decision-making process, and participation in the communities. The social dimension is interested in education and the development of competencies for creating the potential of local communities (Saebi et al., 2018).

Environmental Dimension: It is concerned with the environmental footprint of business operations-that is, waste reduction, resource conservation, and production sustainability (Parida et al., 2022). Companies need to use strategies that

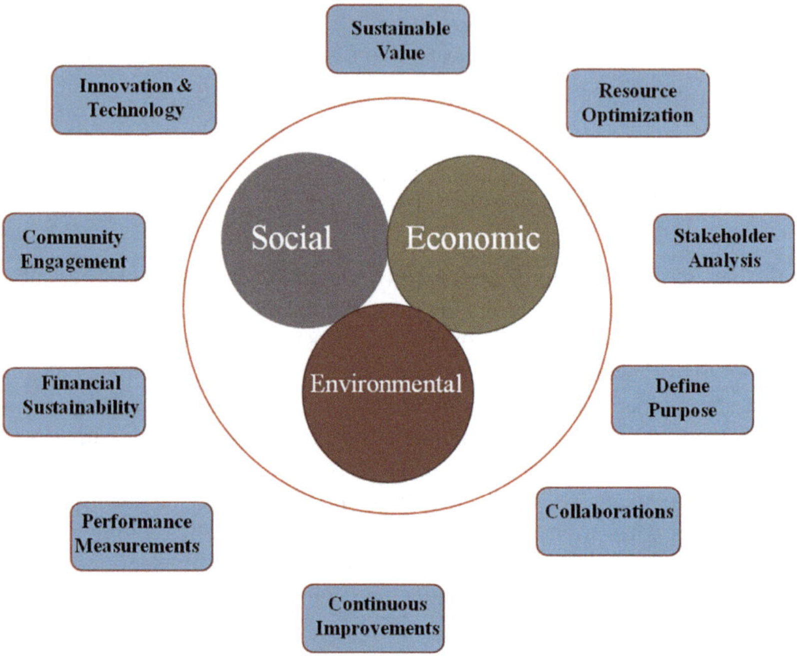

Fig. 1 Sustainable business model framework

reduce carbon footprints, promote renewable energy, and minimize non-renewable resources.

As Ooi & Nasiru, 2015 and Olurina et al., 2023 say, around the core other steps or strategies that would promote the successful implementation of sustainable entrepreneurship are:

1. Stakeholder Analysis: The needs and interests of employees, consumers, investors, and the community are crucial for business practices to be aligned with societal and environmental goals.
2. Resource optimization: Proper use of the resources like energy, water, and raw materials reduces waste generation along with minimizing the environmental effects through saving costs and ensuring better operation.
3. Community involvement: Good relations between businesses and local communities combined with integrating the feedback of communities while taking business decisions bring excellent social benefits and also give a strong base to social initiatives.
4. Financial Sustainability: Develop and maintain robust financial plans that support the bottom line while supporting sustainable development objectives, where businesses compete and promote positive social and environmental impact.

Every step is one which will ensure that sustainable entrepreneurship flourishes with all its economic, social, and environmental dimensions interlocking and mutually supporting within this integrated model. Figure 1 indicates Sustainable Business Model Framework.

3 Mechanisms of Sustainable Entrepreneurship in Inclusive Growth

3.1 Waste Reduction and Resource Management

Sustainable entrepreneurship is intrinsically linked with the so-called model for the circular economy whereby resource reduction and optimum utilization become top focus themes in production and consumption chains (Shane & Venkataraman, 2000). In sharp contrast, unlike the linear economy whereby one lives by "take, make, dispose," here with this alternative, closed-loop way exists about producing again and again and renewing while recycling a product so to speak to minimize more direct extraction of raw material reduce wastes and minimize a variety of impacts on environmental condition (Parida et al., 2022). One of the key principles of the circular economy is designing for sustainability. Products should be designed to be long-lasting, repairable, and recyclable so that they can be reintroduced into the production cycle rather than ending up as waste (Olurina et al., 2023). Also, resource optimization strategies include efficient production processes, eco-friendly inputs, and green logistics in distribution (Onakoya, 2013). Such companies take strategic measures to focus on waste minimization through production and transportation, thereby reducing carbon footprint and conserving energy. It also needs to encourage sustainable consumption and a repair culture. The consumer is becoming more interested in buying things that are repairable and reusable, and business can aid this by providing repair services and supporting sustainable behavior (Da Silva et al., 2022). Collection, recycling, and upcycling create a robust system of reusing products efficiently, and remanufacturing adds more life to products by refurbishing them and putting them back into the market (Mondal et al., 2023). However, when these principles are followed as shown in Figure 2, businesses can go a long way in ensuring a reduction of environmental degradation and resource depletion toward the sustainable utilization of resources.

3.2 Social Responsibility and Community Development

Another key aspect of sustainable entrepreneurship involves CSR activities that are helpful in building the community and ensuring social equity (Filser et al., 2019). CSR activities are more than generating profit (Nayak & K.R.P, 2022). These have

Fig. 2 Circular economy in practice

more to do with creating social and environmental value. Nowadays, several businesses are adopting CSR within their business models. Business operations are now aligned toward goals that are supportive to the community, the environment, and other broader societal requirements (Markard et al., 2012). These involve skill development and education, and employment generation, especially among marginalized and underprivileged sections of the population ("Innovating Education and Educating for Innovation," 2016).

For instance, investment in skill development programs may help build local capacity by empowering the people with the expertise required to participate in sustainable industries. In addition to providing valuable job opportunities, this creates a skilled workforce that can contribute to green and circular economy sectors (Malhotra et al., 2024). Companies like Patagonia and Unilever have been at the forefront of CSR efforts, implementing community-based programs that support environmental education, skill-building, and fair labor practices. Moreover, sustainable entrepreneurship helps in sustainable growth through the job opportunities brought about by industries focusing on a greener world, be it renewable energy, or waste management and sustainable agriculture (Malhotra et al., 2024). This kind of development promotes economic benefits besides uplifting society by providing dignified employment means and improving their social well-being (Da Silva et al., 2022). In other words, it is in the synergy between CSR programs and community development policies that business prosperity and social well-being mutually reinforce and contribute to long-term inclusive growth.

These sustainable entrepreneurial mechanisms lead to economic prosperity but in a way that promotes an equitable and resilient society.

4 Role of SMEs, Start-Ups, and Innovation Ecosystems

SMEs and start-ups are essential for the development of sustainable entrepreneurship, particularly through job creation and green innovation. Most SMEs are resource-constrained and cannot invest in large-scale sustainable technologies (Olurina et al., 2023). However, their agility and flexibility allow them to implement niche, eco-friendly innovations that meet local markets and specific environmental challenges (Mahato & Jha, 2023). They contribute substantially to employment in the local economy, especially in rural settings (Da Silva et al., 2022). They also rely on the government schemes for financing and sustenance (George et al., 2014). Large corporations are better positioned to invest in emerging sustainable technologies and be within the regulatory frameworks because they have more financial resources as well as economies of scale (Bhattacharjee & Mukherjee, 2023). While they generate large-scale employment, their impact on local job creation is often diluted by outsourcing practices. Large enterprises can also spearhead industry-wide sustainability initiatives, making use of the large market influence and their massive R&D departments. However, both SMEs and large enterprises face different kinds of challenges and opportunities depicted in Table 1: SMEs focus on local innovation, and start-ups serve as precursors for broader innovation system-wide transitions toward sustainability.

Table 1 Comparative analysis of SMEs and large enterprises in sustainability adoption

Aspect	SMEs	Large enterprises
Resource efficiency	Often face resource constraints, which limits their ability to invest in sustainable practices	Have greater financial resources and economies of scale to invest in sustainable technologies
Job creation	Contribute significantly to employment in local communities, especially in rural areas	Generate large-scale employment but often outsource jobs, limiting local impact
Innovation	Agile and flexible, SMEs often create niche, sustainable innovations in response to market demand	Large R&D departments lead to large-scale innovations but with a focus on mass production
Access to funding	Limited access to capital for sustainable investments, often relying on government schemes	Easier access to capital, private investors, and can self-fund sustainability initiatives
Regulatory compliance	Often struggle to meet sustainability regulations due to cost constraints	Better positioned to comply with sustainability regulations due to established resources
Market influence	Limited influence due to smaller market reach and production capacity	Significant influence on the market and supply chains, with ability to drive industry-wide changes

5 Policy Frameworks and Government Initiatives

As per M et al., 2023, the government schemes outlined in Table 2 are integral to supporting sustainable entrepreneurship in India, targeting various sectors with specific objectives and benefits. The "Start-up India" Scheme aims to foster entrepreneurship and innovation by providing easier access to capital, tax exemptions, and market access, particularly benefiting technology, manufacturing, and services sectors. The "Atal Innovation Mission (AIM)" focuses on promoting research, development, and sustainable solutions through innovation grants, funding for incubation centers, and public–private partnerships, with emphasis on technology, renewable energy, and manufacturing. The "Make in India" initiative encourages sustainable production processes and innovation in manufacturing and renewable energy, offering investment facilitation and development of innovation ecosystems. The "National Clean Energy Fund (NCEF)" supports clean energy projects, the promotion of energy efficiency, as well as innovation in the area of renewable energy while offering financial assistance and research funding. The "PM Gati Shakti National Master Plan" is improving logistics efficiency and green infrastructure thereby reducing carbon emissions in general through sustainable transport and the infrastructures. Lastly, the "National Biofuels Policy" promotes the use of renewable biofuels and eco-friendly alternatives, offering technology grants and incentives for biofuel production. While these schemes have successfully encouraged sustainable practices across sectors, there are gaps in coherence, interconnectivity between schemes, and alignment with global best practices, which could be addressed through improved coordination, enhanced stakeholder engagement, and monitoring mechanisms to ensure effective implementation and measurable outcomes.

6 Challenges and Opportunities

The challenges and opportunities in sustainable entrepreneurship are closely associated, in that while the several obstacles present obstacles toward growth they also create doors open for innovation and progress, as Table 3 portrays.

6.1 Difficulties

Probably one of the biggest bottlenecks is the availability of capital: many start-ups are just unable to scale their sustainable solution up (Filho et al., 2018). But the Government schemes like AIM, now growing support from venture capitalists, and impact investors as providing grants and funding opportunity research and innovation (George et al., 2014). However, high initial investment will probably be a challenge too—it might deter many business groups from adopting sustainable

Table 2 Key government schemes supporting sustainable entrepreneurship in India

Government scheme	Objectives	Target sectors	Benefits
Startup India scheme	Encourage entrepreneurship and innovation, provide easier access to capital, and foster job creation	All sectors, especially technology, manufacturing, and services	Tax exemptions, funding support, ease of doing business, innovation hubs, and market access
Atal Innovation Mission (AIM)	Promote innovation, research, and development, focusing on the creation of sustainable solutions	Technology, renewable energy, and manufacturing	Innovation grants, funding for incubation centers, mentorship programs, and public–private partnerships
Make in India	Encourage manufacturing and sustainable production processes	Manufacturing, renewable energy, infrastructure	Investment facilitation, ease of doing business, and development of innovation ecosystems
National Clean Energy Fund (NCEF)	Promote clean energy projects, develop renewable energy sources, and support energy efficiency	Renewable energy, clean technologies	Financial support for clean energy projects, research funding, and innovation in sustainable energy
PM gati shakti national master plan	Improving infrastructure for faster transportation and logistics with a focus on sustainability	Logistics, transport, infrastructure	Development of green infrastructure, improved logistics efficiency, and reduction in carbon emissions
National biofuels policy	Promote renewable biofuels, reduce resilience on fossil fuels	Renewable energy, agriculture	Biofuel incentivization, technology grants, and replacing fossil fuels with eco-friendly alternatives

technology (Nayak & K.R.P, 2022). Yet, improvements, including solar power innovation, come to significantly reduce costs toward the achievement of sustainability in various areas. There's another barrier: a limitation of awareness and education that only green building certifications, like LEED, as well as sustainability education programs for consumers, can effectively influence greater demand for such

Table 3 Challenges and opportunities for sustainable entrepreneurship

Challenges	Opportunities	Examples
Lack of access to capital	Government schemes, venture capital, and impact investors are increasingly supporting sustainable start-ups	**Atal Innovation Mission (AIM)** provides grants for R&D and innovation-driven projects
High initial investment costs	Technological advancements reducing costs in renewable energy, waste management, etc	**Solar power innovations** are lowering installation and operational costs for businesses
Limited awareness and education	Awareness programs, sustainability certifications, and eco-labeling can boost demand	**Green Building Certification** programs like IGBC encourage businesses to adopt sustainable building practices
Regulatory hurdles and compliance	Increasing policy support and favorable regulations for green businesses	**Startup India** provides a platform for businesses focusing on sustainable solutions
Supply chain sustainability	Developing resilient supply chains by adopting circular economy principles	**Patagonia** adopting circular manufacturing and take-back programs for used clothing
Competition with conventional business models	Growing consumer demand for sustainable products and services	**Tata Group** integrating sustainable practices across their portfolio from energy to manufacturing

environment-friendly products and services (Mondal et al., 2023). The regulatory hurdles continue, but policy support with initiatives like Start-up India has been building up, which positions the businesses in a much better space to comply with the regulations on sustainability.

6.2 Opportunities

There's always this great impetus for change: Tata Group has now integrated sustainability into all of its portfolio offerings, growing consumer demand for sustainable products is. International support for green financing and businesses like Patagonia embracing circular economy principles add to this, turning challenges into opportunities for sustainable growth and innovation (Markard et al., 2012). As the business world navigates through these challenges and seizes the new opportunities, the pathway toward a more sustainable and profitable future becomes more apparent than ever.

7 Public–Private Collaboration for Inclusive Growth

Public–Private Collaboration Model for sustainable entrepreneurship is a dynamic ecosystem, where different stakeholders play intertwined roles in driving environmental and social change (Filho et al., 2018). The center of this model is "Sustainable Entrepreneurship" as the central goal; from this hub, key stakeholders include the public sector, private sector, academic institutions, NGOs, and local communities, all contributing and benefiting from the collaborative effort (Markard et al., 2012). The public sector will provide policy frameworks, funding, and infrastructure. The private sector will drive technological innovation and investment (Bhattacharjee & Mukherjee, 2023). Academic institutions will contribute through research and educational programs. NGOs will focus on advocacy and grassroots efforts (George et al., 2014). Local communities will be very important in decision-making and co-creating sustainable solutions (Bartolacci et al., 2019). These stakeholders, through bi-directional relationships, work together in a collective effort to promote a sustainable entrepreneurial landscape, where mutual benefits flow between them and create a balanced and thriving ecosystem (Malhotra & Kiran, 2023). Figure 3 indicates Public–Private Collaboration Models.

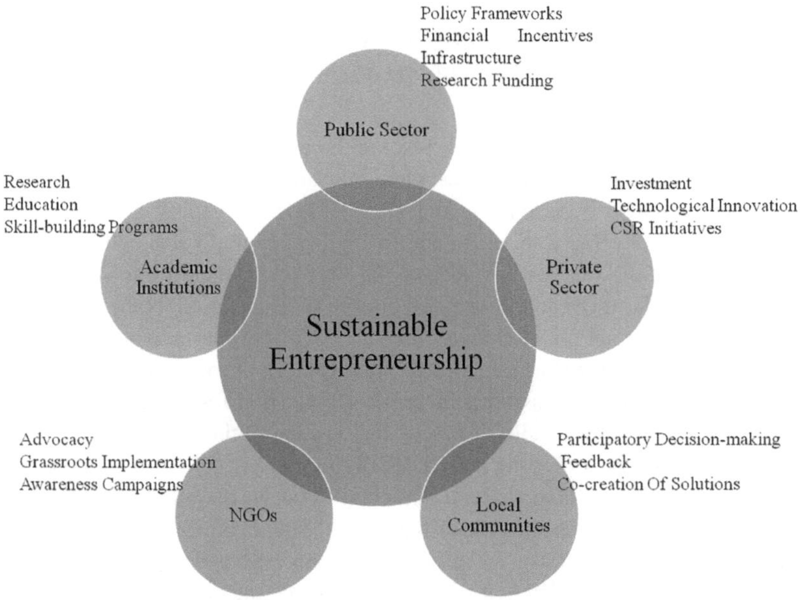

Fig. 3 Public–Private collaboration models

7.1 Importance of Partnerships to Scale Sustainable Innovations

As shown in Fig. 3, Partnerships are key to the scaling of sustainable innovations in that they help businesses and organizations pool their resources and share knowledge (Filser et al., 2019). The only way a business or an organization can scale sustainable innovations through partnerships is by the pooling of resources, knowledge sharing, and access to complementary strengths (Markard et al., 2012). It becomes possible for the scaling of technologies and business models that otherwise would be almost impossible for one entity, especially in renewable energy and waste management. For sustainable innovation to be successful, great upfront investment, access to specialized expertise, and capabilities to navigate regulatory challenges would be needed (Filho et al., 2018). Start-ups and small businesses can gain access to those resources and support by partnering with other stakeholders, whether government bodies, large corporations, research institutions, or NGOs (Bartolacci et al., 2019). Partnerships, in addition, reduce risks. They spread the financial burden, enhance market reach, and improve the efficiency of implementation.

7.2 Case Studies of Successful Collaborations in Renewable Energy and Waste Management

1. **Renewable Energy:** Tesla and Panasonic the best example of successful cooperation in renewable energy is Tesla and Panasonic. Tesla is the electric vehicle manufacturer and Panasonic is the global leader in energy solutions, which combined to develop and manufacture high-efficiency solar cells along with energy storage systems (*New Atlas*, n.d). This partnership has scaled up the creation of solar roofs and Powerwall energy storage solutions for both residential and commercial use (*PV Magazine*, n.d). With this partnership, more people have access to renewable energy and, therefore, contribute to a cleaner energy grid as dependence on fossil fuels decreases, given that Tesla's energy technology is combined with Panasonic's manufacturing capabilities (*WIRED*, n.d.).
2. **Waste Management:** Veolia and SUEZ (*Veolia*, n.d.; *SUEZ*, n.d.) depicts; in the area of waste management, the partnership between two of the world's largest environmental services companies, Veolia and SUEZ, is an example of how partnerships can enhance waste recycling and resource recovery. The two companies have collaborated on different projects to enhance the circular economy, which includes waste reduction, increasing recycling rates, and developing sustainable waste-to-energy solutions. For example, there is Veolia and SUEZ in France who have implemented projects transforming waste into energy and using advanced waste-to-energy plants that convert municipal solid waste to electricity and heat. These collaborations in France have played a crucial role in managing the wastage levels more efficiently and simultaneously contributing toward energy security

with the minimal environmental damage. Veolia and SUEZ brought scalable, effective solutions for the ever-growing global waste crisis together through combining their respective technologies, expertise, and operational networks.

These case studies point out the ways and means through which industry partnerships achieve more effective, innovative, and scalable solutions for sustainability. By sharing their expertise, resources, and risk, sustainable innovations not only reach higher markets but also contribute significantly toward environmental goals.

8 Recommendations

Strengthening of policy framework, education enhancement, and making use of technology to encourage sustainable entrepreneurship along with innovation in specific areas such as renewable energy, waste management, and green technologies, thereby making these sustainable options available to everyone, especially in the underrepresented regions.

1. Policy Incentives for Scaling Sustainable Entrepreneurship Policymakers are the enablers of scaling sustainable entrepreneurship through an enabling policy environment. An important factor to invest in green technologies and sustainable business models is strengthening incentives in policies. This could be tax breaks, subsidies for investments in clean energy, and grants for sustainable innovation. In addition, mechanisms that introduce carbon pricing may spur the growth of the green economy as businesses will be able to receive favorable loans based on sustainable practices. Regulations must be clearer, consistent, and more beneficial to newly emerging entrepreneurs, especially SMEs, which are normally not sufficiently resourced to penetrate complex regulatory compliance requirements.
2. Awareness Education of Sustainability and Entrepreneurship: In an enlightened entrepreneurial ecosystem, there can be success of sustainable business models. Hence, awareness education on sustainability and entrepreneurship should be encouraged particularly in grassroots level. Such areas will involve embedding the sustainability-related curricula within institutions and also through the training programmers as well as other materials for entrepreneurship which are specialized. Business incubators and accelerators focused on sustainable ventures would be developed to provide start-ups with significant support and mentorship as well. Awareness campaigns may engage a community and a consumer at a level in which consumers would understand the value of sustainability and their role in supporting these green businesses.

Technology Leverage for the Rural and Underserved Entrepreneur: It is essential to address the gap between the urban areas and the rural areas concerning opportunities to access sustainable entrepreneurship with technology leverage. The use of digital tools, online platforms, and mobile applications will serve to provide rural and underserved entrepreneurs with access to information, markets, and fundable opportunities.

Small-scale green businesses will be taken to larger markets through e-commerce platforms. Technology-driven solutions in renewable energy, water management, and waste recycling will help scale the business. Micro-loans and crowd funding accessible through mobile-based financial services will allow rural entrepreneurs to have a chance to invest in sustainability. Technology will be realized at the grassroots level, providing impetus for sustainable local economic development, thus creating new vistas in which job opportunities can be produced and environmental improvements can also be made.

Taken together, these policy recommendations help make an even more supporting entrepreneurial ecosystem exist for the successful establishment and scaling of sustainability entrepreneurship throughout business development and scaling. Effective utilization of the better policies toward education will definitely help build a globally significant future through green entrepreneurship with high applicability of effective technology utilization to make such sustainable practice habitual at the end.

9 Conclusion

Sustainable entrepreneurship is increasingly proven to be a transformative driver of inclusive economic growth. The approach differs from other kinds of business models because instead of being only profit-focused in sacrificing social and environmental causes, sustainable entrepreneurship embodies value creation over the long run for all stakeholders involved in its business. They generate economic benefits and do efforts in addressing the present environment issues and the society in achieving equity. For instance, opportunities for such marginalized communities by green-tech businesses, fair trade, and other related eco-products will increase and generate job opportunities, bring down people into poverty, and lead eventually toward a much more resource-use-efficient business ecosystem where a lot less waste takes place and leads to greater redistributions of benefits from this economics. The way to build resilient and equitable societies is by integrating economic, social, and environmental goals. Sustainable entrepreneurship is in inherent synergy with the triple bottom line as it seeks to have its profit, people, and planet balanced. If seen from the perspective of an economic dimension, building a long-term value without resource drain or ecosystem destruction characterizes this sustainability. For a social dimension, fair wages and labor rights plus enhanced empowerment of communities mark attainment in this area. This aspect ensures that business activities impact the local populations positively. It promotes the notion of sustainable entrepreneurship in supporting business models that can solve the issues of gender inequality and economic disparity. The ultimate target of environmental goals includes bringing carbon footprints to minimum numbers, promoting clean energy to increase its consumption, adopting circular waste management, or doing anything else to guarantee that economic growth will be developed without any exploitation of resources for future generations. Together, these integrated goals make up a holistic approach to business where resilience is achieved through diversified strategies adapted to environmental change and equity through equal access to opportunities for all members of society. This will weave the principles of entrepreneurship into the fabric of growth, fostering inclusive and sustainable growth that benefits current and future generations, building a more resilient global economy.

Funding This research did not receive any funding/grant from public, Commercial or private funding agencies.

Conflict of Interest The authors declare that they have no known competing financial interests or personal relationships that could have appeared to influence the work reported in the paper.

References

Bartolacci, F., Caputo, A., & Soverchia, M. (2019). Sustainability and financial performance of small and medium sized enterprises: A bibliometric and systematic literature review. *Business Strategy and the Environment, 29*(3), 1297–1309. https://doi.org/10.1002/bse.2434

Bhattacharjee, P., & Mukherjee, J. (2023). *Sustainable Trends & Practices in Management, Humanities & Social Science Research.* https://doi.org/10.36426/excelleropen/isbn978-93-93734-83-92023/eojul23-04

Da Silva, L., Soltovski, R., Pontes, J., Treinta, F., Leitão, P., Mosconi, E., De Resende, L., & Yoshino, R. (2022). Human resources management 4.0: Literature review and trends. *Computers & Industrial Engineering, 168*, 108111. https://doi.org/10.1016/j.cie.2022.108111

Filho, W. L., Tripathi, S. K., Guerra, J. B. S. O. D. A., Giné-Garriga, R., Lovren, V. O., & Willats, J. (2018). Using the sustainable development goals towards a better understanding of sustainability challenges. *International Journal of Sustainable Development & World Ecology, 26*(2), 179–190. https://doi.org/10.1080/13504509.2018.1505674

Filser, M., Kraus, S., Roig-Tierno, N., Kailer, N., & Fischer, U. (2019). Entrepreneurship as catalyst for sustainable development: opening the black box. *Sustainability, 11*(16), 4503. https://doi.org/10.3390/su11164503

George, G., Rao-Nicholson, R., Corbishley, C., & Bansal, R. (2014). Institutional entrepreneurship, governance, and poverty: Insights from emergency medical response services in India. *Asia Pacific Journal of Management, 32*(1), 39–65. https://doi.org/10.1007/s10490-014-9377-9

Innovating Education and Educating for Innovation. (2016). In *Educational research and innovation.* https://doi.org/10.1787/9789264265097-en

Kumar, R., Goel, R., Singh, T., Mohanty, S. M., Gupta, D., Alkhayyat, A., & Khanna, R. (2024). Sustainable finance factors in Indian economy: Analysis on policy of climate change and energy sector. *Fluctuation and Noise Letters, 23*(2), 2440004–339.

Mahesh, M. K., Aithal, P. S., & S, S. K. R. (2023). Government initiatives and digital innovation for Atma Nirbhar MSMEs/SMEs: To achieve sustainable and inclusive economic growth. *International Journal of Management Technology and Social Sciences, 8*(1), 68–82. https://doi.org/10.47992/ijmts.2581.6012.0256

Mahato, J., & Jha, M. K. (2023). Does financial inclusion promote sustainable livelihood development? Mediating effect of microentrepreneurship. *Journal of Financial Economic Policy, 15*(4/5), 485–499. https://doi.org/10.1108/jfep-05-2023-0134

Malhotra, S., & Kiran, R. (2023). Examining the relationship between entrepreneurial perceived behaviour, intentions, and competencies as catalysts for sustainable growth: An Indian perspective. *Sustainability, 15*(8), 6617. https://doi.org/10.3390/su15086617

Malhotra, S., Kaur, T., Jain, K., Pandey, P. K., & Sengupta, A. (2024). Inclusive entrepreneurship ecosystem for PwDs: A reflection on SDGs. *International Journal of Manpower*. https://doi.org/10.1108/ijm-01-2024-0042

Markard, J., Raven, R., & Truffer, B. (2012). Sustainability transitions: An emerging field of research and its prospects. *Research Policy, 41*(6), 955–967. https://doi.org/10.1016/j.respol.2012.02.013

Mondal, S., Singh, S., & Gupta, H. (2023). Green entrepreneurship and digitalization enabling the circular economy through sustainable waste management—An exploratory study of emerging economy. *Journal of Cleaner Production, 422*, 138433. https://doi.org/10.1016/j.jclepro.2023.138433

Mohanty, R., Samal, A., & Kumar, R. (2024). Green finance supports sustainable environment: Empirical approach towards applicability in tourism sector. *Theoretical and Practical Research in Economic Fields, 15*(1), 114–126.

Nayak, K, R. P., R. (2022). Sustainable entrepreneurship research in emerging economies: An evidence from systematic review. *Journal of Entrepreneurship in Emerging Economies, 16*(3), 495–517. https://doi.org/10.1108/jeee-03-2022-0099

New Atlas. (n.d.). *Tesla and Panasonic partnership in renewable energy*. Retrieved November 25, 2024, from https://newatlas.com

Olurina, J. O., Gidiagba, J. O., Ehiaguina, V. E., Ndiwe, T. C., Ayodeji, S. A., Banso, A. A., Tula, O. A., & Ojo, G. G. (2023). Engineering innovations and sustainable entrepreneurship: A comprehensive literature review. *Materials and Corrosion Engineering Management, 4*(2), 70–79. https://doi.org/10.26480/macem.02.2023.70.79

Onakoya, A. J. (2013). Entrepreneurship, economic development and inclusive growth. *IEEE Series*. https://ijsse.org/articles/ijsse_v1_i3_375_387.pdf

Ooi, Y. K., & Nasiru, A. (2015). Entrepreneurship education as a catalyst of business start-ups: A study on Malaysian community college students. *Asian Social Science, 11*(18). https://doi.org/10.5539/ass.v11n18p350

Parida, R., Katiyar, R., & Rajhans, K. (2022). Identification and analysis of critical barriers for achieving sustainable development in India. *Journal of Modelling in Management, 18*(3), 727–755. https://doi.org/10.1108/jm2-11-2020-0288

Pradhan, R. P., Arvin, M. B., Nair, M. S., Hall, J. H., & Bennett, S. E. (2021). Sustainable economic development in India: The dynamics between financial inclusion, ICT development, and economic growth. *Technological Forecasting and Social Change, 169*, 120758. https://doi.org/10.1016/j.techfore.2021.120758

PV Magazine. (n.d.). *Tesla and Panasonic: Collaboration on solar energy and storage solutions*. Retrieved November 25, 2024, from https://pv-magazine.com

Ramadani, V., Agarwal, S., Caputo, A., Agrawal, V., & Dixit, J. K. (2022). Sustainable competencies of social entrepreneurship for sustainable development: Exploratory analysis from a developing economy. *Business Strategy and the Environment, 31*(7), 3437–3453. https://doi.org/10.1002/bse.3093

Saebi, T., Foss, N. J., & Linder, S. (2018, August 14). Social entrepreneurship research: Past achievements and future promises. *Journal of management, 45*(1), 70–95. https://doi.org/10.1177/0149206318793196

Satar, M. S., Aggarwal, D., Bansal, R., & Alarifi, G. (2023). Mapping the knowledge structure and unveiling the research trends in social entrepreneurship and inclusive development: A bibliometric analysis. *Sustainability, 15*(7), 5626. https://doi.org/10.3390/su15075626

Schot, J., & Steinmueller, W. E. (2018). Three frames for innovation policy: R&D, systems of innovation and transformative change. *Research Policy, 47*(9), 1554–1567. https://doi.org/10.1016/j.respol.2018.08.011

Shane, S., & Venkataraman, S. (2000). The promise of entrepreneurship as a field of research. *Academy of Management Review, 25*(1), 217–226. https://doi.org/10.5465/amr.2000.2791611

Sharma, A., Das, N., & Singh, S. P. (2023). Causal association of entrepreneurship ecosystem and financial inclusion. *Heliyon, 9*(3), e14596. https://doi.org/10.1016/j.heliyon.2023.e14596

Spigel, B., & Harrison, R. (2017). Toward a process theory of entrepreneurial ecosystems. *Strategic Entrepreneurship Journal, 12*(1), 151–168. https://doi.org/10.1002/sej.1268

SUEZ. (n.d.). *SUEZ's advanced waste management solutions.* Retrieved November 25, 2024, from https://www.suez.com

Veolia. (n.d.). *Veolia's role in waste-to-energy and recycling innovations.* Retrieved November 25, 2024, from https://www.veolia.com

Vig, S. (2022). Sustainable development through sustainable entrepreneurship and innovation: A single-case approach. *Social Responsibility Journal, 19*(7), 1196–1217. https://doi.org/10.1108/srj-02-2022-0093

WIRED. (n.d.). *Tesla's vision of integrated renewable solutions with Panasonic.* Retrieved November 25, 2024, from https://www.wired.com

Green Nanotechnology Innovations for Achieving India's Sustainable Development Goals by 2030

Monu Bhardwaj, Namrata Prakash, and Rupa Khanna Mahlotra

Abstract India's ambitious Sustainable Development Goals (SDGs) for 2030 require a convergence of science, technology, and policy. This research explores the transformative role of green nanotechnology in addressing critical challenges related to clean energy, sustainable agriculture, water purification, environmental remediation, and healthcare. By integrating green synthesis methods, eco-friendly nanomaterials, and innovative applications, this study evaluates how nanotechnology can accelerate progress across multiple SDGs. Key findings include the development of bio-inspired nanoparticles for efficient water treatment, capable of removing up to 98% of heavy metals and organic contaminants; nano-enabled fertilizers and pesticides that enhance agricultural yields by 30% while minimizing environmental toxicity; advanced photovoltaic nanomaterials that increase solar energy efficiency by 20% with minimal carbon footprints; and cost-effective nanomedicines that improve the reach and affordability of healthcare in rural regions. The study identifies critical enablers such as public–private partnerships, targeted funding, and capacity-building initiatives to scale these technologies. Green nanotechnology innovations have the potential to significantly contribute to India's SDGs, particularly in clean water and sanitation (SDG 6), affordable and clean energy (SDG 7), sustainable cities and communities (SDG 11), and climate action (SDG 13), offering eco-efficient solutions, creating green jobs, and reducing dependency on non-renewable resources. However, challenges include high initial development costs, limited technical expertise and infrastructure in rural areas, ecological risks associated with nanomaterial accumulation, and difficulties in scaling laboratory innovations to industrial applications. This research provides a roadmap for leveraging green nanotechnology to achieve India's SDG targets by 2030, underlining the need for interdisciplinary approaches and sustainable innovation.

M. Bhardwaj (✉) · N. Prakash
School of Management Graphic, Era Hill University Dehradun, Dehradun, Uttarakhand, India

R. K. Mahlotra
School of Commerce Graphic Era Deemed to Be University Dehradun, Dehradun, Uttarakhand, India
e-mail: head.comm@geu.ac.in

R. Kumar and S. N. Mohanty (eds.), *Sustainable Economy Models in the Age of Industry 5.0*, https://doi.org/10.1007/978-981-96-4104-8_11

Keywords Green nanotechnology · Sustainable Development Goals (SDGs) · Climate action · Circular economy · India 2030

1 Introduction

Green Nanotechnology Innovations for Achieving India's Sustainable Development Goals by 2030 highlight the relevance of green nanotechnology to meet the sustainable development demands of rapidly growing India. The chapter discusses the adaptations and contributions of nanotechnologies to serve mankind without causing environmental or health damages. The chapter reviews the Indian sustainability imperatives and identifies green nanoinnovations and strategies that are exclusively designed and developed to satisfy these sustainability needs. Nanoparticles and their technologies are growing at an unprecedented scale. These novel technologies have been aptly named nanotechnology. Continuous ongoing research in the field has proliferated an increasing list of possibilities where they can deliver extraordinary benefits to various activities of human life. Special features of nanomaterials also raise concerns over any resultant hazards to human health and safety or to the environment. Additionally, energy and natural resource requirements in their production, the reality of pollution at production sites, and relevance in relation to present strategies for reducing environmental threats may all be judged as potentially limiting factors for technological advancement (Sahu et al., 2021).

1.1 Background and Significance

Over the years, India has transitioned toward a developed economy, targeting annual growth rates of 8% to fulfill the needs and requirements of its expanding 1.3 billion people. Rapid urbanization and industrialization are adding to environmental degradation, causing climatic changes, natural resource depletion, increasing pollution, and biodiversity loss, which restrict future well-being. India's pledge of the Paris Agreement advocated a large number of policy measures and compromised solutions for numerous stakeholders, striving to achieve the country's nationally determined contribution, formulated on important elements such as low-carbon economic growth, energy efficiency, and the adoption of larger portions of renewable energy sources. The Sustainable Development Goals indicate concrete measures that define consensus about the world we want to leave behind for future generations (Dioha and Kumar, 2020).

Nanotechnology is associated with economic reforms and disruptive science-led technology, which are manufacturing eco-friendly materials, reducing resource consumption, adopting intellectual knowledge across all aspects of human activities, and learning unique properties and attributes of 1–100 nm sizes. The promising potential applications range from green buildings and renewable energy to water recycling and purification, waste management, and pollution control. Substituting traditional technology and innovative components with a profitable, unusual system needs to resort to a scientific laboratory base to visualize how the assembly of atoms and particles then impacts the prospective performance. As nanotechnology is counted among the key areas for generating future opportunities, investments shifted toward the design and development of smart business practices where benefits exceed costs, guaranteeing a sustainable and qualitative life for all people. As developing countries hold limited capability and resources, the structure and scope of successful research programs in fostering green nanotechnology applications for fulfilling environmental challenges become a crucial topic in need of immediate deliberation (Khan, 2020).

1.2 Purpose of the Study

Nanotechnology is the science, engineering, and technology related to nanoparticles, which are particles with at least one dimension of less than 50 nm. These nanoparticles have unique properties due to their small size, high surface area, and very high reactivity. Nanoparticles exhibit the properties of strength, high electrical and thermal conductivity, and the optical properties of the materials when bonded to polymers, ceramics, or metals. Nanotechnology has huge potential, and its related innovations can be applied to a wide range of applications in the areas of electronics, medicine, energy production, and consumer products, among others. It has been declared by many countries as a key general-purpose technology of the twenty-first century that will pave the way to reach economic prosperity, societal benefit, and well-being among its citizens (Haleem et al., 2023).

Nanotechnology stimulates innovation. It is predicted to be the foundational business for approximately US$ 3.1 trillion by 2015. Hence, nanotechnology holds huge potential in terms of economic benefits to producers of such innovative products. The large perceived potential of nanotechnology could face public reactions if its potential safety, health, environmental, regulatory complexity, and ethical issues are not properly managed. Hence, there is an imminent need for nanotechnology innovation, responsible development, and stewardship exhibiting sustainable growth. This study has been conducted to understand nanotechnology innovation and to assess its potential for being utilized to achieve India's Sustainable Development Goals by the year 2030. The objectives are to list various nanotechnology innovations in different sectors, understand the status quo of select nanotechnology innovations, and assess their impact on the achievement of India's SDG, so that India can meet its targets by 2030.

2 Understanding Sustainable Development Goals (SDGs) 2030

2.1 Overview of SDGs

United Nations General Assembly Resolution 1725, adopted in 2015, is a set of seventeen worldwide objectives for sustainable development that will remain in effect until 2030.The SDGs are part of the 2030 Agenda. A holistic and coordinated strategy for environmentally friendly growth is required to achieve the lofty and interconnected goals. The 17 goals include 169 targets covering a broad range of sustainable development issues. These include ending poverty and hunger, improving health and education, making cities more sustainable, combating climate change, and protecting oceans and forests (Tsalis et al., 2020). The SDGs are interconnected and should be addressed together in harmony. Governments are encouraged to turn to the goals when making decisions on activities to work on at the domestic and international levels. The SDGs are a reported improvement on the Millennium Development Goals (MDGs) initiated in 2000 and concluded in 2015.

The principal outcome of the 2012 United Nations Conference on Sustainable Development was the agreement by member states to launch a process to develop a set of SDGs that have been establishing the post-2015 agenda. The open working group proposal for Sustainable Development Goals has been commended for the breadth, the focus on disparities, the coverage of the goals and targets, and the recognition of the critical role of culture as a fundamental driver of sustainable development. Some of the changes were characterized as big steps in some areas with a view to correctly reflect funding sources and with the absence of 'develop' in many of the goals and targets. The new SDGs carried more political weight than the MDGs, as they were all considered by all countries, rich, large, and small alike, unlike the MDGs which were focused more on the poorer nations.

2.2 Relevance to India

This research reveals that green nanotechnology innovations are well-aligned with targets set by India under its Sustainable Development Goals and SNDP-2030 framework. The 12th Five Year Plan of India recognized that nanoscale science and engineering are growing very fast and hold particularly strong potential for India. Given our core strength in certain critical sectors and niche technologies, we need to develop designated areas of strategic applications for nanoscience, particularly in healthcare, agriculture, water, food, and energy, and then focus on their rapid development, leading to application. India has potential inherent advantages including its pluralistic culture, flexible scientific capabilities, demographic advantage, and strong IT industry; hence we can seize this opportunity for rapid advancement and sufficiency in strategically important areas by making substantial investments in this sector & on National Nanotechnology Initiative..., 2020).

This is an endorsement of the view expressed here that nanotechnology can play an important role in overcoming many major challenges created by growing populations, aging infrastructure, and the need for sustainability—within energy, water, the environment, and the potential risks of climate change anytime beyond 2030—as well as economic and social development. Our current academic discoveries, as well as increasing focus on risk governance and regulatory approaches to nanotechnology, combine aspects of material and device research with current research in economics and social equity. After four years of close consultation and consensus with state governments, the private sector including the Indian industry, and expert groups, it is more than fortuitous that the National Institution for Transforming India worked together for 15 months with 13 dialogue forums to put forth its 2030 framework on 23rd September 2015 for preparing India's SDGs, and that the data required to track the achievement of the national targets against SDGs indicators will be launched at the 1st SDGs India Index and Dashboard on 25 December 2018 with the presence of the Vice Chairman of NITI Aayog.

3 Nanotechnology and Its Applications

Nanotechnology is the manipulation of matter on an atomic, molecular, and supramolecular scale. A novel approach through green nanotechnology is to develop clean, safe, non-toxic, and eco-friendly treatment technologies. Green nanotechnology refers to the use of processes that are environmentally friendly, meaning they are harmless to the environment and also involves the use of nanotechnology to enhance the environmental performance of processes in a particularly wide range of materials and applications. The term green in green nanoscience or green nanotechnology lends its sense to the area that is dedicated to allowing the development of new nanoscale materials, particles, and devices that have features which are eco-friendly Aithal and Aithal (2021a, 2021b).

Green nanotechnology exists to address problems associated with conventional, unmodified, or unscaled-up synthesis approaches that make toxic byproducts and generate significant unit-specific environmental, toxicological, and health concerns. In general, processes can be considered green if, simultaneously or not, they contribute to environmental quality, ecological sustainability, safety, and equity. Green nanotechnology works on the basic concept of the three R's of sustainability, 'Sustainable Nanotechnology,' also known as the dematerialization concept. The first green nanotechnology that was developed is the 3-nm Bio-Cluster for oral delivery of insulin and the environmental remediation of heavy metals. The concept of a white biotechnology sector builds upon the current developments and large impact in the fields of biotechnologies in industries tasked to minimize the environmental and ecological externalities of heavily labeled activities (Khan, 2020). In fact, a number of currently available commercial applications have already been demonstrated to be environmentally friendly in the sense of being safe and low impact on the environment.

3.1 Definition and Basics

The concept of green nanotechnology has been derived from the principles of green chemistry. Green nanotechnology involves the design, production, and use of nanotech applications that reduce the use and generation of harmful substances such as heavy metals and have potential health and environmental implications. Green nanotechnology encompasses both pollution prevention and creates green engineering approaches within a broad research field. Hence, together with the importance of understanding and effectively utilizing the developed nanotechnologies to prevent their adverse effects on human health and the environment, there are also noticeable adverse effects caused to life and the environment by the rapidly growing field of nanotechnology itself (Ciambelli et al., 2020). Scientists should always be aware of the needed development of environmentally friendly and sustainable nanotoxicological devices. Their development should be planned simultaneously and with the engagement of society, as the proposed risk-benefits of the developing nanotechnology-based field mostly demand and are supported by society.

The applications of green nanotechnology include medication, the food industry, filtration/desalination of water, operation of prototypes in the energy field, printing, lifetime expanders for tests in vivo, and public health, reactive and shiny dimension inks, antimicrobial floorings, and windows that block the amount of light traveling through them when the building becomes very hot. However, the scientific development in the field of green nanotechnology still has a long way to go. Only through the recalibration of green nanotechnology could nanotechnology achieve a good social virtue upswing by society. Moreover, nanotechnology could only achieve a good social virtue upsurge through the incorporation of principles such as sustainable innovation and responsible research.

3.2 Current Applications in Various Sectors

Currently, tapping sunlight as solar energy is one of the most popular applications of green nanotechnology. Other areas include the storage of solar energy in fuel cells, polymer electrolytes, photocatalysis of water, photovoltaic cells, and dye-sensitized solar cells. In membrane technologies, high performance and separation of gases in hydrogen fuel cells, desalinization, fuel cells, and low-fouling membranes are achieved with green nanotechnology. Various types of nanosensors, such as optical sensors, DNA sensors, and capacitor-based sensors, can be used for on-site monitoring of environmental parameters. For molecular imaging in biology, nanomaterials can be used as MRI contrast agents to track the labeled cells and hyperthermia. In carbon sink technology, particularly the vanadium nanoclusters efficiently dissolve carbon dioxide and catalyze the conversion of carbon dioxide to fixed organic compounds with the use of ultraviolet light Assad et al., 2022). Balancing air moisture for enhanced agricultural production and water management in agriculture are important economic applications. However, these applications need to consider the criticality of watersheds and the environment.

Development in textile nanotechnology to create ultraviolet protective and antibacterial or antifungal properties has progressed to the stage of commercial products. In the area of energy conservation, the use of nanomaterials for enhancing the insulating properties in polymer films is another triumph. Magnetic nanoparticles and nanorods for cancer destruction, drug delivery, and targeted therapy are probably the most debated applications in medical research at present. Prevention-focused medical research is also utilizing the ultrasound-induced heating effect of gold nanospheres at tumor sites. Green nanotechnology innovations are leading to both prevention and cure of diseases. In the area of nanospheres, mesoporous silica is being replaced by carbon nanospheres. These are produced in the laboratory for nucleation and growth of silica and zeolite nanoparticles. Catalytic applications have replaced this with the creation of rapidly obtained ordered mesoporous carbon nanoparticles. Since carbon nanospheres can be functionalized with various groups on the carbon nanosphere surface, the resultant nanosphere could find potential application in the pharmaceutical or catalysis industry (Nasrollahzadeh et al., 2020).

4 Green Nanotechnology

Society is advancing in the areas of science, technology, and manufacturing of nanomaterials for various applications. Likely, while achieving the sustainable goals in the future, it would be necessary to stop the release of hazardous nanomaterials into the environment, particularly in its embryonic stage. In this chapter, we have included some nanomaterials which would not only have zero effect on human health and the environment but also clean it.

In order to overcome future toxicological problems arising from the production of the above nanoparticles, nanotechnology should be applied in the greenest way possible by reducing the hazardous effect associated with it. So, it is necessary to apply the principles of green technology to avoid polluting the environment and to carry out the detoxification process simultaneously. The phrase 'green nanotechnology,' called green manufacturing, aligns principles of green chemistry with the unique benefits of nanotechnology. Green nanotechnology aims to reduce the potential of nanotechnology to generate nanomaterials and products that are not harmful to humans or the environment. This can be achieved by designing nanomaterials that provide beneficial nanotechnology to sustainable development objectives. In the development of Science, Technology, and Innovation Policy following the Sustainable Development Goals objectives for 2030, nanomaterials are recognized for their implications toward environmental and human safety and health. Regulations, guidelines, and standardization have also been established. The main positions for the sustainable development objectives are given in a table. In order to link green nanotechnology with sustainable development, a brief account of the different properties of nanomaterials is described, which are related to achieving equality objectives. Only a few reports are available on the sustainable development goals' green nanotechnology approaches. What was found is an emphasis on the photocatalytic activity of nanoparticles and a few environmental application descriptions that occur naturally from nanoparticles. To truly promote the concept of green nanotechnology and its association with the other six characteristics, more research is needed. With this in mind, we invite all researchers to extend their results according to the following different administration backgrounds. The relationship between green nanotechnology and ensuring the achievement of sustainable goals by 2030 is supported (Larrañaga-Tapia et al., 2024).

4.1 Concept and Principles

The core philosophy of green nanotechnology revolves around protecting the environment and human health by governing the findings right from the very beginning of any project. Green nanotechnology is efficiently utilized by rational design approaches that greatly reduce waste materials. Using green nanotechnology for environmental observing systems brings the two fields closer together. The beneficial capabilities of nanomaterials become distinctive when nanostructures are at play. One of the key principles of green nanotechnology is the incorporation of microorganisms, such as phytoplankton or bacteria, to complement nanoelements' functional properties, enabling a wide range of pollutants' detection and filtering (Bahrulolum et al., 2021).

4.2 Advantages and Challenges

From boosting the efficiency of solar cells made from bay berries and oranges to the treatment of contaminated water with fruit peels and packaging using nano-ink or nanocomposites made from natural resources, the world of nanotechnology offers a wide range of opportunities to fulfill a variety of sustainable development goals for 2030. Goals for 2030 include that everyone has access to clean water and affordable, sustainable energy Aithal and Aithal (2021a, 2021b). Food and waste streams should be recycled efficiently, the climate should be kept under control, and ecosystems should be managed in such a manner that sufficient land and water are available for the needs of future generations. Additionally, strong goals are set to improve public health and decrease resource conflicts. India's share toward the world objectives is also demonstrated. The inclusion and exclusion criteria for applying these sustainable development objectives in the Indian context are recommended.

5 Role of Green Nanotechnology in Achieving SDGs

Nanotechnology offers a variety of strategies that can contribute to positively influencing the achievement of the United Nations Sustainable Development Goals. Strategies have been discussed that use nanoparticles intended to help solve specific problems, such as nanoparticles for water decontamination, carbon dioxide sequestration, noise reduction, and nanosensors to monitor the environmental impact of industry (Pokrajac et al., 2021). An improved understanding of technical, social, political, and economic conditions as enabling or supporting responsible innovation is of great value to society in general. Maximizing environmental performance and benefits is considered when innovating graphene and other nanomaterials in emerging photocatalysis and nanomedicine, among other highly promising applications, which represents a fundamental challenge. India's rapid transition and bright future need an innovative and transformative approach that takes a global opportunity—a green nanotechnology—to rapidly develop and disperse advanced nanomaterials with superior properties, utilizing the unique characteristics of nanomaterials to achieve sustainability and people-centered innovation combined with circular economy principles to optimize materials management, clean manufacturing, and regeneration in commercial real-world products and sustainable solutions. Given that many nanomaterials can be described as natural-based, biologically driven, and environmentally sound, India has not only the opportunity but also the responsibility to safely develop and utilize products for people and our planet.

5.1 Overview of SDGs 2030

Since 2015, India has redoubled its efforts to meet the Sustainable Development Goals (SDGs) by 2030. Green nanotechnology offers novel solutions to accelerate the speed and success with which we can meet this goal. A few of the targets under the 17 different SDGs, with emphasis to be achieved by 2030, are: SDG-3 (Good Health and Well-Being), on issues related to health; SDG-7 (Affordable and Clean Energy), to ensure access to affordable, reliable, and sustainable energy for all; SDG-9 (Industry, Innovation and Infrastructure), on building resilient infrastructure, promoting sustainable industrialization, and fostering innovation; SDG-11 (Sustainable Cities and Communities), ensuring sustainable and resilient cities by providing access to safe, affordable, accessible, and sustainable transport systems for all, improving road safety; SDG-12 (Responsible Consumption and Production), and encouraging countries to promote the use of cleaner and more efficient green and innovative solutions in production and consumption; SDG-14 (Life Below Water), focusing on conserving and sustainably using the oceans, seas, and marine resources for sustainable development; and SDG-15 (Life on Land), on sustainably managing forests, combating desertification, halting, and reversing land degradation.

In order to achieve these objectives of the SDGs by 2030, green nanotechnology research should focus on designing sustainable and green solutions that ensure meeting the SDGs—methods that are efficient, affordable, and cost-effective, but at the same time, minimize the depletion of natural resources and reduce the overall impact on the economy and society. We believe a circular economy model followed by the Regenerative Development Framework could be adopted by conscious policymakers and researchers in green nanotechnology and allied disciplines to meet the UN SDG objectives of 2030. Furthermore, green nanotechnology-based innovations should be adequately funded, and healthcare policy should be reshaped for India to have a better chance of achieving the 2030 goals (Khajuria et al., 2022).

5.2 *Key Areas of Impact*

The field of nanotechnology for greening purposes is quite vast, with a very broad range of applications. With regard to a country, besides its contributions to the SDG goals, it also creates ample opportunities for its public and private R&D efforts to be steered in the right direction so that the national benefits can be maximized. At this juncture in time, India has its own list of national priorities and accordingly, there is a need to identify areas where the application of Green Nanotechnology (GNT) can help meet those national objectives Aithal and Aithal (2021a, 2021b). In the backdrop of India's identified priorities and coinciding with the nation's aspirations for becoming a world leader in certain sectors and its scope to provide solutions based on the country's proficiency in IT and software, the following are some of the key areas where GNT can lead to innovations: (i) GNT in food processing for increased storage life via non-toxic nanocoatings (ii) GNT for energy generation and storage (iii) GNT in water treatment technologies (iv) GNT in healthcare as alternatives to conventional avenues (v) GNT in hydrogen generation technologies and delivery (vi) GNT in cotton viscose processing and nanosensors for textiles (vii) ICTs that benefit from GNT (viii) Green Zero Energy Light Emitting Smart Cities approach to smart buildings.

6 Research Methodology

6.1 Research Design

This study will employ quantitative methodologies to comprehensively explore the role of green nanotechnology in achieving India's Sustainable Development Goals (SDGs). The research design is structured into the following phases:

6.1.1 Exploratory Phase:

- A detailed literature review will analyze existing green nanotechnology innovations and their applications in sustainable development.
- Case studies of successful implementations worldwide will be studied to identify scalable solutions for India.

6.2 Data Collection Methods:

6.2.1 Secondary Data:

- Published journals, conference papers, and patents related to green nanotechnology.
- Reports from Indian government initiatives, including the Ministry of Science and Technology, NITI Aayog, and international organizations like the UNDP and WHO.

6.3 Analytical Tools and Techniques

6.3.1 Qualitative Analysis:

- Thematic analysis for interviews and focus groups to identify trends and barriers.
- SWOT analysis to evaluate strengths, weaknesses, opportunities, and threats of green nanotechnology in India.

6.3.2 Policy Analysis:

- Examination of existing Indian policies and their alignment with green nanotechnology innovations.
- Comparative analysis of international best practices to recommend strategies for India.

6.4 *Success Stories in India*

A few success stories in the area of green nanotechnology, involving the application of some of the above-finished products from industries for achieving SDGs in India, are of interest. They were translations of basic research demonstrating to the industry, provided by the scientists of the Department of Atomic Energy-funded Nano Science and Technology Initiative. Successful technology transfers in this journal are based on the results of commercial activities of industries, which are described briefly.

In the cases of the fabrics, it was observed that the grey and treated water did not show any presence of very low-level nanoparticles, while the conventional fabric processing industries generate significant amounts of such particles in water during the removal of metal ions. The case of antibacterial paint is another interesting one. Even the TDS, which includes metal-oxide-containing nanoparticles, if present in water, is reported to induce some biological response. In the water treatment sector, a report concludes by mentioning that based on results, the bioaccumulation of Ag from several engineered nanomaterials in the aquatic environment might contribute to potential adverse effects (Uddin, 2021).

6.5 *Global Best Practices*

Nanotechnology lies at the forefront of the converging technologies agenda and is an area of intense global interest. The global nanotechnology R&D expenditure share of individual nations from 2001 to 2015 highlights the significant investment by countries, particularly the USA, during this period. Important strides in nanotechnology R&D are being made. The 3 + 3 best practices for nanotechnology research and the National Nanotechnology Safety Infrastructure training provide the best practices for enhancing the contribution of nanotechnology to society in a responsible manner. A portfolio of health and safety guidelines, practices, and communication tools for nanotechnology and nano-enabled products will be necessary to enhance society's acceptance and benefit from them.

The companion Best Practices Documents establish the technical details for establishing and enforcing specific expectations frequently associated with green nanotechnology by the broad community (Gottardo et al., 2021).

7 Policy and Regulatory Framework

Advances in green nanotechnology have provided an unprecedented opportunity in advancing development to benefit both present and future generations. In order for it to play this transformative role, it is important to create an environment that fosters the technology alongside identifying, monitoring, and addressing potential externalities or risks that could hinder realization of the same. Achieving this balance, while bearing in mind the importance of reducing social disparities, will require a coordinated and coherent policy framework, a flexible investment climate, and coherence across national and local governing infrastructure like the knowledge and capacities of existing regulations, intellectual property rights strategies, and a nexus of previously or contemporarily established lawful norms. This chapter presents, albeit qualitatively, policy, regulatory, and actual cases of stimulating and incubating enabling green nanoinnovations that could be deployed to augment India's ability to achieve critical sustainable development goals by 2030.

Among the conclusions reached, a targeted incentive mechanism that fosters the development and use of green nanotechnology for those sectors that mostly stunt sustainable development will be critical in realizing its full potential. Such incentive mechanisms could be crafted to encourage environment-centric private financing alongside local government support for risky ventures, promote behavioral change among potential end-users or investors, develop academic infrastructure to support the trickle-down commercialization model, or even incentivize onshoring Aithal and Aithal (2022). The 2030 strategic position will require prudent utilization of public funds not only in terms of percentage of interventions but also in terms of the scale of intervention, the degree of direct usage, and the nature of intervention.

7.1 Current Policies in India

India has several policies to address sustainable development and emphasizes its industry, technology, and development capacity. This includes three major initiatives in six key technology areas aimed at achieving its sustainable development goals through appropriate policies (Narwane et al., 2021). These pertain to: the National Action Plan on Climate Change, the National Solar Mission, the National Environment Policy, the National Missions on Enhancing Energy Efficiency, the National Mission on Sustainable Habitat, and Ganga rejuvenation. The National Action Plan on Climate Change is organized as eight distinct 'National Missions,' including renewable energy, energy efficiency, sustainable agriculture, and water, with the other missions focused on aspects such as sustainable strategies for Himalayan ecosystems.

The goal of the National Action Plan on Climate Change is to address the identified issues and implement sustainable solutions. The National Missions on Enhancing Energy Efficiency are oriented toward carbon trading and certifying energy-efficient industries. The National Mission on Sustainable Habitat aims at the development of smart cities which will incorporate low-carbon and resource-efficient development. Ganga rejuvenation is a program for the internal nourishment of rivers through five key components: Integrated River Conservation Program, which prepares the Ganga River Conservation Master Plan Alagirisamy and Ramesh (2022).

7.2 International Collaborations and Agreements

Aspects of the green nanotechnology innovations proposed in this study are intimately linked to international agreements and diverse research and economic relationships between India and other countries at the governmental, industrial, national laboratory, and academic levels. Prominent treaty-type international agreements that underlie India's aspirations to meet the UN SDGs by 2030 include the Paris Agreement and the Minamata Agreement. (JAMEEL & Ahmed, 2021).

The first week of November 2017 was designated as World Green Nanotechnology Week when international research in universities and businesses was showcased in expert presentations, student seminars, meetings with industry, and science outreach programs. Ideally, India, due to its large population and high energy usage, should be the focus of maximum attention for cooperative research leading to green nanotechnology innovation in clean energy and quick mitigation of adverse environmental effects on human health and the available ecosystems caused by air, water, and soil pollution.

8 Challenges and Future Directions

India's Aspirations for SDGs by 2030:

- A green approach is essential to achieve the Sustainable Development Goals (SDGs).
- Innovative thinking and alternative development strategies are necessary.

Challenges with Nanomaterials:

- Many challenges and unknown risks are associated with nanomaterials.
- Thorough assessment of these risks is crucial for sustainable development.

Avoiding Greenwashing:

- The focus should not overly highlight the negative aspects of nanomaterials.
- Innovation in green nanomaterials must avoid superficial sustainability claims.

Government Role in Oversight:

- Agencies should monitor and report nanomaterial usage transparently.
- Public reporting should balance transparency and protection of business-sensitive information.

Progress and Vision for Nanotechnology in India:

- Near-term milestones to track progress have been initiated.
- Much remains to be done to achieve long-term goals.
- A sustained pursuit of an ambitious vision is required to address global challenges.

8.1 Technological Barriers

Existing nanotechnological innovations that deliver green or other sustainable development benefits are relatively small in number compared to the overall nanotech innovation output available. In particular, the contribution of Indian nanotechnological innovations is rare. Consequently, the benefits of nanotechnology, which aid the leadership of the United Nations to deliver on its promise of a better future through Sustainable Development Goals, are at stake. Since India's stated enduring technological vision-objective is to become self-reliant in critical technologies by enhancing R&D intensity to 2.0% of GDP, utilizing its large working population strategically to its advantage by transforming education, skills, talent, and intellect from 'demographic dividend' to 'demographic advantage.' The same may be expected to have a demonstrable positive impact on increasing the human element of knowledge, pushing the national economy into innovation-driven growth, besides preserving the high yields of national biotechnology penetrations and conversions to multiple ventures that are finally meant for a sustainable planet with poor and rich living in harmony without stark extremes of inequality.

8.2 Economic Implications

One of the potential economic consequences of adopting green nanotechnologies is the use of nano-structured materials in various applications, leading to a transition from resource-based economies to knowledge-based economies. India should recognize the potential importance of nanotechnology innovation in its strategy to achieve sustainable development, engage in the development of an institutional process that can steer and promote innovation in this area, so that India remains in the vanguard of this technology and fully exploits its potential benefits. This also presents numerous trade opportunities for India as an early entrant into a market that has infinite possibilities. India can be a big player in the production and use of engineered nanomaterials and also play a vital role in the global market to dominate and steer it on a sustainable path. Fast tracking of green nanotechnology will help bring down energy demand and reduce environmental pollution. It will help in sustainable growth and achieve sustainable development goals by 2030 to bring inclusiveness and a higher standard of living to its people all around India and also help the global community work for a better future.

9 Conclusion and Recommendations

The Nano Mission provides opportunities for innovations and newer methods to address many of the important issues concerning India's challenges associated with climate change and the goals of sustainability. There is a huge potential for advancements in the field of nanotechnology leading to the creation of green nanotechnology. Importantly, it is about creating value for society and contributing toward the goal of inclusive and sustainable growth. The study revealed that the focus areas encompassing climate change, environment, and energy align with the priority areas identified by the nano-related ministries and departments. Based on the above, the following are the overall suggestions to leverage the opportunities and realize the potential of nanotechnology for addressing India's challenges associated with climate change: There is a need to enhance translational research activities for creating nanomaterials that offer dual benefits, where nanomaterials, devices, and systems not only possess small carbon or ecological footprints but also have the ability to mitigate the existing problems associated with climate change. The research landscape can move further in the value chain by creating products that are carbon-free or carbon-negative, thereby offering benefits of direct and indirect carbon footprint reduction. There is a need to address public concern when dealing with the potential growth of these materials to realize the full benefits and minimize adverse impacts. All these actions would have a major positive effect on society. The challenge is how to reap the many benefits of nanomaterial development while also minimizing potential unintended consequences. There was a lack of firm answers to this challenge inherent in nanoscience while displaying the potential to significantly affect how human beings live and how they prepare themselves to conduct the business of nanoscience and technology.

9.1 Summary of Key Findings

This study is aimed at achieving and accelerating the intended purposes of the adoption of green nanotechnologies for environmental protection and sustainable development by both policy and scientific approaches. Nanoscience and nanotechnology have the profound potential to attain social, economic, environmental, and national security benefits for the entire community compared to any existing technology of a similar level of research, even though policy inadequacies may lead to a faux pas regarding these potential gains. Such potentials are realized through predictability, control of nanoscale structures, and the unique properties and novel applications these materials offer. These special features are greatly important in our efforts to create a truly sustainable, well-off, and high quality of life for the public—both today and tomorrow—within the framework of green nanotechnologies. It is globally recognized, however, that the expectation of these benefits is deeply associated with the potential risks that such emerging capabilities bring to the ecosystem, as we also noted in this study. This has made the integration of nanotechnology, environment, health, and safety research a priority for society and policymakers.

Academic and policy research into the environmental and sharing issues of nanotechnology has been well celebrated internationally, which has even called for advances in nanotechnology to meet national problems and societal needs in areas such as water, energy generation and conservation, pollution prevention, and better biotechnology applications; hence, the birth of green nanotechnologies. In this study, we have disclosed and identified how reengineered green nanocomposites could be potentially crucial and effective materials for addressing social challenges of Sustainable Development Goals, and inclusive and transformative solutions for populations seeking sustainable well-being and high quality.

9.2 *Policy Recommendations for India*

India has taken numerous initiatives to promote nanoscience and nanotechnology through investments and policy interventions at different levels. Numerous policy considerations could be taken into account in mainstreaming green nanotechnology toward sustainable development in India for a greater transformative impact: Consider the challenges of complementing national and regional innovation systems by promoting win–win initiatives such as bilateral or multilateral co-funding of public–private research partnerships. Work on priority themes such as water, energy, health, food, and the environment to achieve mission-oriented goals for social welfare, health, and protection of the environment. This will receive popular acceptance. Track funding through periodic reports on investment in selected identified priority themes of sustainable development for financial clarity. This will also foster budgetary transparency and contribute to potential IT-enabled India's leadership in nanotechnology and green nanotechnology research and applications. Design efficiency through specific product designs in addition to improvements in process efficiency, including the development of integrated life cycle assessment tools and indicators for green nanotechnology. Identify and promote areas of advantage such as traditional knowledge like Ayurveda, handicrafts, and the traditional knowledge of tribals in order to benefit from comparative advantages such as local production through local consumption concepts. This includes areas such as traditional Ayurvedic medicine and plant extracts, which could prove helpful in reducing the cost of medicine.

Funding and Conflict of Interest Statement A "Funding and Conflict of Interest Statement" is a declaration within a research paper or proposal where authors explicitly disclose any financial sources that supported their study, alongside any potential conflicts of interest that could influence the research findings, such as personal affiliations, financial gains, or relationships with companies or organizations directly related to the study topic.

References

National Academies of Sciences, Division on Engineering, Physical Sciences, National Materials, Manufacturing Board, Committee on National Nanotechnology Initiative, & A Quadrennial Review. (2020). A quadrennial review of the national nanotechnology initiative: Nanoscience, applications, and commercialization. [HTML]

Aithal, S., & Aithal, P. S. (2021). Green nanotechnology innovations to realize UN sustainable development goals 2030. *International Journal of Applied Engineering and Management Letters (IJAEML), 5*(2), 96–105. supublication.com

Aithal, S., & Aithal, P. S. (2021). Green and eco-friendly Nanotechnology–concepts and industrial prospects. *International Journal of Management, Technology, and Social Sciences (IJMTS), 6*(1), 1–31. supublication.com

Aithal, P. S., & Aithal, S. (2022). Opportunities and challenges for green and eco-friendly nanotechnology in twenty-first century. In *Sustainable nanotechnology: Strategies, products, and applications* (pp. 31–50). researchgate.net

Alagirisamy, B., & Ramesh, P. (2022). Smart sustainable cities: Principles and future trends. In *Sustainable cities and resilience: Select proceedings of VCDRR 2021* (pp. 301–316). Springer Singapore. researchgate.net

Assad, H., Kaya, S., Kumar, P. S., Vo, D. V. N., Sharma, A., & Kumar, A. (2022). Insights into the role of nanotechnology on the performance of biofuel cells and the production of viable biofuels: A review. *Fuel, 323*, 124277. [HTML]

Bahrulolum, H., Nooraei, S., Javanshir, N., Tarrahimofrad, H., Mirbagheri, V. S., Easton, A. J., & Ahmadian, G. (2021). Green synthesis of metal nanoparticles using microorganisms and their application in the agrifood sector. *Journal of Nanobiotechnology, 19*, 1–26. springer.com

Carlsen, L., & Bruggemann, R. (2022). The 17 United Nations' sustainable development goals: A status by 2020. *International Journal of Sustainable Development & World Ecology, 29*(3), 219–229. [HTML]

Ciambelli, P., La Guardia, G., & Vitale, L. (2020). Nanotechnology for green materials and processes. In *Studies in surface science and catalysis* (Vol. 179, pp. 97–116). Elsevier. [HTML]

Dioha, M. O. & Kumar, A. (2020). Exploring the energy system impacts of Nigeria's nationally determined contributions and low-carbon transition to mid-century. *Energy Policy, 144*. [HTML]

Gottardo, S., Mech, A., Drbohlavová, J., Małyska, A., Bøwadt, S., Sintes, J. R., & Rauscher, H. (2021). Towards safe and sustainable innovation in nanotechnology: State-of-play for smart nanomaterials. *NanoImpact, 21*, 100297. sciencedirect.com

Haleem, A., Javaid, M., Singh, R. P., Rab, S., & Suman, R. (2023). Applications of nanotechnology in medical field: a brief review. *Global Health Journal, 7*. sciencedirect.com

Jameel, A., & Ahmed, W. (2021). Sustainable development goals and India's commitment to gender justice. *Society & Sustainability, 3*(2), 68–86. riiopenjournals.com

Khajuria, A., Atienza, V. A., Chavanich, S., Henning, W., Islam, I., Kral, U., ... & Li, J. (2022). Accelerating circular economy solutions to achieve the 2030 agenda for sustainable development goals. *Circular Economy, 1*(1), 100001. sciencedirect.com

Khan, S. H. (2020). Green nanotechnology for the environment and sustainable development. Green materials for wastewater treatment. [HTML]

Larrañaga-Tapia, M., Betancourt-Tovar, B., Videa, M., Antunes-Ricardo, M., & Cholula-Díaz, J. L. (2024). Green synthesis trends and potential applications of bimetallic nanoparticles towards the sustainable development goals 2030. *Nanoscale Advances, 6*(1), 51–71. rsc.org

Narwane, V. S., Yadav, V. S., Raut, R. D., Narkhede, B. E., & Gardas, B. B. (2021). Sustainable development challenges of the biofuel industry in India based on integrated MCDM approach. Renewable Energy, 164, 298–309. somaiya.edu

Nasrollahzadeh, M., Sajjadi, M., Soufi, G. J., Iravani, S., & Varma, R. S. (2020). Nanomaterials and nanotechnology-associated innovations against viral infections with a focus on coronaviruses. *Nanomaterials, 10*(6), 1072. mdpi.com

Pokrajac, L., Abbas, A., Chrzanowski, W., Dias, G. M., Eggleton, B. J., Maguire, S., ... & Mitra, S. (2021). Nanotechnology for a sustainable future: Addressing global challenges with the international network4sustainable nanotechnology. acs.org

Rajneesh, S. (2023). *Strategy of Karnataka to achieve aspirant Sustainable Development Goals 2030*. Agricultural Sector in India. [HTML]

Sahu, T., Ratre, Y. K., Chauhan, S., Bhaskar, L. V. K. S., Nair, M. P., & Verma, H. K. (2021). Nanotechnology based drug delivery system: Current strategies and emerging therapeutic potential for medical science. *Journal of Drug Delivery Science and Technology, 63*, 102487. [HTML]

Sobti, R. C., & Sarin, A. (2024). Science, technology and innovation for achieving sustainable development goals. In *Role of science and technology for sustainable future: volume 1: Sustainable development: A primary goal* (pp. 3–13). Springer Nature Singapore. [HTML]

Tsalis, T. A., Malamateniou, K. E., Koulouriotis, D., & Nikolaou, I. E. (2020). New challenges for corporate sustainability reporting: United Nations' 2030 Agenda for sustainable development and the sustainable development goals. *Corporate Social Responsibility and Environmental Management, 27*(4), 1617–1629. researchgate.net

Uddin, F. (2021). Environmental hazard in textile dyeing wastewater from local textile industry. *Cellulose, 28*, 10715–10739. [HTML]

An Adaptive Framework for Empowering Marginalized Communities Through Human-Centric Artificial Intelligence (HCAI) for Decision-Making in Poverty Reduction Using Regression Techniques

R. Vidya, P. Deepan, and N. Arul

Abstract This paper presents an adaptive decision support framework utilizing Human-Centric Artificial Intelligence (HCAI) to aid poverty reduction efforts among marginalized communities. The model dynamically incorporates real-time data inputs such as income distribution, welfare metrics, and poverty headcount ratios. By integrating community feedback loops, the system continuously adjusts to the needs of local populations, providing contextually relevant insights and tailored policy recommendations. Leveraging machine learning algorithms, including regression, this approach achieves high predictive accuracy and actionable, region-specific insights for decision-makers. Findings suggest the model's potential for scalable, ethically aligned poverty alleviation. Different regression model techniques were applied to the model. XGBoost outperforms other models, achieving an accuracy of 63% and the least error of 12% so it can be considered the optimal model. The residual plots have confirmed the performance of the XGBoost model for capturing the complex relationship of the data more effectively.

Keywords Regression techniques · Poverty · XGBoost · Random Forest · Decision Tree

1 Introduction

Since poverty is cut across the world, poverty reduction remains one of the fundamental global issues, especially for vulnerable groups vulnerable to economic, social, and structural vulnerabilities. Customarily, poverty-related decision-making has been a traditional logic in which top and down dominate and thus, fail to consider the

R. Vidya (✉) · P. Deepan · N. Arul
Department of CSE (AIML), St. Peter's Engineering College, Hyderabad, Telangana, India
e-mail: vidyarajesh23@gmail.com

R. Kumar and S. N. Mohanty (eds.), *Sustainable Economy Models in the Age of Industry 5.0*, https://doi.org/10.1007/978-981-96-4104-8_12

necessities of such communities. The human-centric artificial intelligence (HCAI) approach is proposed as an effective way to introduce changes that will involve the perception of the needs and experiences of disenfranchised groups of the population. It is for this reason that this paper develops a data-driven approach with HCAI to inform targeted approaches to poverty eradication. From the poverty dataset, the model extracts the trends from the available economic variables, such as the Gini coefficient, headcount ratios, and different types of welfare.

Even most current AI solutions do not contain the human-oriented aspects vital for effective interventions in poverty challenges. In this paper, the framework of Human-Centric Artificial Intelligence (HCAI) is built to incorporate local knowledge, ethics, and culture into poverty reduction models. Poverty is among the world's most serious challenges; billions of people are living in poverty across the globe. Currently, the World Bank reveals that about 9% of the world's population lives under $1.90 or the poverty line. In addition to material recessions, poverty entails inadequate provisions of important and basic needs in the community including health, education, and employment. Such challenges are not only intensified by systemic exclusions such as discrimination of women, or geographical and political unrest but they erect hurdles that act as impediments to the improvement of the social welfare of the otherwise socially excluded groups.

As for the conception of poverty alleviation, the approaches have been macroeconomic policies, governmental welfare, and international aid. Though these have achieved some fruitful results, these efforts are generally insufficient in terms of comprehensiveness to respond to the corresponding community-level poverty which is a multi-dimensional phenomenon. In particular, many of the interventions are non-targeted and do not consider the geo-socioeconomic characteristics, culture, and requirements of the targeted communities. However, the disparities between policy and implementation led to inefficiency, lost resources, and flawed solutions that do not help the neediest population.

With the recently emerging technologies in AI and machine learning, there is a new and unique chance to remodel the fight against poverty. AI supports choosing the right priorities and funding by analyzing resource distribution; on the other hand, by predicting poverty levels and detecting populations at high risk, it strengthens the concept of efficient policymaking. Although the use of artificial intelligence in this field has been on the rise, there are certain prospects associated with it. Current AI solutions are not sensitive to culture or ethics and could thus reveal unfair results which will only add to heightened unfairness.

Author (Usmanova et al., 2022) discussed the different aspects of artificial intelligence for predicting poverty levels. The authors perform a brief literature search on the AI methods such as machine learning and data mining that have been successfully applied to predict poverty indicators. This paper shows how AI aids the analysis of key socioeconomic drivers of poverty and how these can be used by policymakers. In addition, the authors describe some limitations of the field like data quality, ethnic aspects, and the requirement of increased cooperation between all the interested parties to bring the best of AI to eliminate poverty. The study (Zhao et al., 2019) systematically measures poverty estimation in Bangladesh using diverse data sets. To illustrate their work, the authors demonstrate how satellite imagery socio-projective databases and other relevant datasets can help improve poverty diagnosis and measurement. The study proves how random forest regression adopted in poverty evaluation can translate proven strategies of volatile influence for general poverty evaluation when official data sampling techniques are hard to administer. The results underline the potential for adopting the multi-source data approach in analyzing poverty dynamics and guiding the potential actions of policymakers and other stakeholders. The authors in (Hassan et al., 2024) use machine learning models to assess the 2020 Somali Demographic and Health Survey (SDHS) and determine the poverty determinants in Somalia.

In this case, the authors intend to enhance the performance of the poverty estimation by using different machine learning techniques; moreover, the authors find out the socio-demographic and health characteristics of poverty in the region. The implications of the findings are useful for specific poverty interventions because they provide policymakers with useful information about specific groups of people in need. The poverty rate and open unemployment rate are analyzed using nonparametric regression techniques in (Andriyana et al., 2019). This paper illustrates the nonparametric models for analyzing and forecasting poverty levels as a function of unemployment data. This approach is expected to allow better data modeling of poverty factors since some data do not fit the traditional parametric models' assumptions and could be a better tool for policymaking. In this paper (Rajasekaran & Priyadarshini, 2021) using various regression analyses and optimization methodologies, the authors identify factors regarding product returns, a major concern to e-commerce business. By using regression algorithms, a measure of the level of association between various factors like customer characteristics, type of product, and return rates can be established. Considering all the possible applications of regression techniques, we focus on applying regression analysis to predict the poverty reduction rates for empowering marginalized communities.

2 Literature Survey

Reviews of poverty reduction strategies already present in the literature indicate that context and inclusiveness of policy options are critical. There are several relevant research works where technology is employed to overcome poverty and its issues. The article (Mhlanga, 2024) focuses on how big data reshapes financial technology to deepen access to financial services for more people. This paper explores the contribution of big data analytics in increasing the economic status of semi and low-income earners and other vulnerable groups. The author explains that one application of it is that they help in the proper analysis of credit scores of the applicants, efficient design of products, and cutting costs of operations, thereby achieving effective financial inclusion. The study also substantiates issues like data privacy and regulation, which are significant for effectively utilizing big data in the finance industry.

The study (Mohammed et al., 2024) aims to examine the role that unconventional banking plays in poverty alleviation with special attention to the Hail region. With the help of regression analysis, the authors evaluate the positive impact of microfinancing, community-based credit services, etc. on the economy's performance for the targeted clients grouped within the low-income category. These nontraditional banking methods are explored in terms of accessibility and usage and policy implications for facilitating financial access in these regions are presented based on findings from the research study.

The authors in (Zegeye et al., 2022) have looked into the impact that employing agricultural technologies has on poverty in rural areas in 2022. Demonstrating the beauty of data analysis, the researchers employed a multinomial endogenous switching regression model to 2336 farm households to assess factors that determine the use of technology and its influence on poverty in the households. This study reveals that the use of several agricultural technologies has a greater impact on increasing consumption levels. Educational level, receipt of remittance income, farmer's visits by extension workers, and availability of credit are some of the influential parameters that help shape adoption levels. Technology adoption is being recommended through strategic organizational efforts to extend the fight against poverty in Ethiopia.

The authors (Zhao & Herencsar, 2022) in analysis show the Positive Impact of Big data in Mobile Networks concerning poverty reduction through logistic regression. They use mobile data analytics to pinpoint such populations and the distribution of these resources. This offers significant information on such economic status and lifestyle, therefore improving the targeted aims of poverty elimination. Recognizing the strengths of big data in improving accuracy in poverty targeting, the authors find value in employing big data in supporting social welfare policymaking. The authors in (Ha et al., 2021) are integrating the affecting factors of urbanization and poverty reduction in Vietnam. In turn, quantitative data analysis proposed by the study explores how poverty reduction results from urbanization, employment, economy, and living standards. The authors note that the growth of urbanization increases opportunities to access better infrastructure, education, and health, which positively, enhances the human quality of life of individuals in urban areas. But they also talk of square socioeconomic implications of fast urbanization inclusive of inequity and pressure on the environment explaining the necessity of sound urban development strategies.

There has been much discussion of AI in development recently, there are already many cases of using AI to improve resource allocation, health, and education in underserved communities. However, a dearth of scholarly work investigates the use of HCAI within the sphere of poverty reduction and how relevant stakeholders especially in deprived communities are engaged in the process.

The authors (How et al., 2020) present human-centered AI that infuses ethics and extends AI technology to future-deprived communities. They describe how using AI for predictive modeling of creditworthiness, risks, and financial needs when infused with human-centered design, catalyzes access to fair and quality financial solutions. The application of this approach can enhance the prospects of providing financial services that will foster economic inclusion to the several segments of the population who are locked out of the banking systems on a standard basis.

3 Proposed Methodology

3.1 Data Preprocessing

Data is collected and preprocessed from the provided global poverty and inequality dataset (2022), which includes key economic indicators from various countries, and the entire process, is depicted in Fig. 1. The data preprocessing and initial exploratory data analysis (EDA) reveal several insights. The preprocessing includes data cleaning, handling the missing values, data conversions, and standardization. This is the initial step in preprocessing where the inaccuracies, inconsistencies, and errors in the dataset are identified and corrected. There were a total of 108 columns in the dataset and some missing values were found. The columns with more than 50% of the missing data were removed, and the columns with missing values were filled with mean value and mode value for the categorical data. The compatibility of the analysis is ensured through data conversions converting categorical variables to numerical ones. Standardization is crucial for machine learning algorithms, where we ensure the numerical features with a mean of 0 and a standard deviation of 1.

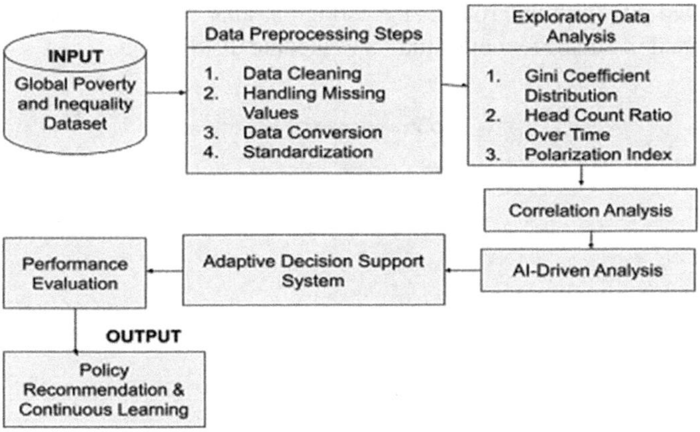

Fig. 1 Block representation of developing an adaptive decision support system

3.2 Exploratory Data Analysis

The dataset used in this study comprises several poverty indicators. It includes key metrics such as the Gini coefficient for measuring income inequality metrics, headcount ratios at various income thresholds for international, lower-middle, and upper-middle, income shares, and polarization indices for social polarization and inequality is measured with metrics like Palma ratio, and S80/S20 ratio P90/P10 ratio. These indicators provide a comprehensive view of income inequality and poverty distribution across different socioeconomic groups.

Gini Coefficient: The box plot in Fig. 2 visualizes the Gini coefficient across various countries, highlighting income inequality. Some countries show higher levels of inequality than others. The histogram shows the distribution of the Gini coefficient, which measures income inequality. The plot includes a density curve (KDE) to highlight the overall trend.

Headcount Ratio over Time: The line plot in Fig. 3 shows how the headcount ratio at the international poverty line varies over time for different countries. This can provide insights into poverty trends across regions. This line plot shows the trends of headcount ratios across various countries over the years. The individual lines represent the headcount ratio based on the international poverty line for each country, with slight transparency to avoid overcrowding.

Polarization Index: One metric for determining the level of polarization in a population or dataset is the Polarization Index. It usually measures how wealth or income is concentrated into discrete groups, suggesting possible divides within a community, in the context of socioeconomic study and income distribution.

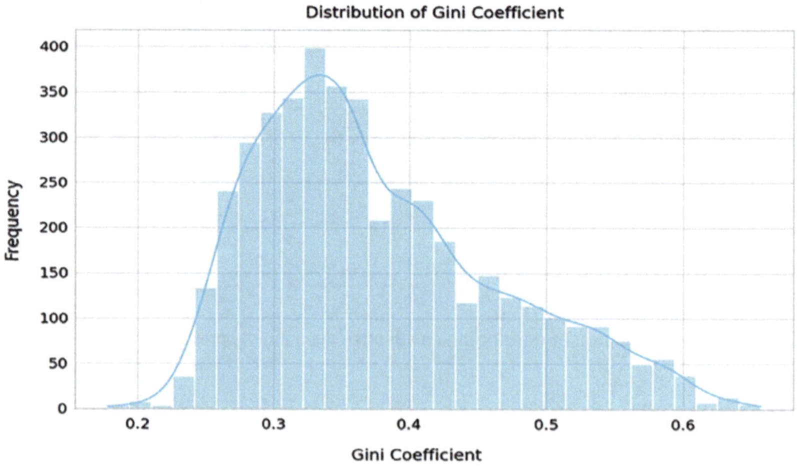

Fig. 2 Gini coefficient distribution graph

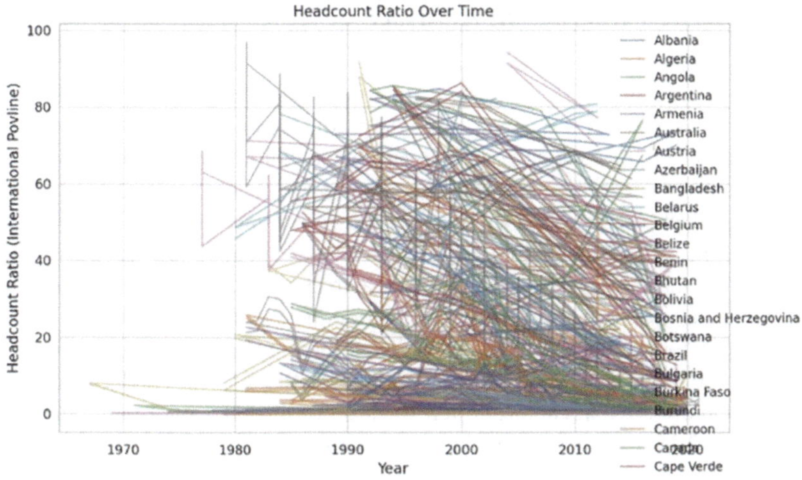

Fig. 3 Sample visualization of headcount ratio over time

Correlation Matrix: The heat map in Fig. 4 shows the correlation between key economic indicators such as the Gini coefficient, headcount ratio, Palma ratio, S80/S20 ratio, and P90/P10 ratio. The color intensity indicates the strength of the correlations, providing insights into the relationships between these metrics. Strong correlations may indicate relationships between income inequality and poverty levels.

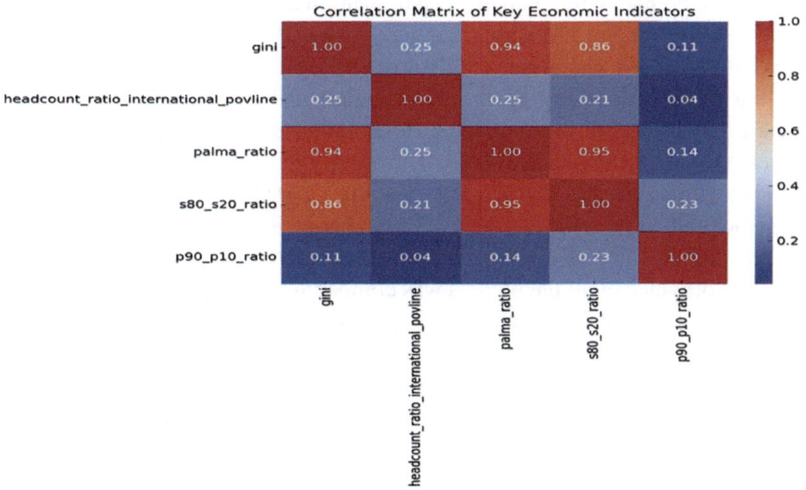

Fig. 4 Correlation heat map

3.3 AI-Driven Regression Analysis

We apply fundamental data analysis techniques such as regression and clustering for different purposes. Regression is used for predictive modeling involving analysis of the relationship between a dependent variable and one or more independent variables. Clustering is used for exploring the data where it groups linked data points into clusters based on their features, without requiring predefined labels. We apply regression to find the income inequality, welfare type, and headcount ratio and clustering to segment the data based on country and year. Regression models are employed to estimate the headcount ratio based on specific economic characteristics such as the Gini coefficient, income, and type of welfare. We apply a simple Linear Regression (LR) model to predict the effect of the headcount ratio, a Random Forest (RF) model that is robust and effective in capturing non-linearity of the headcount ratio, and Ridge Regression (RR) is applied to tackle the problem of multicollinearity by adjusting linear regression and control over-fitting, especially when multicollinearity issues arise with adjusting alpha value to 0.1 controlling regularization. Gradient Boosting (GB) Regression is applied in which an ensemble is learning model where several strong, learned models are built up step-by-step focusing on the residuals of previous models learned.

Support Vector Regression (SVR) is a non-linear model but the kernel trick is to do a regression. Feature scaling is required at first and is done using the Standard Scalar technique. Lasso Regression (LR) is another linear model, but instead of applying L2 regularization to shrink the coefficients, it applies L1 to further increase the model's simplicity by setting some of the coefficients to zero. Elastic Net Regression (ENR) is a combination of L1 and L2 regularization techniques which is also applied to the proposed model. Decision Tree Regression (DT) divides the data set into groups based on the decision rules. XGBoost (XGB) is an efficient gradient-boosting algorithm strongly focusing on the performance and the use of tabular datasets. Standard Scalar is used in this context to standardize the features which is quite significant for algorithms such as SVR. The accuracy and the error rate of the applied regression techniques are depicted in below Fig. 5. MSE describes the error rate of each model and a lower MSE value is always better. R^2 denotes the variance in the data and a higher value of variance is always considered better.

The performance of all the models was compared with the hybrid model in terms of Mean Squared Error (MSE) and accuracy (R^2). MSE measures the average squared difference between the actual and predicted outcomes and R^2 measures how best the model fits the data. The performance optimization is also carried out using hyperparameters like alpha in the case of Ridge Regression to optimize the performance of the model.

The results of the regression techniques are analyzed where Ridge Regression obtains an MSE value of 632.64 and R^2 of -0.92. It states that the Ridge Regression produces poor performance with a very high error rate and a negative accuracy value. It clearly shows that it unfits the data where negative R^2 performs worse than a horizontal line that is the mean of the data. In the case of Lasso Regression, the

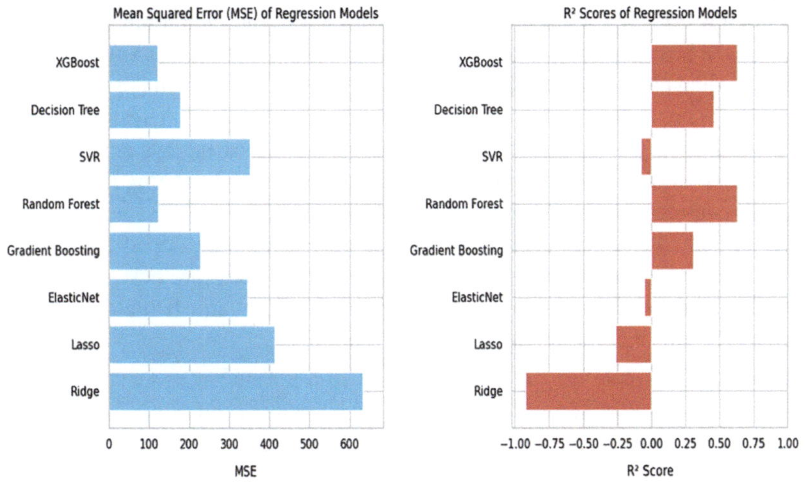

Fig. 5 Graph representing performance analysis of different regression models

obtained MSE value is 414.64 and R^2 is −0.26. Lasso regressions performance is also poor but comparatively better than Ridge. The regularization has a moderate positive effect when compared to Ridge. It struggles with the datasets evidenced by the negative R^2 value. In the case of Elastic Net Regression, the obtained MSE value is 346.63 and the R^2 score is −0.05.

These results show that Elastic Net performs better than Lasso and Ridge regression techniques with a lower MSE value and R^2 value which is close to zero, where there is no variance. The Gradient Boosting model obtains an MSE of 227.49 and R^2 of 0.31. Gradient Boosting improves the performance significantly when compared to the above models. Here R^2 value of 0.31 starts capturing the variance in the data even though it is not a still ideal model. Support Vector Regression obtains an MSE value of 351.87 and R^2 of −0.07. SVR's performance is not satisfied with a negative accuracy value and lacks good performance because of the sensitivity to outliers and data scaling. The negative R^2 value denotes poor fit. Decision Tree Regression obtains an MSE of 176.88 and an R^2 value of 0.46 which is moderately well, but always an accuracy of above 60% is a good score count to be considered. Random Forest Regression results with an MSE of 123.23 and R^2 of 0.63, which is the best accuracy score compared to the above-obtained models. A higher value of R^2 (0.63) and a lower value of MSE is an optimal choice for the given dataset since it balances both accuracy rate and generalization. XGBoost performs similarly to Random Forest with an MSE of 120.46 and R^2 of 0.63. This is also another best model equal to Random Forest in terms of performance. So we can conclude that RF and XGB are out-performing models that handle non-linearity and interactions in a better way when compared to Ridge, Lasso, Elastic Net, Gradient Boosting, Decision Trees, and SVR where they fail to capture the complex interactions.

We extracted a few more visualizations to understand and compare the performance of the best-performing regression models using residuals as shown in Figs. 6 and 7. Residual plots show the differences between the observed and predicted values for each outperformed model. The best model should have their residuals that are scattered randomly around zero. It is usually used in accessing systematic patterns. The residuals should be randomly distributed around zero indicating that the model very well captures the trends and patterns of the dataset.

The predicted and actual plots are compared for Random Forest and XGBoost and the points close to the diagonal line indicate the models are best. The more the points are close to the predictions the more perfect they are. If the deviations are more, it indicates that the model is over-predicting.

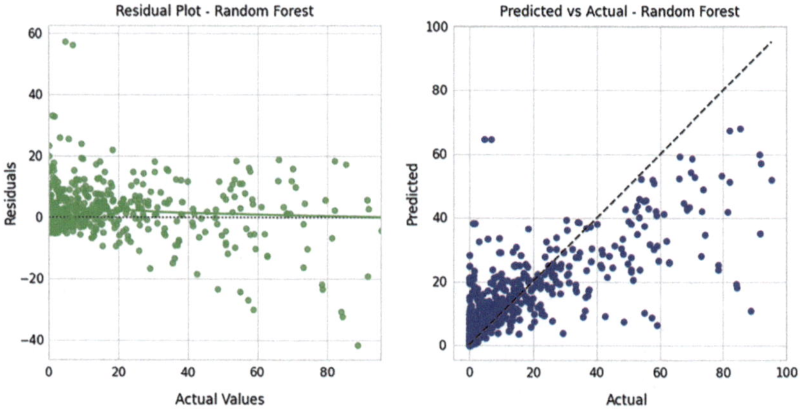

Fig. 6 Residual plots representing residuals in Random Forest

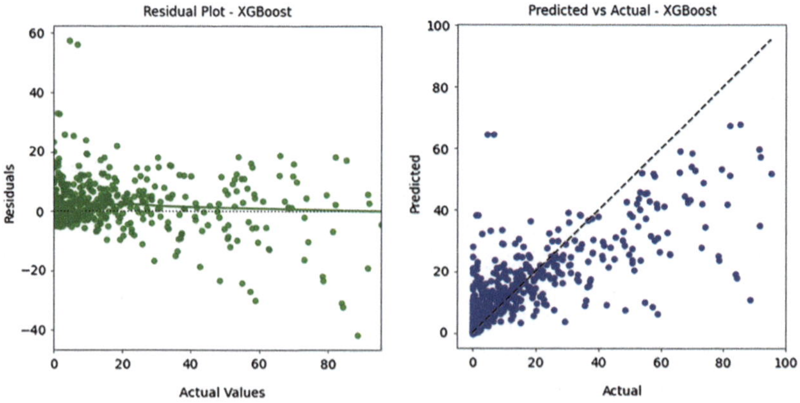

Fig. 7 Residual plots representing residuals in XGBoost

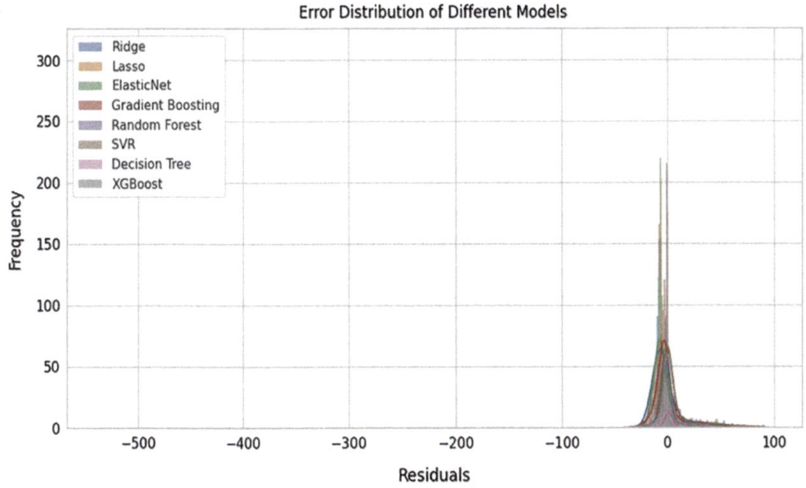

Fig. 8 Error distribution graph of different models

The histogram, represented in Fig. 8, illustrates the residuals as a distribution of the prediction errors from the dataset. The normal distribution in the graph denotes that the model is well-fitted with its peak at Zero.

3.4 Adaptive Decision Support System

The model adapts to real-time data inputs from local communities, using continuous feedback to adjust recommendations. As the AI processes new data on income, inequality, and welfare, it provides updated strategies, ensuring the system evolves in response to changing socioeconomic conditions. By collecting current data from nearby communities, the Adaptive Decision Support System (DSS) can change the recommended advice based on ongoing input. Therefore, policy suggestions remain highly effective and salient in the ever-changing socioeconomic environments of the system. Since the policy recommendations of the DSS are based on poverty headcount, inequality, and welfare indicators, they employ machine learning algorithms. Every suggestion is changed if new data and feedback are entered. Most specifically, the proposed Adaptive Decision Support System (DSS) is capable of changing recommendations depending on the real-time data obtained from neighboring communities. Consequently, policy propositions remain persuasive in addition to being relevant given the continually evolving state of socioeconomic environments. The DSS forms policies related to living people in poverty using machine learning algorithms. The DSS uses machine learning algorithms to produce policy insights based on welfare metrics, income disparity, and the number of people living in poverty. Each proposal is updated when new information and input are incorporated.

The objective is to minimize the poverty headcount PH_t and IE_t is formulated as in (1).

$$OPR_t = \min(PH_t, IE_t) \tag{1}$$

where OPR_t is the optimized policy recommendation, α_t is the real-time input data at time t. α_t is the predictive model updated at time t to provide the policy insights. CF_t indicates the Community Feedback at time t and is used to update the insights of the model, with adjustment to different parameters on welfare distribution, the income redistribution, and ensure relevance to adapt the recommendations. Poverty reduction and inequality reduction are balanced through the optimization function as shown in (2).

$$\min_{OPR_t}(PH_t\alpha_t, CF_t + \omega \cdot IE_t(\alpha_t, CF_t)) \tag{2}$$

where ω is the weighting factor. Community representatives are involved in training and validation for the DSS, a model training system that aligns decision-making with local requirements. It produces insights to help policymakers find ways that effectively reduce poverty, like social security programs, targeted expenditures in education, and equitable income distribution. In order to effectively meet the needs of marginalized communities, feedback loops make sure the system stays current and relevant.

The DSS is developed based on the history data set. The update module polishes up the model along with the real-time data and feedback trickling in from the community. Then the relevant insights are generated based on the real-time updates that are transmitted to the server. In a way to ensure recommendations are conscious of the Get Name model parameters are constantly modified from feedback from the local populace. The system generates differentiated recommendations based on model alterations for example increasing social aid where poverty levels are projected to increase. The verdict and recommendations mean the model improves with time and allows the DSS to adapt to the evolving needs of the socioeconomic landscape. The decision-making process is still consistent with the conditions of deprived regions as the DSS is based on local concerns. Because the community members participate in the training and validation of the models used to enhance recommendations, feedback loops originating from such communities ensure that the system remains contextually relevant. DSS generates information that reveals best practices in designing poverty elimination strategies to inform policymakers. Redistribution strategies are one of these suggestions for fair income distribution in areas with significant income inequality. Specific suggestions for funding education in areas with lower average incomes are suggestions for welfare assistance in areas with high rates of poverty.

4 Results

The performance of the model is evaluated using Accuracy and Mean Squared Error (MSE) metrics. The scores in Table 1 indicate the model's high accuracy in predicting poverty-related metrics and generating reliable recommendations. XGBoost is considered the better choice with 0.3% reduced MSE and is computationally efficient with flexibility in tuning. The Gradient Boosting Approach in XGBoost captures the complex non-linear patterns.

The poverty gap index is predicted using XGBoost Regressor as XGBoost is the optimal fit achieving a good accuracy and low error. The dataset is split into 80% for training and 20% for testing. XGBoost is trained using the training data. The obtained accuracy of 0.99 and 0.11 of MSE indicates an excellent fit as shown in Table 2.

The accuracy explains the 99.79% of variance in the target and MSE indicates the closeness of the prediction to the actual values. The insights into the performance

Table 1 Performance metrics results

Model	MSE %	R^2%	Description
Ridge Regression	63.3	−92	Poor fit with negative accuracy
Lasso Regression	41.5	−26	Little better than ridge but the accuracy is negative
Elastic net	34.7	-5	MSE reduction indicates better performance than Ridge and Lasso but still the predictive accuracy is negative
Support vector regression	35.2	−7	Poor fit due to negative accuracy
Gradient boosting	22.7	31	Catches variance in the dataset but still not good enough to handle complex interactions in the dataset
Decision tree	17.7	46	Moderate fit model with positive accuracy, but an accuracy of above 60% is ideal
Random Forest	12.3	63	Better performing model handling complexity and non-linearity effectively with balanced MSE and accuracy
XGBoost	12	63	Ideal model with optimal accuracy and minimal error

Table 2 XGBoost performance metric outcome

Metrics	XGBoost performance score
Mean squared error	0.11822442777676707
R^2 score	0.997891109442394

and the patterns within the data for the proposed adaptive decision support system are visualized in Figs. 9, 10, 11, and 12.

Table 2 depicts the outcome of XGBoost performance using accuracy and error metrics. XGBoost is used to predict the poverty gap index based on the headcount ratio at the international poverty line if it is above 30%, the income gap ratio above 0.2%, and the mean income value is less than $5000 based on the regions, and if none of the above is identified, the monitor is on progress.

The findings in Fig. 13 offered here offer specific advice on the best strategies for each region based on its SECTOR analysis of that area's socioeconomic demographics to combat poverty and income inequality. What emerges from this analysis would be country characteristics that show that poverty analysis cannot be made in

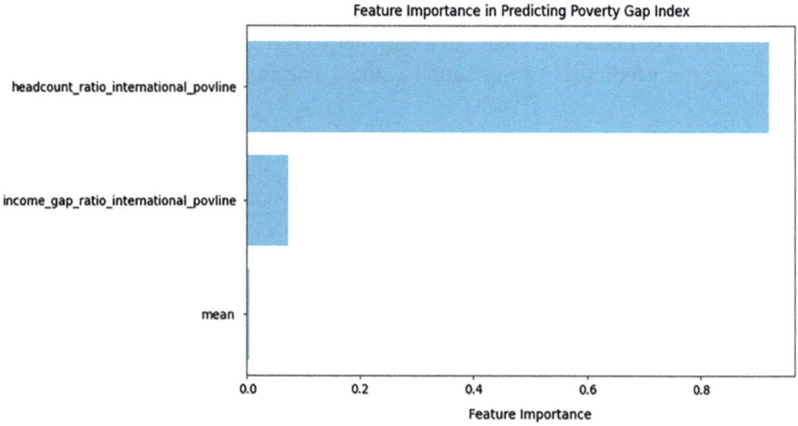

Fig. 9 Feature importance plot of the trained model

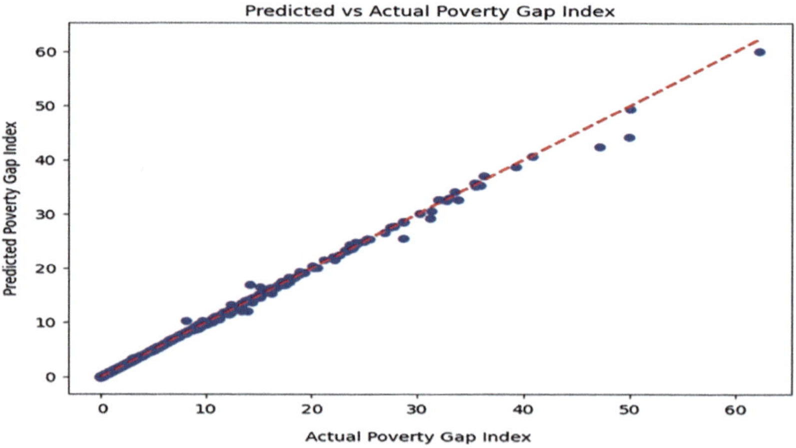

Fig. 10 Chart representing the difference of predicted and actual poverty gap index

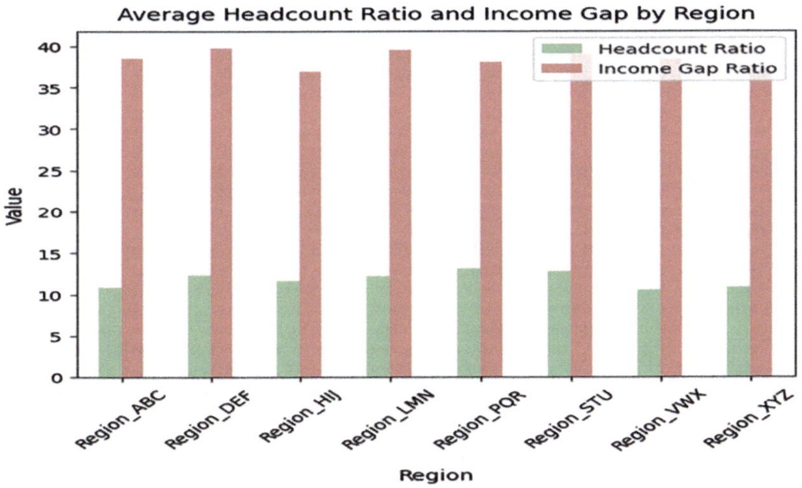

Fig. 11 Region-specific insights visualization

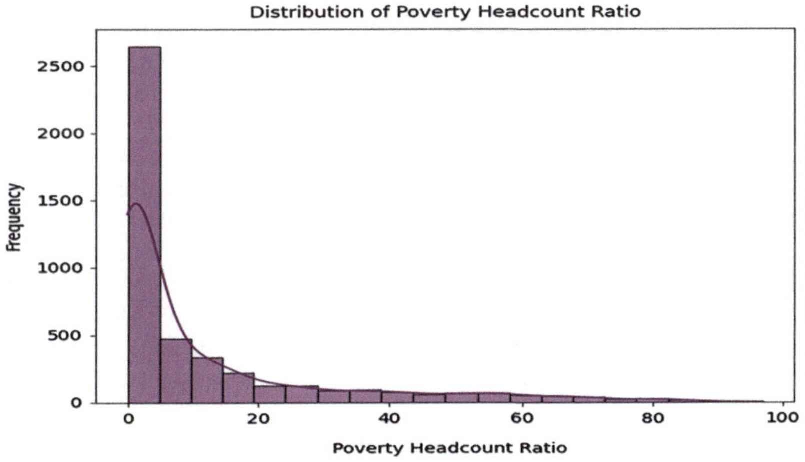

Fig. 12 Poverty Headcount Ratio distribution chart

isolation of country-specific factors. For instance, headcount ratios and inequality indicators variably changed in some countries such as Albania from 1996 to 2012. Cross-sectional estimates for the previous year showed a headcount ratio of the international poverty line at 0.92 in 1996 but have fallen to 0.31 by 2008 indicating improvements in poverty eradication. However, the Gini coefficient, the chief measure of inequality, could be a problem and called for more effectiveness in social regulation. In this work, the AI model was able to determine factors that gave rise to poverty reduction in various countries. Cross-validation procedures were used

```
In Region_LMN: Consider increasing welfare support as poverty headcount is high (above 30%).
In Region_LMN: Economic growth strategies are needed due to low average income (below $5000).
In Region_PQR: Consider increasing welfare support as poverty headcount is high (above 30%).
In Region_LMN: Consider increasing welfare support as poverty headcount is high (above 30%).
In Region_PQR: Consider increasing welfare support as poverty headcount is high (above 30%).
In Region_ABC: Consider increasing welfare support as poverty headcount is high (above 30%).
In Region_PQR: Consider increasing welfare support as poverty headcount is high (above 30%).
In Region_DEF: Economic growth strategies are needed due to low average income (below $5000).
In Region_PQR: Consider increasing welfare support as poverty headcount is high (above 30%).
In Region_LMN: Consider increasing welfare support as poverty headcount is high (above 30%).
In Region_VWX: Economic growth strategies are needed due to low average income (below $5000).
In Region_DEF: Consider increasing welfare support as poverty headcount is high (above 30%).
In Region_LMN: Consider increasing welfare support as poverty headcount is high (above 30%).
In Region_XYZ: Economic growth strategies are needed due to low average income (below $5000).
In Region_XYZ: Consider increasing welfare support as poverty headcount is high (above 30%).
In Region_VWX: Consider increasing welfare support as poverty headcount is high (above 30%).
In Region_VWX: Economic growth strategies are needed due to low average income (below $5000).
In Region_DEF: Economic growth strategies are needed due to low average income (below $5000).
In Region_ABC: Consider increasing welfare support as poverty headcount is high (above 30%).
In Region_VWX: Economic growth strategies are needed due to low average income (below $5000).
```

Fig. 13 Sample output of the recommendation model

to test the efficiency of the model which predicts poverty trends using XGBoost. The community-based validation indicated that incorporating the feedback originating from excluded groups helped make the proposal more relevant and useful. When communities chose to interpret the insights derived from the AI-generated models, there were better proposals for interventions that were more suitable for the local context. The adaptive framework proposed in this paper elaborates on the possibilities of HCAI can ensure poverty reduction, especially in the framework of marginalized communities. We complete the circle of AI analysis and decision-making by empowering communities to be the focus of what drives the decision process and helps to identify the most relevant approaches to meet the identified needs. Also, this approach eliminates the problem of bias that arises due to algorithms and the unethical treatment of already vulnerable groups of people. Thus, the findings of the research are following when using the concept of multi-dimensional poverty indices. Income growth and economic redistribution are important but should be connected with policies on social justice and the labor market by giving people education services, health care, and social protection.

The proposed model offers community-related data for income distribution, enhancing the welfare of society and reducing the inequalities in the communities that support sustainable economic development. Feedback loops are integrated to develop continuous changes in socioeconomic policies. The findings of the model support the policymakers to extract insights for solving poverty issues.

5 Conclusion

In this paper, an adaptive architecture was proposed to introduce Human-Centric AI into poverty reduction decision-making. In this paper, we utilized a large-scale and rich poverty dataset to show that AI offers a mechanism for offering focused

medical tourism enables the developing nations like India in enhancing the healthcare amenities, better set-ups, increasing income and employment prospects on the other hand. However, achieving this status is not possible with proper branding of the medical tourism offered by the country. Marketers say that "Branding is the art and corner store of marketing" (Philip Kotler). The facilities provided by these hospitals stand ahead as best in the sector worldwide, and they use medical tourism as the core concept in attracting patients all around the world. In fact, these hospitals are vitally creating a strong brand name in the market and reiterate the brand concept through various ways. People develop the general impression that brands and branding are a part of manufactured products only. But today, most of the marketable products are services, which are clear from the contribution of the service sector in GDP from many of the developing countries of the world, including India. In recent times, the tourism sector has emerged in the global map, contributing 8–9% yearly, increasing employability and leveraging world GDP. Presently the figure of international travelers' arrivals is expected to grow from 1 billion to 1.8 billion in 2030, nations wide-reaching progressively and adopted an evolving tactic concentrating on tourism. Among the different categories and types of tourism, what is more fascinating and fetching large number of travelers is the medical and wellness segment (Kumar et al., 2024). So, health care, which is treated as a service-oriented product gradually, gets converted into a marketable commodity by creating brand preferences through generic strategy of differentiation. In India, healthcare providers can differentiate their service from service providers in other countries through the differentiation mechanism of healthcare tourism, as India has unparalleled competitive strength as compared to other countries of the world.

2 India and Medical Tourism

Dash (2021), in his paper, mentioned that medical tourism across globe is shaping with the growth rate of 21.1% till 2027. Medical tourism has got very high potentiality in India due to various factors. Even long back India had been a destination of patients around the world and of late, the trend achieved fast pace due to development of infrastructure, transport and communications, availability of good hotel facilities and presence of expert doctors (Bagga et al., 2020). At present the international value of medical tourism is estimated around USD 54.4 billion and projected to reach USD 207. 9 billion by 2027, recording a CAGR of 21.1% growth forecasted for the duration between 2020 to 2027). The exponential growth of Indian Medical Tourism is reported to grow incrementally to more than US$ 9 billion from 2020 onwards. However, while doing intense comparison between Indian Foreign Tourist Arrivals (FTAs) and Inbound Medical Tourist Arrivals (IMTAs) validates the increased footfalls in its global foreign tourist influxes the equal hop is missing in the case of inward medical tourist influxes (Baiev et al., 2020). It is supported by multiple evidences that India's Medical Tourist Arrivals are decreased by 7% of the entire inward-bound tourist arrivals from different countries. Although as per the record of 2019 approximately foreign medical tourist arrivals were near to 6,97,453, though by 2020 India expected nearly 1 million Medical Tourist Arrivals (Grant Thornton Report 2015).

The major factors that boosted the concept of medical tourism in India are low cost as equated to many leading countries of the world. The cost is less than 20% of the total expenses incurred by a patient on an average in the USA for the similar treatment. For example, a heart value replacement in the US costs approximately $200,000. But it can be conducted in India at a cost of $12,000 to $14,000 only. The cost advantages can be best utilized by the hospitals to appoint experts and provide high-quality service (Ghasemi et al., 2021; Guiry & Morgan, 2021; Parekh et al., 2021). This is vital for developing a strong brand name in the market, especially dental, cosmetic, cardiac and non-conventional therapies. In countries like Europe, a patient has to wait long time to get an appointment with a doctor whereas the waiting time is very low in India. The gestation time is quite low so that the patient will get adequate treatment in right time. This helps that segment of patients who are not covered by the insurance facilities. World-class quality and facilities with world-class medical infrastructure are provided by the leading hospitals approved by Joint Commission International (JCI). The facilities provided by hospitals such as Apollo, Escorts, Wockhardth and Max Healthcare are the best in the world. The medical education in India attracts professionals as the teaching centers select candidates on the basis of merits rather than money. The higher education opportunities are very high, less costly and hence affordable to the common man who has got intellectual capability and professional acumen and hence the medical system is considered to be safer and more reliable. The rich cultural heritage of the country with its vivid cultural backgrounds, lifestyle, topographical conditions, weather, etc. has attracted tourists around the world for decades. These peculiarities can be used to attract patients from various parts of the world under the name of medical tourism. Absence of language barrier as English education is much prevalent. Usually, language is a major hindrance when people travel across the world. Since English is a common language around the world and it is widely used in India, majority of the people who visit the country need not have to be concerned about the language barriers. After liberalization and globalization in 1991, the government has been giving much attention to build up world-class infrastructure facilities. This has paved to the development of transport and communications, road systems, public utility facilities, etc. Leading healthcare providers that offer medical tourism in India as per government official list (hci pom.gov) are:

a. Medanta, Gurugram
b. Kokilaben, Mumbai
c. Fortis Memorial Research Institute, Gurugram
d. Indraprastha Hospital, New Delhi
e. Artemis Hospital, Gurugram
f. GGHC, Chennai
g. Nanavati Super Specialty, Mumbai
h. Manipal Hospital, Bengaluru

i. Appolo, Chennai
j. l Max, Delhi

3 Research Design

The methodology adopted in present research is case study methodology of brands providing medical service in India to foreign travelers. In past, also case-based methodologies were used as research approach where an extensive literature based on published data was used to make cast study analysis to strategies medical tourism in emerging or developing nations (Eisenhardt, 1989; London & Hart, 2004). The advantage of this method is that it allows researchers to do in-depth analysis of the selected cases so as to understand the strategies adopted by them to attract tourism and design future marketing strategies to prolong their business interest. Hence, author adopted qualitative research approach to extract possible factors leveraging sustainable growth among low-income groups in India. Evaluation of multiple cases permits scrutiny of questions relevant framed in research objectives like what are the diverse innovation strategies considered in business model that healthcare organization can adopt to deliver inclusive services and how precisely it contributes in economic development of host country (Yin, 1992). The secondary data was collected from published report, article and information provided on creditable sources was extracted to decode the research questions. List of the top medical services providers was taken from the official list provided by the government of India.

4 Medial Tourism and Its Impact on Economic Growth of Low-Income Groups

Various published reports mention the fact that the cost of medical services in few cases like dental, plastic surgery and angioplasty is quite cheaper in developing or underdeveloped countries in comparison with developed nations (Gopalan et al., 2021). Presently, several tour and travel agencies are making arrangements that substantiates medical services with pleasant vacation attracting the attention of many travelers and service providers, the organizations promoting medical tourism, health analysts and policymakers. The national certified agencies and policy originalities are now required to stimulate and endorse medical tourism by host countries. Several authors identified three main drivers of medical tourism across globe, namely source to generate foreign exchange revenue (Klijs et al., 2016; Lautier, 2008), boosts revenue from tertiary sectors like retail, hospitality industry, transportation sector and real estate industries influencing employment opportunities for host citizens (Beladi et al., 2019; Hopkins et al., 2010; Lautier, 2014).

Literature in the past provided sufficient evidence of tourism having instrumental influence in driving economic advantage to host nations by generating sizeable opportunities of employment, increase in income among community members and several benefits to governing bodies. In various arguments over the collaborative advantage between economic development and poverty elevation to attain inclusive evolution, World Bank and other agencies like OECD are now foraying toward UN Sustainable Development Goals (SDG-8). Nevertheless, increased development remains extremely disputed explicitly when it comes to role tourism plays. These issues are, however, still not discussed in length to measure the real influence of tourism in economic development. The present chapter underwrites the argument by cross-examining tourism-based inclusive growth using evidence from developing economies of South Asia like India. The study is raising the fundamental query to understand whether tourism-based economic growth can be inclusive in the short to medium terms, drawing on evidence from groundwork done in India. While travel can generate employment, this can be risky and poorly paid, limiting openings for poverty mitigation (Jeyacheya & Hampton, 2020).

5 Possible Strategies to Build Market for Medical Tourism to Support Economic Development

5.1 Branding of the Leading Hospitals in Medical Tourism

Many hospitals are land in a dilemma when they think about branding. According to Chris Bevelo, construction of a brand name is crucial for positioning and differentiating a service provider. Healthcare branding strategy attempts to attract new patients, increase physician referrals and partnerships. The question is whether to brand the hospital or brand the doctors or to brand the tourism facilities. Building brand name is imperative in medical tourism as the targeted population is attracted from different parts of the world. When a person selects a hospital for treatment, the highest priority is the facilities and the reputation of service. In service reputation, we need to add the quality of the personnel who are working there. Hence the priority always lies in promoting the facilities and services provided by the hospital. The branding for medical tourism comes as secondary as it is a secondary option in many cases. But building a strong brand name is imperative to attract the attention of the target customers which requires careful planning, long-term commitment and a creatively designed and executed marketing. All the leading players in the market rely on the generic strategy recommended by Michael E. Porter in his famous work "Competitive Strategy: Techniques for Analyzing Industries and Competitors" (1980). According to Porter, in a highly competitive world, a firm can rely on generic strategies such as low cost, differentiation, focused low cost and focused differentiation. Each hospital wants to develop a brand name based on the special healthcare service offered by them so that the organization can differentiate it from others. So, these players rely on country differentiation with focused differentiation through healthcare tourism by branding it in the name of the hospitals. From the point of view of the leading global markets, the cost also is low, especially for patients from Europe, Middle East and other richest countries of the world. So, in fact, in building up a strong brand name, the strategy recommended by Michael E. Porter is totally utilized by these organizations. Considering the cases of these hospitals in leveraging their medical services for foreign travelers are showcased through the details showing how these hospitals differentiate their services:

1. **Apollo Hospitals**: One of the major healthcare organizations in India is one of the best hospitals. The hospital is specialized in dealing of disease related to heart, spine, orthopedics, cancer, etc. taking critical care and preventive measures.
2. **All India Institute of Medical Science, New Delhi**: AIIMS a multi-specialty hospital listed under medical tourism is worldwide acknowledged in the field of medical education, research and specialized treatments in various fields of medical sciences. There are around 25 clinical departments apart from the specialty centers. The various specialty centers include: Cardio-Thoracic Sciences Center; Centre for Dental Education and Research; Neurosciences Centre; Trauma center; Cancer Hospital; Drug Dependence Treatment Centre; and Ophthalmic Sciences.
3. **Aravind Eye Hospital, Madurai:** The hospital has well-equipped specialty clinics with comprehensive support facilities. The major facilities at Aravind Eye Hospital include Fluorescein Angiography; Electroretinogram (ERG); Ultra-sonography (3D Ultrasound); Laser: ROP (Retinopathy of Pre-maturity); Photo Dynamic Therapy (PDT); Transpupillary Thermo Therapy (TTT), etc.
4. **B.M. Birla Heart Research Center, Kolkata**: It has built reputable image as India's most progressive heart center, with highly specialized exhaustive care units which are specially tailored to cater to the range of patient needs and are equipped with state-of-the-art computerized system.
5. **New Delhi Escorts Heart Institute and Research Center**: The hospital has the most advanced laboratories for investigative tests in the fields of Nuclear Medicines; Radiology; Hematology; Transfusion Medicine and Microbiology. EHIRC is unique in the field of preventive cardiology with a fully developed program of Monitored Exercise, Yoga and Meditation for life- style management.
6. **Fortis Hospital, Chandigarh**: Fortis Hospital is a JCI-accredited multi-specialty hospital and super specialty in heart which uses latest medical technology, setting up international standards in providing patients with highest degree of personalized services. It is affiliated to Partners Healthcare System Inc. (PHS), USA, and has received the Best Design Award by the American Institute of Architecture in 1999.
7. **Jaslok Hospital, Mumbai**: Jaslok hospital and research center has links with Stanford University and is a center for seminars and conferences on new and modern techniques. The hospital is known for IVF procedures and is also one of the leading facilities in India for organ transplants and has links with hospitals in the USA to pioneer new frontiers in medical and diagnostic techniques. The IVF department is one of the most modern in the world, allowing per screening of embryos for parents who are careers for genetic diseases. The surgical facilities

also utilize minimally invasive techniques to reduce risks and make the whole recovery process easier.

8. **Mallya Hospital, Bangalore**: "Individual care to deliver best results – that is what hospital focuses at every step." Here the patients experience the concept of Total Quality Management (TQM), and each staff is driven by the vision of quality and commitment in giving individual attention for maximum treatment outcome. It is the first multi-specialty hospital in the country to receive the coveted ISO-9002 certification award.
9. **Narayana Hrudayalaya, Bangalore**: Narayana Hrudayalaya is known as one of the world's largest heart hospitals for infants, with all super-specialties that the medical world offers. The group performs largest heart surgeries on children in the world providing cardiac care to children from 73 countries.

5.2 *How Branding Will Help the Hospitals*

Strong brands control its own existence; if the brand is good and branding is effective, it ensures customer loyalty. Clear brand ensures positioning between physicians, staff and customers. It ensures consistent communication between the organization and the stakeholders. Branding supports multi-channel and social media initiatives. Internal branding activities will help the organization to implement total quality initiatives. However, before strategizing this, one should be prepared to answer the limitations. There are certain barriers in branding health services in India. The law pertaining to the nature of medical practice, accreditation and standards, immigration, etc. at times doesn't favor smooth functioning. Cost escalation in using imported equipment, skilled doctors and costly infrastructure considerably, India will lose the status as a low-cost attractive destination for health care in future. It is quite difficult to build up a strong link between the various participants in the local system such as medical, tourism, hospitality, insurance and airlines to fulfill the major goal. There is no proper tracking of the requirement of the targeted customers by understanding their needs and requirements from time to time by many of the service providers.

5.3 *Promotional Initiatives*

The Medical Tourism Association has developed various initiatives to help international hospitals build a strong brand and image so as to enable them to become leaders in the global healthcare industry. It has developed a guide (MTA Hospital Destination Guide) that mentions the health care offer of a specific hospital and represents a powerful tool to market its services to target customers. It is represented by attending World Travel Marts and World Medical Fairs and tie up with various agencies abroad involved in medical tourism. It conducts continuous medical education for doctors, on-campus patient education programs, free health checkups for doctors, etc. It regularly has ad campaign in leading international magazines, new media, TV, etc. and ties up with various embassies, free camps in various countries, etc.

6 Conclusion

Findings of the study showcase that medical tourism in India is growing at its fullest heavily augmenting inward tourist influxes. At present, most the international medical tourism is determined by the economic value, quality of services it offers to the travelers. Based on these benefits, India can successfully endorse these double-folded rewards possibly increase the footfalls and foreign exchange earnings. Economies like USA and UK collectively compose around 23% of the Indian inward travel market share and rank second and third after Bangladesh in Inward tourism (Statistics 2021). Medical tourism, if branded strategically in India, can bring travelers in all season, 365-day destination irrespective of the on/off season divide and shield inbound tourism from the devastation of economic meltdown and recession. Additionally, medical tourism can contribute remarkably toward employment, enhanced foreign exchange revenues, exponential surge in various investments like infra in healthcare institutions and also result in reverse medical trench of medical professionals and support staff to India. Careful marketing in source markets will permit India being placed as the furthermost reasonably priced global quality medical tourism destination. While considering the branding of healthcare service providers, we need to assess certain ethical considerations in a service industry. Health care, in fact, is a service industry and service motivated, but now it is given revenue generating business concept. So many experts opine that the service orientation is undergoing rapid transformation toward materialism which may create so many issues and problems in the society. One of the major criticisms is that there exists the possibility of unhealthy competition in the market and the health service may orient toward an oligopolistic market system in future. When reputed brands are created and the brand name is stabilized in the market, it may lead to oligopolistic tendencies which may hamper the local stakeholders' interests adversely. But as business is becoming global and heavy investment is taking place for modernization of hospitals, they need to develop marketing strategies to attract the customers and thereby recover the investment profitably.

7 Limitations and Future Recommendations

India is a premier medical hub in Southeast Asia continent and maintains its competitiveness and international reputation, it is now crucial to have a thorough awareness of the expectations and views of medical tourists. The mounting competition in the global medical tourism sector has forced destination marketers to implement innovative tactics meant to continuously draw foreign medical tourists and cultivate repeat business for the economic development for the host nation and its citizen especially representing low-income group. India must continue to hold its position as a top medical tourism destination globally in this fiercely competitive environment. In order to position, further, the study's limitations include its small sample case size and the narrow range of activities examined.

Furthermore, to classify more on the variables contributing toward economic progression of medical tourism in India, future research should broaden its geographic scope to include a wider range of locales. For the benefit of the global medical tourism business, a more thorough analysis of cross-national differences in medical tourists' attitudes, behaviors and perceptions would be extremely beneficial. It is anticipated that these studies will produce useful information that influences industry strategies and governmental laws, allowing for the creation of specialized programs to successfully draw in and serve medical tourists from particular nations.

Funding Statement Authors have not taken any funding support from any private or government agencies to publish this paper.

Conflict of Interest We authors of the paper declare that all the points mentioned below are true to my knowledge current research work is original, and has not been published in any form prior to this submission. Also, there is no conflicts of interest.

References

Bagga, T., Vishnoi, S. K., Jain, S., & Sharma, R. (2020). Medical tourism: Treatment, therapy & tourism. *International Journal of Scientific & Technology Research, 9*(3), 4447–4453.

Baiev, V. V., Bakhov, I. S., Holovach, N. V., & Zgalat-Lozynska, L. O. (2020). The economic determinants of the world medical tourism industry development. *Journal of Environmental Management and Tourism, 10*(6), 1392–1398.

Beladi, H., Chao, C. C., Ee, M. S., & Hollas, D. (2019). Does medical tourism promote economic growth? A cross-country analysis. *Journal of Travel Research, 58*(1), 121–135.

Bookman, M. Z., & Bookman, K. R. (2007). Introduction to medical tourism. In *Medical tourism in developing countries* (pp. 1–19). Palgrave Macmillan US.

Cham, T. H., Lim, Y. M., Sia, B. C., Cheah, J. H., & Ting, H. (2021). Medical tourism destination image and its relationship with the intention to revisit: A study of Chinese medical tourists in Malaysia. *Journal of China Tourism Research, 17*(2), 163–191.

Chaulagain, S., Pizam, A., & Wang, Y. (2021). An integrated behavioral model for medical tourism: An American perspective. *Journal of Travel Research, 60*(4), 761–778.

Dash, A. (2021). Exploring visit intention to India for medical tourism using an extended theory of planned behaviour. *Journal of Hospitality and Tourism Insights, 4*(4), 418–436.

Eisenhardt, K. M. (1989). Building theories from case study research. *Academy of Management Review, 14*(4), 532–550.

Ferreira, F. A., & Castro, C. (2020). Medical tourism in Portugal—A potential niche market. In *Advances in tourism, technology and smart systems: Proceedings of ICOTTS 2019* (pp. 615–625). Springer Singapore.

Ghasemi, P., Mehdiabadi, A., Spulbar, C., & Birau, R. (2021). Ranking of sustainable medical tourism destinations in Iran: An integrated approach using fuzzy SWARA-PROMETHEE. *Sustainability, 13*(2), 683.
Gopalan, N., Mohamed Noor, S. N., & Salim Mohamed, M. (2021). The pro-medical tourism stance of Malaysia and how it affects stem cell tourism industry. *SAGE Open, 11*(2), 21582440211016836.
Guiry, M., & Morgan, A. (2021). Mexico's medical tourism healthcare service quality: A content analysis of international patients' online testimonials. *International Journal of Leisure and Tourism Marketing, 7*(2), 105–135.
Gupta, P. (2018). Bangladeshis top the list of foreigners visiting India for medical tourism.
Hopkins, L., Labonte, R., Runnels, V., & Parker, C. (2010). Medical tourism today: What is the state of existing knowledge? *Journal of Public Health Policy, 31*(2), 185–198.
https://www.hcipom.gov.in/docs/medical%20tourism.pdf
Jeyacheya, J., & Hampton, M. P. (2020). Wishful thinking or wise policy? Theorising tourism-led inclusive growth: Supply chains and host communities. *World Development, 131*, 104960.
Klijs, J., Ormond, M., Mainil, T., Peerlings, J., & Heijman, W. (2016). A state-level analysis of the economic impacts of medical tourism in Malaysia. *Asian-Pacific Economic Literature, 30*(1), 3–29.
Kucukusta, D., Hudson, S., & DeMicco, F. J. (2019). Medical tourism: Strategies for quality patient/guest experiences. *Journal of Hospitality and Tourism Insights, 2*(3), 221–223.
Kumar, N. (2020). *Over 50% medical tourists to India are from Bangladesh.*
Kumar, R., Goel, R., Singh, T., Mohanty, S. M., Gupta, D., Alkhayyat, A., & Khanna, R. (2024). Sustainable finance factors in indian economy: Analysis on policy of climate change and energy sector. *Fluctuation and Noise Letters, 23*(2), 2440004-339.
Kumar, S., Gowreesunkar, V. G., Kumar, S., & Kumar, V. (2022). Delving medical tourism in India: Insights and challenges. *Journal on Tourism & Sustainability, 6*(1), 24–98.
Lautier, M. (2008). Export of health services from developing countries: The case of Tunisia. *Social Science & Medicine, 67*, 101–110.
Lautier, M. (2014). International trade of health services: Global trends and local impact. *Health Policy, 118*(1), 105–113.
London, T., & Hart, S. L. (2004). Reinventing strategies for emerging markets: Beyond the transnational model. *Journal of International Business Studies, 35*, 350–370.
Lunt, N., Horsfall, D., & Hanefeld, J. (2016). Medical tourism: A snapshot of evidence on treatment abroad. *Maturitas, 88*, 37–44.
Oberoi, S., & Kansra, P. (2019). Factors influencing medical tourism in India: A critical review. *SAMVAD, 17*, 9–16.
Olya, H., & Nia, T. H. (2021). The medical tourism index and behavioral responses of medical travelers: A mixed-method study. *Journal of Travel Research, 60*(4), 779–798.
Parekh, J., Jaffer, A., Bhanushali, U., & Shukla, S. (2021). Disintermediation in medical tourism through blockchain technology: An analysis using value-focused thinking approach. *Information Technology & Tourism, 23*, 69–96.
Rahman, M. K. (2019). Medical tourism: Tourists' perceived services and satisfaction lessons from Malaysian hospitals. *Tourism Review, 74*(3), 739–758.
Rai, A., & Rai, A. (2019). Medical tourism: An introduction. In *Medical Tourism in Kolkata, Eastern India* (pp. 1–41). Springer International Publishing.
Sag, I., & Zengul, F. D. (2019). Why medical tourists choose Turkey as a medical tourism destination? *Journal of Hospitality and Tourism Insights, 2*(3), 296–306.
Yin, R. K. (1992). The case study method as a tool for doing evaluation. *Current Sociology, 40*(1), 121–137.
Zarei, A., & Maleki, F. (2019). Asian medical marketing, a review of factors affecting Asian medical tourism development. *Journal of Quality Assurance in Hospitality & Tourism, 20*(1), 1–15.

Futuristic Challenges in Circular Economy: Transforming Challenges into Opportunities in Industry 5.0

Gayatri Rathi and Samta Kathuria

Abstract Sustainability through a circular economy using artificial intelligence within the context of Industry 5.0 is therefore a viable line of development. Industry 5.0 focuses on the placement of people alongside smart technology to work together and best fit circular economy guidelines, including waste reduction, better resource utilization, and recycling. This study also explores whether AI has a role in making linear industries circular by offering cutting-edge technological applications for waste and resource management, as well as supply chain traceability utilizing predictive modelling, self-acting sorting, and digital replicas. The research methodology of the study is done after reviewing high-quality research reviews and articles to draw evidence from existing literature and case studies to establish the use of AI in the circular economy to overcome these challenges. Further, the absence of clear legislation to address the use of AI in circular models hinders the advancement of circular economy applications, particularly in developing nations that have poor technological environments. To overcome these challenges, this study proposes a behavioural systems approach. Finally, outlining the state-of-the-art advancements of AI, the paper appeals to industries, policymakers, and communities for cooperation to realize AI's potential and pursue the Industry 5.0 goals for developing a sustainable and resilient economy worldwide.

Keywords Circular Economy (CE) · Industry 5.0 · Artificial Intelligence (AI) · Waste management · Resource optimization

G. Rathi
Faculty of Law, Swami Vivekanand Subharti University, Meerut, India
e-mail: gayatrirathi950@gmail.com

S. Kathuria (✉)
Parul Institute of Law, Parul University, Vadodara, India
e-mail: samtakathuria14@gmail.com

R. Kumar and S. N. Mohanty (eds.), *Sustainable Economy Models in the Age of Industry 5.0*, https://doi.org/10.1007/978-981-96-4104-8_14

1 Introduction

This has led to the globalization of sustainable development, which tasks our economic systems to progress from the earlier model of linear economy to circular economy. The circular economy can be viewed as an innovative model making people's effort to curb waste and achieve the efficient utilization of resources in stark opposition to the so-called linear economy (Adel, 2022).

Circular economy deals with an effort to bring the product life cycle closer or eliminate gaps in it by redesigning products, recycling, or reintroducing some materials as shown in Fig. 1. However, as the economies become more developed, there are so many difficulties in the implementation of circular economy principles. Circular economy has long encountered certain challenges, and in the recent past, the emergence of Industry 5.0 has presented new openings through which obstacles of circular economy have been addressed by economies. Industry 5.0 is a new form of industrial revolution that combines artificial intelligence, robotics, smart automation, blockchain, and the decision-making capabilities of IoT with the creativity of the human mind as shown in Fig. 2.

India, like most of the other developing nations, is at the crossroads in implementing such innovations. Due to the large population and a rapidly developing

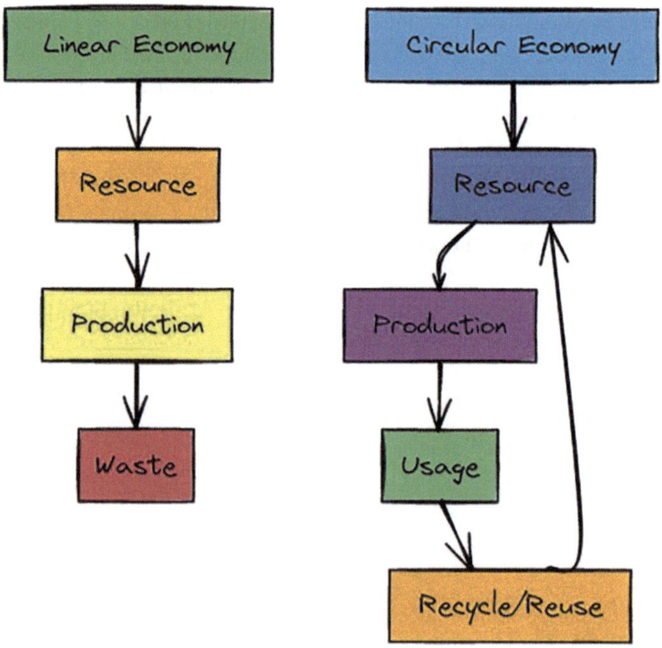

Fig. 1 Comparison of linear and circular economy models

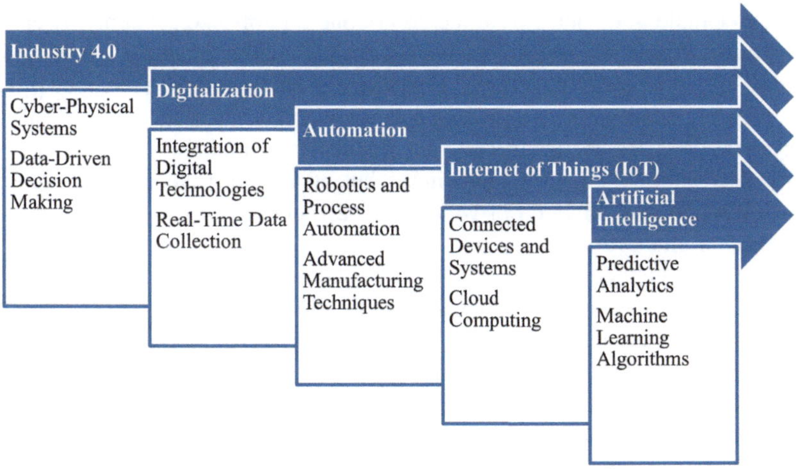

Fig. 2 Detailed evolution from Industry 4.0 to Industry 5.0

economy, it has specific difficulties in waste disposal, resource utilization, and environmentally friendly industrialization. Industry 5.0 presents a clear design of how these challenges can be turned into opportunities in embracing the circular economy. However, implementing a model that does not link institutions to a clearing organization is a relatively recent development that cannot be achieved without technology, legislation, policies, and awareness. Industry 5.0 might transition the industry between the extremes of technological efficiency and sustainability by providing organizations with more efficient techniques for waste reduction and employment generation with environmental conservation as a priority in addition to efficiency.

The focus of this research article is to analyse whether and how Industry 5.0 can improve the circular economy to meet its challenges and constraints. We will examine how these two models could be blended in terms of the capacity of smart technologies with the capacity to transform waste minimization, product lengthening, and resource recycling. The article will also discuss various risks of Industry 5.0 in the Indian context, like high costs inherent in technological adoption, limited technical skills, and conformity to globally acceptable normative measures required to make these innovations competitive at the international level. By critically analysing these factors, it will be possible to establish several areas where Industry 5.0 could leverage circular economy's existing challenges to sustainable development opportunities.

1.1 Overview of Circular Economy

As a circular economy is a new concept proposed for implementation, it has been defined as a closed-loop economy where the use of products, components, and materials is optimized to serve its highest value at any one time. Originally known as

Cradle to Cradle, it deviates from the do-nothing-end-of-pipe economy by encouraging reuse, repair, refurbishment, and recycling of the products and materials. Circular economy is based on three key principles: eliminating the scrapped product and unregulated emissions, prolonging product and material usage, and replenishing ecosystems. With the change in the linear model to the circular model, industries can greatly decrease their environmental impact but, at the same time, enhance economic growth through the collection of waste materials and the creation of new products from them (Rame et al., 2024).

While the economic model studied in schools has a clear and definite life expectancy while ending in the discard dump, the circular economy seeks to have loop circuits within industrial processes. It minimizes the use of scarce natural resources, emissions, and resource wastage and carries the general notion of efficiency in energy utilization. Circular economy has not been without some challenges, such as technological setbacks, market factors, and new regulations required for circular economy organizational changes. Nevertheless, the usage of circular economy design is emerging all over the world, including countries such as India, where the pressure on natural resources and pollution is rising rapidly. Industry 5.0 offers the chance to address some of these challenges by integrating technologies into circular economy practices.

1.2 Emergence of Industry 5.0

Industry 5.0 is the next level to Industry 4.0, which seeks to integrate the use of smart technologies in industries with an increased focus on the integration of human beings. Whereas Industry 4.0 was all about the integration of smart technologies in production to enhance automation, Industry 5.0 is all about human-centric. As such, it aims to use the combination of humans and artificial intelligence, robotics, etc. to offer solutions that not only work but are courteous to society and the environment as well. From the point of view of the circular business model, there are many splendid opportunities that Industry 5.0 brings to improve sustainability. For instance, the recent development of AI can work for supply chain integration, minimizing the wastage of production lines, and predicting the overall maintenance of necessary machinery to support the principles of the circular economy. In the same way, there is a possibility to predict more accurate amounts of resources for manufacturing and improve the efficiency of recycling and upgrading of products through the use of robotics and IoT. This symbiosis between man and machine is essential for managing new multi-layered industrial environments in which it is sometimes hard to find harmony between the need for economic development and environmental protection.

The current move to Industry 5.0 is set to redefine the industrial environment in India. Indeed, there is one compelling argument that has come out of the implementation of sustainable development strategies, such as through this new industrial revolution: the prospects for it. With sustainability still a far-off reality in India, the advent of Industry 5.0 can help in overcoming the challenges of the current circular economy systems that are proven to be unfit for the future, which nudges the country to look beyond the current systems and create a more sustainable and robust economy suitable for the future (Liu et al., 2024).

1.3 Research Objectives

The main research question of the study is: How can Industry 5.0 enhance the circular economy to mitigate the challenges faced by the strategy, especially in the Indian context? This research will focus on clarifying important technological trends and policy actions that might turn environmental and economic sustainability threats into new opportunities for development. Specific topics of the article will include waste minimization, resource optimization, and product life-cycle elongation; however, it is equally important to note that many challenges have to be overcome to unlock those opportunities, such as a lack of adequate technological and legislative knowledge.

1.4 Industry 5.0 and Circular Economy

The penetration of Industry 5.0 into the topical circular economy environment presents a paradigm shift in how industries will address sustainability, material management, and manufacturing. For that reason, industrial culture is diminishing and the conflict between people and the environment is growing continuously as the environmental problems are supposed to be growing with time, and such problems include but are not limited to resource scarcity, climatic change, and pollution. Industry 5.0, which is characterized by increased cooperation and collaboration between humans and different smart technologies such as AI, robotics, and blockchain, offers the enabling keys for furthering the tenets of the circular economy (Adel, 2022).

Industry 5.0 makes smart and sustainable mechanisms of service delivery that change traditional industrial structures into ecologically friendly ones. This change is very relevant in developing countries, for example, India, given the increasing industrialization, population, and urbanization, which compounds the environmental stress. These were treated challenges by Industry 5.0 in terms of innovation and human-centricity for a sustainable circular economy path.

1.5 Evolution from Industry 4.0 to 5.0

The industry transition from Industry 4.0 to 5.0 is a considerable leap to new practices in the industry characterized by human integration and customization. Originally, Industry 4.0 was focused on the digitalization of the production and manufacturing industries, which included automation, AID, and IoT. These technologies fostered efficiency and precision in manufacturing processes or accuracy in manufacturing processes while paying less attention and human participation/sustainability. Industry 4.0 established a strategic emphasis on efficiency and effectiveness that made organizations and companies productive and cost-cutting-oriented; however, there is a lacuna in environmental managerial concerns, especially recycling and resource management.

Industry 5.0, on the other hand, is derived from the prior Industry 4. Innovative development, where the focus is on humans and sustainability. According to the primary definition, the use of human and machine synergy is the primary concept behind Industry 5.0. In contrast to the powerful promotion of the complete substitution of human labour by the use of various technologies, Industry 5.0 aims at the harmonization of a human and a tech workforce. This makes it easier to identify the potential for more client-specific and therefore more efficient production systems, which are in line with circular economy thinking. While Industry 5.0 presupposes the developments of Industry 4.0, it goes far beyond the economic purpose and focuses on the sustenance of growth by encouraging innovations in aspects like waste minimization, resource usage enhancement, and product life-cycle elongation. For Indian industries, which are witnessing a process of fast industrial transition, it is important to assess the shift to Industry 5.0 as a chance to improve sustainability without the loss of competitiveness in the global environment (Turner et al., 2022).

1.6 Key Technologies in Industry 5.0

The following are some key technologies that underpin Industry 5.0 and its vision of the integration of people and technology: Today, some of the crucial technologies in this new industrial revolution are artificial intelligence, robotics, blockchain, and IoT. Implementing such technologies helps also to improve the principles of the circular economy by increasing productivity, reducing waste production, and improving recycling. The main driver of change in Industry 5.0 is AI due to its capability to process large amounts of data and provide quality solutions for enhancing decision-making. Relative to the circular economy, AI has advantages for the effective management of the supply chain, minimization of energy consumption, and identification of machinery's need for maintenance, thus enhancing the longevity of the machinery and management of any waste. For example, AI in predictive maintenance means knowing when equipment is about to fail and thus repairing before a full failure results in breakdowns and subsequent demands for new components. This at the same time reduces wastage while, on the other hand, also helps in reducing the utilization of resources by such elements as extending the time taken before industrial assets again become useful.

Robotics is another of the main pillars of Industry 5.0 and contributes to making the recycling and waste processing phases more efficient. AI robots can sort and process waste faster and more effectively than human employees, which promotes the improvement of the efficiency of recycling processes. In manufacturing industries, robotics can also be used in the development of modular products with the ease of disassembling and recycling, hence the goals of the circular economy.

It is becoming possible to use blockchain technology with high levels of security and transparency to follow the life cycle of products and materials from production to consumption, reuse, and disposal. This knowledge is particularly crucial for a circular economy in which system cycles depend on the tracking of assets. Blockchain can guarantee that the material is properly recycled or returned into circulation so that sustainability standards against wastage are met. In India and many developing countries where informal systems of waste management are common, blockchain can provide a formal base to those systems, making new opportunities for sustainable development possible (Ghobakhloo et al., 2024).

This is because IoT, which connects many intelligent devices, enables supervisors to monitor the production processes and resource adoption in real time. Through metering flow into energy, material, and waste IoT systems enable the acquisition of information on how to improve efficiency and minimize environmental footprint. For instance, smart sensors are capable of identifying areas of energy suboptimality or material loss and then making corrections that augment resource utilization. In a circular economy, IoT can also improve organizations' management of their products throughout their life cycle since it is possible to monitor when these products are ready to be used again, repaired, or recycled.

These technologies in Industry 5.0 could bring drastic change in different sectors like manufacturing units, farms, and industrial waste sectors in the Indian context. However, the integration of these technologies is not without some challenges, as this paper is about to unveil. They surveyed that there are challenges such as high expenses, few skilled professionals, and the demand for policies. Nonetheless, the chances for Industry 5.0 as the model to shift the Indian economy towards a more circular economy system are vast.

1.7 Circular Economy Framework

In remaking the consumption systems from the perspective of Industry 5.0, sustainable systems are the driving force provided by the circular economy framework. While linear economics has been around in which raw materials are extracted, subsequently used, and then disposed of, the circular economy aims to design back resources into the loop and keep them in perpetuity. Industry 5.0 builds on this framework by integrating the use of smart technologies to ensure minimized utilization of resources and maximum material recycling and reuse.

A major characteristic of the circular economy is that the design of products and systems avoids waste generation. In Industry 5.0, this principle is even more amplified and supported by robotics and artificial intelligence to make designs of products for durability and modularity and easier disassembly for recycling. Such design means that industries can minimize how much waste they create in product production and consumption phases since they can be taken apart and used in other ways. Furthermore, by using AI, it is also possible to determine where in the production process there are issues of waste and then make some corrections in real time (Tavares et al., 2022).

Closed-loop systems form the core of the circular economy, and it is Industry 5.0 that gives the proper tools to implement it. In closed-loop supply chain management, the material is used severally in its original form by recycling, re-manufacturing, or re-conditioning. The flow of material in such systems can be tracked and managed through technological solutions like IoT and the blockchain of Industry 5.0. IoT on a product can enable a company to know the condition of the product or component and replace or recycle it before hitting the dump. Blockchain helps industries track the life cycle of materials and therefore guarantees efficient reusing of resources.

In India, the transition to a circular economy associated with Industry 5.0 technological solutions can become a real way to mitigate significant environmental issues in the country today that are concerned with more waste management and resource deficiency. Through developing improved systems of production, India will be in a position to decrease the exploitation of physical resources while at the same time fostering growth. But to get here, it will need technological progress with the backing of regulation, advertisement of the concept, and development of education and training that prepares industries and workers to embrace this new industrial revolution.

2 Challenges in Implementing Circular Economy in Industry 5.0

The inclusion of Industry 5.0 technologies in the circular economy model raises numerous futuristic issues that are as much technological as they are social. Despite the numerous promising opportunities of Industry 5.0 as part of further industrial development for increasing resource efficiency or reducing the environmental burden, the process of achieving these goals faces various challenges. The problems of introducing Industry 5.0 solutions to promote sustainable development are particularly acute in such countries as India, as their industries today are incredibly diverse. All of these issues are of a technological, regulatory, economic, and cultural nature, which indicate vast difficulties that need to be overcome to get towards integration. When envisioning a shift to a world where the circular economy process is complexly interwoven with artificial intelligence, robotics, and blockchain, it is aptly crucial to study these barometers of deep-sea challenges in detail and convert them into opportunities for development and innovation (Bhat & Parvez, 2024).

2.1 *Technological Challenges*

The major issue arising when attempting to integrate the circular flow of goods and services within a system specifically designed for Industry 5.0 is the level of technology. As we look at developing circular economy solutions, the integration of new-age technologies like AI, robotics, and IoT is incredibly challenging. These technologies hold the prospects of automating the management of waste or of improving the efficiency of resource utilization and product longevity, but their adoption is knowledge-intensive. In the Indian context, where industrial development is comparatively less advanced than in developed countries, such technological intricacy poses a problem. A lot of industries in the country, especially SMEs, do not have the technological clout and environment that will enable them to adapt to AI applications or implement IoT technology into their operations.

The adoption of these technologies is also costly, which aggravates the situation. AI and robotics remain expensive technologies, which, although they might be simpler to obtain at the current stage, require considerable initial investment into the required hardware and software, as well as specially trained specialists. For many companies, particularly those on low-profit margins, the cost of integrating Industry 5.0 might seem too steep. Further, sustaining and upgrading these modern technologies to continue the delivery of top-notch technologies, given the fast rate at which they are being developed, poses another cost factor. The consistency of the technological framework is also different across different sectors and regions, which hampers integration. While industries in urban locations might be better placed to obtain digital tools and human capital, industries in rural and semi-urban zones face the challenges of accessing and obtaining reliable Internet connections, electricity, and other relevant technologies.

In addition, the technology used in AI/robotics is complex; hence, expertise to control the machines may involve training. India is experiencing a deficiency of qualified workers familiar with these peculiar innovative technologies, which creates a deficit in the human capital that can adapt to Industry 5.0 causes. Thus, the lack of skills is not only a reason for the slow implementation of circular economy strategies but also a factor contributing to the poor performance of implemented circular economy projects. Neglecting infrastructure with all efforts to bring more and more efficient technological solutions will remain the key constraint in the shift towards the circular economy practices of Industry 5.0 (Bhat & Parvez, 2024).

2.2 Policy and Regulatory Challenges

Besides those technological barriers, the absence of transparent policy initiatives and calibrated norms and regulations represents another major impediment to the deployment of circular economy strategies in Industry 5.0. Technology advancements such as AI robotics and IoT have grown exceptionally fast, but regulatory frameworks governing their application have not been developed concurrently. Such a regulatory relaxation is even more alarming in the context of developing countries like India, within which the regulatory conditions for emerging technologies remain insufficient, fragmented, and uneven. Lack of clarity leads to the formation of policies and regulations that are unclear themselves and which create conflicting regulatory frameworks within industries that slow down the development of new sustainable technologies.

For instance, despite India's progress in the regulation concerning some sections of the digital economy, namely data protection through the Personal Data Protection Bill, there also remain unsolved or insufficient legal positions regarding AI, robotics, or blockchain. Such issues arise because the use of these technologies is not adequately regulated, which is a key challenge to businesses in the current Industry 5.0 Industrial Revolution era to advance circular economy practices. Often, such technologies are not implemented into use by the companies until they have legal guidelines that they can follow related to some parameters, including but not limited to liability, security, and ethical use. However, this lack of regulatory guidance not only stifles the uptake of these technologies but also raises the stakes of using such technologies (Masoomi et al., 2023).

Additionally, there are no set international best practices for circular economy endeavours and Industry 5.0 technologies, which further complicate these regulatory hurdles. Currently, only a few countries have begun the process of developing guidelines for implementing circular economy policies; there is no international cooperation in this regard. However, this lack of international harmonization is an issue for India because it is a country that is heavily involved in international supply chains. Due to these diverse standards that the trading partners set, Indian companies may find it extremely challenging to embrace Industry 5.0 technology that provides sustainability.

To address these issues, the Indian government must come up with structured policies that will guide the integration of AI and robotics as well as other industry 5.0 into the circular economy system. On the other hand, it involves designing policies that encourage the use of sustainable technologies by organizations, implementing structures that define the ethical legalities of the technologies, and assigning the roles of international organizations to create universally acceptable standards for circular economy practices.

2.3 Economic Challenges

Economic challenges also emerge as another strain that acts as resistance to the integration of circular economy principles within the industry 5.0 environment. The shift from linear models that have been in practice to circular economy models entails considerable capital outlay both in technology and system change. In India, many industries, along with manufacturing and agriculture, basically where resource costs comprise a large portion of total expenses; this transition may prove costly. Such industries are typically characterized as capital-intensive, which means their financial resources to provide the initial capital for implementing Industry 5.0 technologies such as AI, IoT, and robotics are significantly affected (Gamberini & Pluchino, 2024).

In addition, many formats associated with circular economy practices may not necessarily generate high ROI within short periods. While adopting circular economy models offers advantages in the future like cutting costs on materials, less impact on the environment, and better brand image, the costs of making these changes will limit organizations from doing so. In countries such as India, where many organizations work on slim margins, the risks of implementing new technologies often seem too high.

Research and development are also vital for progressing the circular economy because everybody needs to invest in it, but due to the financial constraints faced by most Indian companies, especially SMEs, everyone needs to invest in R&D. Also, there is a lack of financing for the use of green technologies and following sustainable processes. While the Indian government has proposed multiple policies to support sustainability—for example, the "Atal Innovation Mission" and the multiple green financings—there is still a great need for specific funding mechanisms for industries willing to push the adoption of Industry 5.0 enablers for the circular economy. Eliminating these economic losses will require policy incentives, private, and/or outside investments, as well as innovative financing mechanisms for the circular economy transition to be profitable for industries in India (Danish & Senjyu, 2023).

2.4 Cultural and Behavioural Challenges

Regardless of the technological, regulatory, and economic enablers and barriers, culture and behaviour are key influencers to circular economy implementation within the Industry 5.0 environment. In that respect, one of the main challenges is the implementation of change. A lack of willingness to change is a common issue for many industries, might be clearer as especially traditional ones that must abandon the linear economy thinking that is typical for many industries. This resistance mostly stems from the existing flow and continuity of organizational processes enhanced by little knowledge of the implications of adopting circular economy practices in the long term.

India, for example, is characterized by dissimilarities between different industrial areas and numerous zones; this makes the problem of cultural entrenchment complex. Some industries, especially in regions that may not be so developed technologically, may not be conversant with the circular economy strategy or the advantages of Industry 5.0. This lack of awareness is then attributed to broader social amnesia where short-termism dominates over long-termism. Lastly, the concept of circular economy is sometimes challenging to advocate because it takes more time for businesses to reap profits from circular economy practices than from linear economy ones (Malik et al., 2024).

They also apply to the consumers' attitudes that show disturbed and often unpredictable behaviour. Consumers have a major role in demand sourcing in circular economy systems since they directly demand sustainable products and services. However, in many of the regions across India, the customers continue to be poorly sensitive to the effects of accommodation on the environment. It is crucial to make consumers change their behaviour and suggest choosing products with a longer time existence, using products with built-in production for their further consumption, and supporting companies that use resources as efficiently as possible. This is a process that cannot be solved, concerning education and awareness campaigns only, but with a change of culture in society to a sustainable one, from an environment-friendly point of view.

To address such cultural and behavioural barriers, there is a need to come up with a specific drive to sensitize people on circular economy business practices and Industry 5.0 technologies. Education and training for managers, employees, and customers; promotion of sustainability values by the government in the business and societal management.

3 Transforming Challenges into Opportunities in Industry 5.0

Many possibilities are opening up with the transition to Industry 5.0 to address the issues that have plagued circular economy uptake. The application of new technologies, including artificial intelligence and blockchain, as well as digital twins of industries, opens up opportunities to dramatically increase resource management, improve waste handling, and bring sustainable production into focus. Indian conditions that face even now the problems of waste or inadequate usage of resources are completely compatible with the application of aforementioned technologies as tools that can radically change the contours of the Indian industry and bring it to a circular economy paradigm. Industry 5.0 perspectives of human and machine integration in problem-solving enable the transformation of conventional problems hindering productivity into opportunities for future growth, innovation, efficiency, and environmental sustainability. The uses of all these technologies range from waste, supply chain, product life cycle, and material recycling transparency, which are capable of building a strong economy (Van Erp et al., 2024).

3.1 Leveraging Artificial Intelligence for Waste Management

AI is a crucial enabler for waste management challenges, especially in a circular economy. In the production line in traditional industrial systems, waste is usually poorly disposed of, and most of it is dumped into the environment, especially into a landfill or incinerated, thus worsening the environmental status. These inefficiencies can be addressed by the integration of AI technologies such as predictive modelling, sorting, and improved resource recovery. Among all the fields and areas where the application of AI is most beneficial, there is the prospect of the use of AI for predictive modelling for industries that can predict the tendencies of waste generation, or, in other words, entrust industries to work on the preservation and reduction of waste before it has been generated. Much as with the predictions of when or where a traffic jam may occur, AI analytical tools can use historical data to anticipate where and when waste may happen and prepare in advance for it. In India, where waste management infrastructure is still a growing effort, AI in conjunction with predictive modes can assist municipal governments and industries in the optimization of their waste collection and recycling, thereby enabling efficient waste management with little impact on the landfill as shown in Fig. 3 (Arcot, 2023).

There is another promising area where AI tools can bring a lot of positive changes—automated sorting. Machine vision systems deployed in automated robots can sort waste materials with a higher degree of effectiveness than human workers separating recyclable waste from other waste items. It is also important to note that this technology can assist in the task as the waste sector in India remains informal

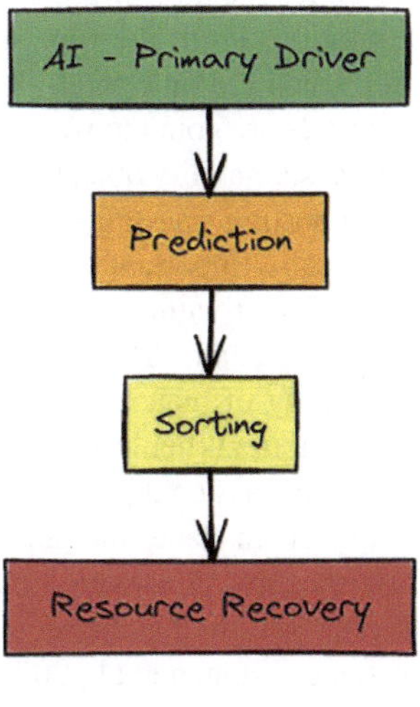

Fig. 3 AI-driven waste management funnel

with manual sorting of the waste conducted. With the help of such an assisted mechanism and by applying AI, it is possible to enhance the recycling features of such operations and make the proportion of the overall recycled articles in the market higher and subsequently the quantity of the buried articles to be minimized.

Resource recovery is also an important function of AI since useful materials are recovered from waste products for reuse in manufacturing. Using machine learning techniques, one can learn the best ways to recycle material from waste using chemical, biologic, or mechanical means. Hence, such AI-driven resource recovery in the growing manufacturing sector in India stands as a potential solution to diminish the dependence on virgin material, drawing a large rationale that comes under circular economy sustainability efforts. Successful case looks at the benefits of applying AI in the European waste management industry and will help India revolutionize waste management through the use of AI, making the process more efficient while lowering costs and promoting environmental sustainability (Dwivedi et al., 2022).

3.2 AI and Supply Chain Optimization

It is no less transformative as far as the application of the circular economy concept within the context of supply chain incorporation of artificial intelligence is concerned. Business supply chains, especially those of large production organizations, are usually convoluted, which may sometimes create poor efficiency that causes wastage of resources and environmental degradation. By boosting the circularity of supply chains through increasing traceability and efficiency and minimizing waste, AI can help circular economy strategies thrive.

A crucial barrier when operating in circular economies heavily remains the ability to extend the economic flow of materials and products. AI can improve the system that tracks materials from the supply chain and guarantee that finished products are either reclaimed, recycled, or remanufactured once they reach their life-cycle expiration. From the time the materials enter one point of the supply chain until they reach the other end, AI algorithms will help a company identify areas where resources are going to waste and help fix them. It also assists companies in achieving their sustainability objectives, in addition to making the customers have an assurance of supporting a sustainable business chain.

It is also important to understand the specific ways in which AI enhances resource management within supply chains. Recuperating information at different stages of the supply chain, AI systems can detect patterns of resource usage and better arrange the supply of materials and products. This is especially significant in domains of production, particularly manufacturing and agriculture, where the issue of scarcity of resources is contiguous. AI can be of great help to companies in India where most supply chains face challenges due to infrastructural hindrances. In moments, the AI furnishes live information that the companies can use to organize their resources in the best way so that the required supplies are utilized to the maximum with minimal wastage (Kannan et al., 2024).

In addition, AI can reduce waste since manufacturing processes can be optimized, and there will be little generation of waste. By using AI, it is possible to predict when equipment is likely to break down and therefore need repair before this leads to wastage of material, which is often the case with most breakdowns. Implemented well, AI can help maintain machines in top working order to minimize the possibility of product delays and decrease the amount of wasted material in production. AI analyses of supply chains in the rapidly emerging industrial sector in India can provide a roadmap for more sustainable ways of production and lower effects of industrial enterprise on the environment.

3.3 Blockchain for Circular Economy

Despite its relative infancy, blockchain tech has the potential to offer a level of profound improvement in circular economy credentials by increasing the solidity of material flows and the accountability of recycling credentials. Closed-loop is the key to the circular economy since all cycles must loop continuously and the resources be recycled. Nevertheless, effective tracking of material throughout the life cycle may often prove to be difficult or even next to impossible given the complex upstream and downstream chain of supply. New trends such as blockchain address this challenge as they ensure clear records of all materials that are involved in the production and disposal process.

Based on the use case of recycling, we can thereafter establish a chain of process blocks where every material has a blockchain. This also makes it possible for recyclable materials to be well recycled and also makes consumers and business persons able to tell whether the products they use are sustainable or not. For example, a manufacturer can apply a blockchain system, which tracks the process of creating a product and recycling or reusing it when it becomes useless. Such openness also establishes a level of trust between the consumers and the commercial entities because they can verify sustainability engagement with third parties.

In India today, most people rely on informal waste recycling bins; hence, the blockchain solution will formalize these systems and enhance the efficiency of the recycling process. Blockchain enables a centralized paper trail for records of waste material so that recyclables are correctly sorted and waste handling adheres to environmental laws. Furthermore, by enabling a record of resource flow, blockchain inspires the improved collaboration of the actors in the circular economy, ranging from producers to waste managers. Circling back, as India evolves towards a sustainable circular economy for clarity will significantly help in promoting transparency and optimizing the processes associated with recycling in the country aims to achieve (Hassan et al., 2024).

3.4 Digital Twins and Predictive Analytics

Another pair of tools that can change resource utilization and sustainable production in the context of Industry 5.0 are digital twins and predictive analytics. A digital twin is a mirror image, in the form of a digital model, of an actual asset or system to track performance data. Industries can build digital replicas of manufacturing processes, individual equipment, or even entire factories and thereby gain a better understanding of how their processes are working and in which areas potential optimization for resource utilization and minimizing waste can be achieved.

Inefficiency, digital twins are most appropriate when it comes to finding out which aspect of the manufacturing process is resource-intensive and which one is wasteful. Digital twins are interactive models that can predict how a given manufacturing process will perform in real life; hence, by modelling various production possibilities, it is possible to achieve maximum yield with minimal resources, energy, and material utilization. For instance, a digital twin of a production line can model the flow of material through it and isolate areas that belong to the slow-moving line or to the regions where material consumption is excessive. Thus, actions solely based on these findings make a change in the production process, influencing the consumption of resources at the cost of waste generation, one of which is a principle of the circular economy.

The use of digital twins in connection with predictive analytics enables industries to foresee equipment breakdowns, slowdowns in production, and other factors that can contribute to waste and loss. Using historical information and trends, predictive analysis divides equipment into categories that need maintenance at certain times, and it can tell when something is likely to go wrong in the production line. This helps firms to prevent further occurrence of downtime and, more so, categorically minimizes the waste associated with equipment breakdowns or delays in production. The manufacturing industries in India face a lot of challenges where breakdown is often costly and causes a lot of wastage of resources. This is where the predictive analytics of the industries can play a crucial role in delivering services that can enable the industries to deliver their services as required and at the same time reduce the impact they may have on the environment.

In conclusion, the use of digital twins and predictive analytics seems to be effective in functioning as instruments in the development of the circular economy principles in Industry 5.0. Through feeding real-time data on resource usage and product throughput, these technologies make it possible for an industry to make informed decisions that will help enhance the sustainable production of goods and optimal usage of resources. But as India introduces more digitalization in industrial areas, the use of digital twins and predictive analytics will be very vital to the development of a circular economy and successful implementation to avoid wastage (Preud'homme, 2024).

4 Legislative and Policy Suggestions for AI in the Circular Economy

Due to the increased challenges created by Industry 5.0's technology in attaining a circular economy model in India, legislative and policy measures emerge highly applicable and important. Creating effective regulation, applying specific policy measures, and promoting synergies between industries and communities are critical milestones on the way to achieving a circular economy transition supported by AI and other digital technologies. Of course, all these legislative and policy recommendations are aimed at managing the diverse issues and realizing the latent benefits that can be derived from integrating AI and the circular economy to help India transition towards a more sustainable and resilient economic future (Dhakal et al., 2024).

4.1 Need for Regulatory Frameworks

The development of standardized legal frameworks for AI is a crucial need for the successful adoption of Industry 5.0 and a transition to circular economies. These frameworks need to articulate the best practicable means to incorporate superior technologies as well as the policies to include in using advanced technologies towards the realization of the top overall organizational objectives of sustainable resource management. At the international and national levels, it is found that legal structures act as the general framework for the promotion of innovation and the management of compliance and risks requisite for the implementation of AI and circular economies (Barata & Kayser, 2023).

Internationally, it makes sense to coordinate the regulations to foster a favourable operating environment for Indian enterprises aiming to enter the global market. It can include their compliance with international policies as to data protection, cybersecurity, and the use of artificial intelligence for the promotion of a circular economy, which will give more confidence to India's initiatives on the international level. Globally, India requires certain policy measures to address the peculiar issues that AI integration calls for in circular economies. This involves developing guidelines for the new generation of AI-based waste sorting and management solutions, including records for traceability of the recycled materials using advanced applications like blockchain, and controlling the application of robotics in industries to avoid displacing people while augmenting human-robot symbiosis (Negi et al., 2024).

In addition, policies should encourage investment in sustainable technologies by providing legal encouragement such as tax breaks and incentives in the form of subsidies and grants for firms deploying the use of artificial intelligence and Industry 5.0 technologies that are circular economy friendly. Such incentives can reduce the financial pressure on companies, which in turn will allow agencies and SMEs to invest in requisite technologies that are otherwise too expensive. Also, there are legal requirements that relate to sustainability reporting and disclosure to ensure that firms adopt a circular economy bearing in mind the four Rs of AI (Atif, 2023).

Furthermore, in the same vein, an effective set-up of specific governmental or non-governmental regulatory authorities or enhancing institutional capacities to monitor the application and adherence to AI technologies in circular economies. These bodies can offer sustainable certification, perform sustainable checking or auditing, and ensure that the firms are embracing the preferred forms of sustainability. With a stronger framework, India continues to harness innovation and improvement of technologies and at the same time help in steering a sustainable environment as well as leverage on the potential economic shocks. Therefore, the presented need for nearer regulatory controls inherent in technological advancement is critical to align technological developments such as AI and Industry 5.0 with sustainability agendas, which in turn provides AI with a clear roadmap for integration (Tyagi et al., 2023).

4.2 Policy Interventions in Developing Nations

Technological and financial barriers to implementing circular economy practices boosted by the technology of Industry 5.0: the case of a developing country, India. To overcome these barriers, there is a need for specific policy interventions if there is to be the provision of an enabling environment that supports the change direction towards sustainable and enhanced technological industrial systems. Such policy interventions should be worded and targeted technically, financially, and infrastructure-wise to accommodate a circular economy.

An important policy lever conducive to the advancement of circular economy is the offer of favourable financial incentives and enabling instruments for companies targeting circular economy technologies. This can include funds, affordable credits, and fiscal exemptions initially given for those firms incorporating AI, robotics, and IoT for waste disposal, resource utilization, and sustainable manufacturing. Where governments adopt such policies, they may make it easier for more businesses, especially SMEs, to embrace such circular technologies through the reduction of costs related to the initial investment in such technologies. The other key policy advancement is technology support improvement. To comply with the goals of Industry 5.0, there is a need to have good digital support infrastructures such as the Internet, a credible supply of power, and well-featured production lines and equipment, among others. Many governments worldwide have made efforts to launch training courses and a campaign that is meant to enhance the movement towards the embracement of advanced technologies by the workforce because the lack of skills often leads to the stagnation of such technologies (Mourtzis, 2024).

Moreover, specific physical policies relating to research and development in circular economy technologies are necessary. To this end, financial support for R&D can result in the development of new and better methods needed to manage waste, recycle it, and recover useful resources from the waste. There exists potential for an ecosystem of innovation where governmental, academic, and industrial sectors can work hand in hand to drive the development of the next generation of technologies and the application of those technologies in real life.

Furthermore, it can also be noted that gaining a partnership with international organizations or utilizing international competencies may increase the speed of implementing the principles of the circular economy. Since policy frameworks refer to the means, a consultation of knowledge exchange, transfer of technology, and collaborative projects with global associates can enlighten Indian industries with advanced technologies and outstanding information about circular economy implementation, which in turn increases their ability to implement circular economy models (Martini et al., 2024).

Therefore, on balance, policy-level targeted actions in emerging economies like India are inherently necessary to fix the technological and financial barriers to the adoption of the circular economy models. These involve the provision of financial incentives, improved technology support systems, an encouragement of R&D, as well as the support of international collaboration, which can all help create a conducive policy environment for the drive for a shift towards a green and circular economy based on technology advancement from Industry 5.0 (Kathuria et al., 2024).

4.3 Encouraging Industry and Community Collaboration

From inventory and beyond, and hence it is important to promote the integration of industries, governments, and local communities within the Industry 5.0 transition towards circular economies. Such collaboration means that advancement in technology, performance, quality, accessibility, and more is done in a way that is responsible and cohesive to all and for all parties. This is the case because nurturing this tripartite partnership could translate into better and enhanced use of circular economy strategies and policies, which enshrine sustainable development. The formation of the PPPs is found to be very useful in promoting cooperation. Such joint ventures can fully capitalize on the potential of two types of players: the technological and resource base of private industry and the regulatory and policy of the government sector. For example, PPPs can help in the implementation of AI systems in waste management through cooperation in terms of resources for research, development, and deployment. It could also be used in the replication process of pilot projects, which makes it easier to see that solutions being brought into the market can benefit a greater number of people (Kirtania et al., 2023).

Besides, local communities must be engaged in different circular economy measures to ensure the high effectiveness and sustainability of the measures. Engagement programs at the community level can help to familiarize the population with the advantages regarding the circular economy and enhanced technologies that are applied to hit these advantages. Outreach programs, sensitization and awareness creation, drama, fairs, interactive sessions, and group decision-making can go a long way to ensure that people take personal responsibility for handling waste, recycling, and ensuring efficient use of resources. This is good for the circular economy because all participants will support the actions if the community is engaged, and it results in a positive feedback loop on what has been done (Payer et al., 2024).

Furthermore, industries can also tie up with academic institutions and research organizations to foster innovation and come up with context-related solutions to circular economy problems. It is at such collaborations that new technologies, methodologies, and best practices, specific to the individual needs of these areas and industries, can be developed. Collectively, these stakeholders will help make certain that the solutions formulated are not only good but also sustainable, given the unique problems experienced by industries and people in India (Sulich, 2023).

In addition, most of the government departments have a function of sharing electronic forums to ensure that all the stakeholders are connected (Sakao et al., 2024). Such platforms can be used as a means for knowledge, resource, and best practices exchange that contributes to the enhanced culture of excellence. As a result, these platforms facilitate the mobilization of all stakeholders in industrialization, policies, community members, and technology gurus for synergy towards meeting set circular economy goals (Sarkar et al., 2024).

5 Conclusion

Industry 5.0 entering the circular economy is a revolutionary way of implementing sustainable industrial novelties that improve the usage of resources and development of the economy. As this research shows, technological solutions of Industry 5.0—artificial intelligence, robotics, blockchain, and IoT—may be a way out of the multiple problems that the circular economy paradigm faces today. They can help industries trace, minimize, and reutilize materials and resources to lengthen the life cycle of products to achieve the goals of the circular economy. India in particular, with its ever-expanding population, coupled with the current rising industrialization pressure on the limited resources and strain on waste management systems, the takeaway from Industry 5.0 holds the promise for a sustainable future for the country.

However, the opportunities are enormous, but so is the competition. A major problem area lies in pillars three, four, and five, including technological barriers such as the high costs for technology implication and the lack of telecommunication structures in rural or semi-urban regions. The structure of industry in India, with its increasing emphasis on technological campuses in the major metro areas and its extensive rural base, also makes Integrated Infrastructure Development and Technological Awareness much more specific. Thus, the absence of attempts to develop this capacity constrains the potential of Industry 5.0 to unlock the next level of circular economy activities. For this reason, countries require proper investment in their digital infrastructure and the education of their employees.

Also, the regulatory and policy environment should change depending on the new technology in the market. Uncertainty as to how the principles of artificial intelligence, robotics, and the Internet of Things are regulated and insufficient definition of their connection to sustainability and circular economy measures is a significant challenge. The actions on these technologies are required and must be inclusive policies that control how they are utilized but also champion their usage. One of the issues can be solved by having clear legal rules, which will give the companies the certainty to invest in solutions relying on artificial intelligence and, at the same time, guarantee adherence to the aspects of ethical non-acceptance, including privacy in the usage of data, job loss, and negative impacts on the natural environment. Internationally, it has become extremely pivotal for India to follow the international standards of regulations so that they can compete in the international market, retaining an environment that may not only encourage sustainable and circular systems but mandate them to be followed.

Other external factors equally contribute to the slow advancement of Industry 5.0 technologies in circular economies, including economic factors. Unfortunately, the initial costs of these technologies are relatively steep, and therefore, many organizations, especially SMEs, are finding it hard to justify the investments because the returns are frequent in the long term. In its early phase, these businesses will rely on tax credits, grants, and low-interest loans to enable them to make the transition. Finally, it has also proven effective to engage different sectors to co-fund pilot schemes through public-private partnerships (PPPs) that can apply successful pilot projects at the scale needed, ensured by the development of partnerships and networks.

The last reason is that cultural and behavioural factors increase the challenge of this transition. Various industries and customers are still reluctant to embrace circular economy strategies as they are disruptive or costly. Education for change and revised community perceptions are key factors in the progress of the circular economy strategies. In the same way, including community participation in the decision-making process and guaranteeing that locals will benefit from those programs will increase their interest in sustaining initiatives. Simply put, all of these challenges are rather serious, but they are not impassable. Industry 5.0 is the best and most promising chance to change India from a prestigious industrialized nation to a remarkable industrialized country and an environmentally sustainable industrialized country simultaneously. Industry 5.0's technological, regulatory, economic, and cultural barriers can be overcome through policy frameworks.to establish a circular economy in India. We see here that AI, robotics, blockchain, and IoT are the possible future solutions to counter the circular economy implementation challenges, which we witness as opportunities rather than threats as a society and economy.

References

Adel, A. (2022). Future of industry 5.0 in society: Human-centric solutions, challenges and prospective research areas. *Journal of Cloud Computing, 11*(1), 40. https://doi.org/10.1186/s13677-022-00314-5

Arcot, S. (2023). Industry 5.0 and Society 5.0. In *Industrial Revolution and Metaverse: Industry 5.0 (Chapter 1)*. QUING Publications. https://doi.org/10.54368/qpbc.2023.1.1

Atif, S. (2023). Analysing the alignment between circular economy and industry 4.0 nexus with industry 5.0 era: An integrative systematic literature review. *Sustainable Development, 31*(4), 2155–2175.

Barata, J., & Kayser, I. (2023). Industry 5.0—Past, present, and near future. *Procedia Computer Science, 219*, 778–788.

Bhat, F. A., & Parvez, S. (2024). Emerging challenges in the sustainable manufacturing system: From Industry 4.0 to Industry 5.0. *Journal of the Institution of Engineers (India): Series C*. https://doi.org/10.1007/s40032-024-01046-y

Danish, M. S. S., & Senjyu, T. (2023). Shaping the future of sustainable energy through AI-enabled circular economy policies. *Circular Economy, 2*(2), 100040. https://doi.org/10.1016/j.cec.2023.100040

Dhakal, S., Parvez, A., Kathuria, S., & Chanti, Y. (2024, May). Face recognition technology powered by artificial intelligence for enhanced security measures. In *2024 Parul International Conference on Engineering and Technology (PICET)* (pp. 1–6). IEEE.

Dwivedi, A., Agrawal, D., & Jha, A. (2022). Studying the interactions among Industry 5.0 and circular supply chain: Towards attaining sustainable development. *Computers & Industrial Engineering, 176*, Article 108927. https://doi.org/10.1016/j.cie.2022.108927

Gamberini, L., & Pluchino, P. (2024). Industry 5.0: A comprehensive insight into the future of work, social sustainability, sustainable development, and career. *Australian Journal of Career Development, 33*(1), 5–14. https://doi.org/10.1177/10384162241231118

Ghobakhloo, M., Mahdiraji, H. A., Iranmanesh, M., et al. (2024). From Industry 4.0 digital manufacturing to Industry 5.0 digital society: A roadmap toward human-centric, sustainable, and resilient production. *Information Systems Frontiers*. https://doi.org/10.1007/s10796-024-10476-z

Hassan, M. A., Zardari, S., Farooq, M. U., Alansari, M. M., & Nagro, S. A. (2024). Systematic analysis of risks in Industry 5.0 architecture. *Applied Sciences, 14*(1466). https://doi.org/10.3390/app14041466

Kannan, D., Shamekhi Amiri, A., TarighiShaayesteh, M., Khalili Nasr, A., & Mina, H. (2024). Unveiling barriers to the integration of blockchain-based circular economy and Industry 5.0 in manufacturing industries: A strategic prioritization approach. *Business Strategy and the Environment*. https://doi.org/10.1002/bse.3886

Kathuria, S., Duggal, A. S., & Lande, J. (2024, May). Artificial intelligence automation in advent of human resource management. In *2024 Parul International Conference on Engineering and Technology (PICET)* (pp. 1–5). IEEE.

Kirtania, A., Kathuria, S., Pachouri, V., & Gehlot, A. (2023, August). Artificial intelligence inclusion in landslide monitoring. In *2023 Second International Conference on Augmented Intelligence and Sustainable Systems (ICAISS)* (pp. 47–51). IEEE.

Liu, S., Li, P., Wang, J., & Liu, P. (2024). Toward industry 5.0: Challenges and enablers of intelligent manufacturing technology implementation under the perspective of sustainability. *Heliyon, 10*(15), e35162. https://doi.org/10.1016/j.heliyon.2024.e35162

Malik, S., Muhammad, K., & Waheed, Y. (2024). Artificial intelligence and industrial applications—A revolution in modern industries. *Ain Shams Engineering Journal, 15*(9), Article 102886. https://doi.org/10.1016/j.asej.2024.102886

Martini, B., Bellisario, D., & Coletti, P. (2024). Human-centered and sustainable artificial intelligence in industry 5.0: Challenges and perspectives. *Sustainability, 16*(13), 5448.

Masoomi, B., Sahebi, I. G., Ghobakhloo, M., &Mosayebi, A. (2023). Do industry 5.0 advantages address the sustainable development challenges of the renewable energy supply chain? *Sustainable Production and Consumption, 43*, 94–112. https://doi.org/10.1016/j.spc.2023.10.018

Mourtzis, D. (2024). Challenges and opportunities of the transition from Industry 4.0 to Industry 5.0. *Manufacturing from Industry 4.0 to Industry 5.0*, 97–131.

Negi, P., Singh, R., Gehlot, A., et al. (2024). Specific soft computing strategies for the digitalization of infrastructure and its sustainability: A comprehensive analysis. *Archives of Computational Methods in Engineering, 31*, 1341–1362. https://doi.org/10.1007/s11831-023-10018-x

Payer, R. C., Quelhas, O. L. G., & Bergiante, N. C. R. (2024). Framework to supporting monitoring the circular economy in the context of industry 5.0: A proposal considering circularity indicators, digital transformation, and sustainability. *Journal of Cleaner Production*, 142850.

Preud'homme, G. (2024, August 1). *Industry 5.0: A sustainable revolution at the heart of tomorrow's plants*. Proaction International Blog. https://blog.proactioninternational.com/en/industry-50-sustainability-revolutionizing-plants

Rame, R., Purwanto, P., & Sudarno, S. (2024). Industry 5.0 and sustainability: An overview of emerging trends and challenges for a green future. *Innovation and Green Development, 3*(4), 100173. https://doi.org/10.1016/j.igd.2024.100173

Sakao, T., Bocken, N., Nasr, N., & Umeda, Y. (2024). Implementing circular economy activities in manufacturing for environmental sustainability. *CIRP Annals, 73*(2), 457–481. https://doi.org/10.1016/j.cirp.2024.06.002

Sarkar, B. D., Shardeo, V., Dwivedi, A., & Pamucar, D. (2024). Digital transition from industry 4.0 to industry 5.0 in smart manufacturing: A framework for sustainable future. *Technology in Society, 78*, 102649.

Sulich, A. (2023, September). Model of relationship between circular economy and Industry 5.0. In *IFIP International Workshop on Artificial Intelligence for Knowledge Management* (pp. 220–236). Springer Nature Switzerland.

Tavares, M. C., Azevedo, G., & Marques, R. P. (2022). The challenges and opportunities of Era 5.0 for a more humanistic and sustainable society—A literature review. *Societies, 12*(6), 149. https://doi.org/10.3390/soc12060149

Turner, C., Oyekan, J., Garn, W., Duggan, C., & Abdou, K. (2022). Industry 5.0 and the circular economy: Utilizing LCA with intelligent products. *Sustainability, 14*(22), 14847. https://doi.org/10.3390/su142214847

Tyagi, A. K., Lakshmi Priya, R., Mishra, A. K., & Balamurugan, G. (2023). Industry 5.0: Potentials, issues, opportunities, and challenges for Society 5.0. *Privacy Preservation of Genomic and Medical Data*, 409–432.

van Erp, T., Pereira Carvalho, N. G., Gerolamo, M. C., Gonçalves, R., Rytter, N. G. M., & Gladysz, B. (2024). Industry 5.0: A new strategy framework for sustainability management and beyond. *Journal of Cleaner Production, 461*, Article 142271. https://doi.org/10.1016/j.jclepro.2024.142271

Transforming Circular Economy Through Industry 5.0: A Deep Dive into Sustainable Innovations

Gayatri Rathi and Samta Kathuria

Abstract Industry 5.0 has further implications for taking sustainability forward by linking up with the circular economy as a valuable avenue to advance and reinvigorate circular practices within manufacturing and supply contexts. Industry 5.0 deals with the integration of humans and advanced intelligent machines with the help of intelligent technologies like AI, robotics, and blockchain. Applying it to the circular economy where reduction in resource use, recycling, and reuse matters, Industry 5.0 can catalyze the changes. This paper identifies Industry 5.0 as the solution to current disruptions such as smart manufacturing, digital twins, and a transparent supply chain. For instance, the circular economy popularized in the European Union through a circular economy action plan and Japan's smart factories where these technological advancements have enhanced recycling, wastage, and resource utilization. However, downside risks involving the increased implementation costs, system integration problems, and absence of uniform international standards remain an issue and repercussions are felt most severely in the developing nations. About these challenges, this study suggests policy intervention in the form of technology development endorsement through tax credit schemes and public-private partnerships alongside availing a development program to increase the literacy levels of workers in circular economy models. In the long run, therefore, Industry 5.0 presents a potential way forward for a more sustainable economy globally although the process's challenges cannot be ignored.

Keywords Circular economy · Industry 5.0 · Sustainability · Smart manufacturing · Artificial Intelligence (AI)

G. Rathi
Faculty of Law, Swami Vivekanand Subharti University, Meerut, India
e-mail: gayatrirathi950@gmail.com

S. Kathuria (✉)
Parul Institute of Law, Parul University, Vadodara, India
e-mail: samtakathuria14@gmail.com

R. Kumar and S. N. Mohanty (eds.), *Sustainable Economy Models in the Age of Industry 5.0*, https://doi.org/10.1007/978-981-96-4104-8_15

1 Introduction

In the conditions of the industrialization of society and focus on consumption, the demand for sustainability is greater in terms of economics and geography. A circular economy is the opposite of the current "take-make-dispose" model of consumption and production, which employs sustainable innovation to avoid using up natural resources and subjecting the environment to any further damage. This approach is very important in the present world toward controlling climate change, preserving natural resources as well as practicing and encouraging sustainable development. In a circular economy, these technologies have a critical responsibility of achieving efficiency, using resources to the maximum, and coming up with creative methods of production. For instance, AI can be used to analyze supply chain management and gain insights into the likely resource demand in most cases eliminating unnecessary resource production and consumption. Robotics showed precision in the making and recycling of several products while blockchain provides a platform to track the origin of the material used thus identifying the eligible source and also showing the best way to recycle the products. Altogether these technologies turn the circular economy into an effective model that sustainable natural and resources while creating new added value for the economy (He et al., 2024). As shown in Fig. 1, Industry 5.0 is the next level of Industry 4.0 where there flows harmony between humans and more technologies like AI, robotics, and blockchain. As opposed to Industry 4.0, which emphasizes the importance of automating processes, the 5.0 version of the Industrial Revolution puts Innovation by focusing on human aspects, combined with using artificial intelligence systems (Malik et al., 2024).

1.1 Research Problem and Objectives

The linear economy model that has characterized the production and consumption systems for over half a century is intrinsically unsustainable. They cause unwanted wastage, over-exploitation of resources, and pollution of the environment, all elements implicated in the current climatic change. The drawback of the linear economy becomes quite apparent in developing nations where economic development is seen to occur at the expense of the environment. In these regions inefficient waste management systems, poor recycling services, and over-reliance on finite exhibit the problem. Besides, the industries in such economies are themselves unable to possess the technological capacity that enables the shift from a linear economy to a circular economy. This is where the concept of Industry 5.0 brings a revolutionary solution into the equation. AI, robotics, and blockchain can be added to the existing conventional process to develop new and superior systems with higher efficiency and lesser wastage of resources. However, the changing of this model is not easy, and the transition requires effort. These challenges include the high costs accompanying new technology implementation, calls for enabling policy, and the present and potential

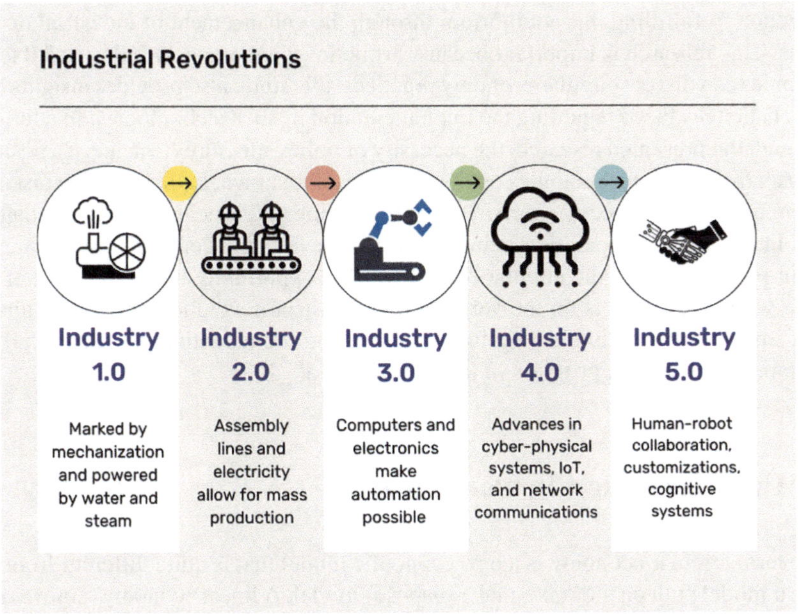

Fig. 1 Evolution of industrial revolutions: from mechanization to AI-driven collaboration

technology divide between the developed and the developing worlds. This research aims to fill these gaps with knowledge of how Industry 5.0 can be used to solve some of the problems of the linear economy model especially in the developing world. The first research questions are therefore: The primary aims of this paper are as follows understand the key technological advancements underpinning Industry 5.0, evaluate how they might affect circular economy efforts, and offer recommendations regarding the issues that surround their application.

1.2 Importance of the Study

Nowadays the demand for innovations that will not harm the environment is more pressing than ever! As the world's population increases, pressures on resources persist, demanding that economies—particularly, emerging ones—devise sustainable strategies for continued stability. In countries like India, where a rapid development of industrial potential is based on the technological growth of industrial production connected with an enhancement of environmental issues, the need for solutions, which promote the economic growth of the country together with persistent requirements for environmental protection, increases. Industry 5.0 provides a direction in fulfilling this equilibrium through the enhancement of industrial operations. This research is important because while investigating how Industry 5.0 technologies can disrupt circular economy practices, this study also provides insights into the challenges faced regarding the implementation of such technologies. In addition, through the presented research, the necessity of policy measures to achieve a positive impact from these technologies on the global level is shown. In this way, the research strengthens the perspective of Industry 5.0, referring to the technological, regulatory, and international cooperation requirements for developing countries to become the main protagonists of the circular economy. This opportunity has come right at the time when the world is on the verge of the industrial revolution and the results of this study could be illuminating for understanding the potential of Industry 5.0 for sustainable economic progress in India (Rame et al., 2024).

2 Understanding Circular Economy

The term circular economy is a new economic model that is quite different from the linear model built on the "take-make-dispose" model. A linear economy removes raw materials from the earth by processing them into goods and then disposing of products after the product cycle is complete. Besides exhausting the natural resources, they are also associated with environmental unfriendliness more so in places where they produce a lot of waste in the process. On the other hand, the circular economy concept revolves around a system that will contain less waste while increasing resource efficiency, and more timeless products. Some of them are cutting down the use of raw materials, using products and components more than once products and components, and reusing materials instead of discharging them to landfill or the environment. Fundamentally the Circular Economy system simply aims to sever the direct link between economic development and resource use as shown in Fig. 2 (Sariatli, 2017).

Three important principles of a circular economy include the following; resource utility which aims at optimizing the use of resources in the production process. This can be done in ways like ensuring the products that are developed are for durable use, increasing the effectiveness of production procedures, and incorporating new techniques that make better usage of certain factors. Moreover, recycling and reuse are other ingredients of the circular economy model. Rather than products with the end of useful life being dumped or buried, they are reused, recycled, or reprocessed with a view of extracting the contained resources and feeding them into the system. This means that there is little requirement for virgin raw materials, less energy is used, and least of all, there is minimal damage to the environment. Also, the circular economy entails product- an as-a-service delivery system in which consumers pay for the right to use a product instead of acquiring ownership of it, thus they force the manufacturers to design long-lasting products that will require frequent servicing and repair (Song et al., 2024).

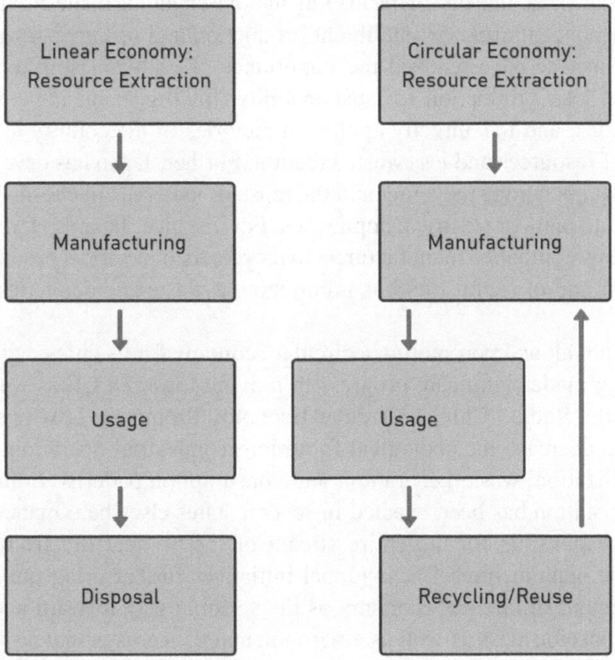

Fig. 2 Comparison of linear and circular economies: While the linear economy follows a 'take-make-dispose' approach, the circular economy emphasizes recycling and reusing resources to minimize waste and extend the lifecycle of materials

2.1 Evolution and Global Adoption of Circular Economy Models

The approach relayed by the circular economy is not novel but its institutionalization and broad implementation have only been trendy for approximately twenty years. Europe particularly the European Union (EU) has been at the forefront of the shift to a practical circular economy through policies and strategies. In 2015 EU launched its Circular Economy Action Plan including a set of measures intended to plan the change in the way goods are developed, procured, and used. This plan focuses on the fragmented management of waste, resource optimization, and the transition toward a circular economy throughout the industries of manufacturing, construction, and agriculture. Based on the EU's circular economy strategy, the management of waste is also a key element that has been targeted as follows; recycling of Municipal waste at a rate of 65% by 2035 and a maximum landfill space of 10% of the total quantity of waste produced.

Japan has also become one of the leading nations engaging in the circular economy shift mainly concentrating on intelligent production and resource-saving technologies. The Japanese have realized the importance of craftsmanship to enhance the efficiency of lean production for sustainability. Intelligent manufacturing, driven by AI, robotics, and IoT directly applies in factories of the country to bring better utilization of resources and less waste creation. Further, Japan has developed policy measures that encourage recycling and the reuse of materials in end-of-life products especially auto parts or electrical appliances. For instance, Japan's Home Appliance Recycling Law compelled manufacturers to recycle their electrical products that have reached their end of useful life thus minimizing e-waste and deliveries of valuable materials.

China is not alone in promoting a circular economy for its cities, with the United States having made significant progress in moving toward a CE as well albeit at a different scale. Rather, China's Circular Economy Promotion Law passed in 2009 endeavors to decrease the ecological footprint of industrial operations in terms of resource utilization, waste generation, and consumption patterns. In the US alone, the EPR legislation has been enacted in several states elsewhere manufacturers are also made responsible for the entire stream of the product life from the recycle bin up to the manufacturer. These global initiatives further bring out the fact that acknowledgment of circular economy as the sensible way forward about resource and waste consciousness as well as environmental concerns is increasingly coming into realization.

2.2 Industry 5.0: Key Technological Advancements

Industry 5.0 is a new concept that is an evolution of the Industry 4.0 framework also known as the fourth industrial revolution which focuses on the automation of manufacturing and the integration of smart technologies such as Artificial Intelligence AI, IoTs, and big data. Industry 4.0 introduced the idea of efficiency and automating systems with as few human interventions as possible, but now, where we are in Industry 5.0 is reversing that notion slightly and reintroducing humans as an essential part of the overall system. Industry 5.0 entails integrating human intelligence and resourcefulness with artificial intelligence and other technologies so that they complement one another for efficiency and efficiency with accountability for responsible industrialization. While Industry 4.0 tends to regard human beings as entities standing outside the closed loop of the automation-based industrial environment, Industry 5.0 takes into account that, in matters that call for flexibility, imagination, and especially ethical considerations, people will remain inevitable. This new phase of industrial development aims not only the increase productivity rates but also the improve the potential of human actors in creating innovative proposals that are socially and environmentally responsible in their impact on guiding industries toward sustainability, especially within a circular economy model (Van Erp et al., 2024).

The differences between Industry 4.0 and 5.0 are also expressed in the purpose of attention. Industry 4.0 sought automation and optimization of processes while Industry 5.0 is the adoption and implementation of top technologies with a focus on human factors. It seeks to help industries adapt to societal importance, such as making better use of resources, fair treatment of workers, and minimizing wastage. Industry 5.0, therefore, signifies the combined attempt of creative integration of human capabilities with the help of artificial intelligence to move from the trend of mass production to slow and sustainable development of goods and services that are economically significant but socially and environmentally responsible. This transformation makes Industry 5.0 especially topical in overcoming the drawbacks of the linear economic model that is dominant at the present stage and aims at the transition to a circular economy that is less resource-consuming and less destructive to the environment.

2.3 Core Technologies of Industry 5.0

At the heart of the processes in Industry 5.0 are a series of technologies and improvements in a circular economy that are interconnected. Artificial intelligence is an excellent tool to analyze big data; therefore, companies can use the result to make optimal decisions to reduce the costs of resources. In a circular economy, AI can help companies better determine where resources are needed most and when to cut down on excess material that could otherwise go unused. AI systems can also help maximize product life cycle by achieving predictive maintenance of machinery, and general equipment to avoid hitches hence prolonging the maximum useful life of assets.

Robotics another core technology enhances AI in the sense that it performs operations that are repetitive such as recycling and production. Robotics enables the timely disassembly of products and enables recovery of valuable material from products at their disposal stage. In smart factories people and robots collaborate to eliminate waste and enhance the efficiency of production, minimizing errors and providing higher accuracy of using resources, which also contributes to the successful implementation of circular economy strategies. In this aspect, Industry 5.0 combines humans and machines to create solutions that improve efficiency cohesively with sustainable resource use in industries (Arcot, 2023).

Digital twin is also another core technology in the industry 5.0 which is a virtual model of physical assets or processes. Manufacturing digital twins provides real-time modeling of products at every stage of the product life cycle, which helps companies avoid problems, minimize the use of resources, and dispose of goods efficiently. Having accurate information on how individual products are used and how they deplete or degrade over time helps firms to create products that are more robust, easier to rectify, and accommodating to circular economy strategies. Applying digital twins in smart manufacturing enables industries to attain enhanced nearness in the

management of resource use, energy intake, and wastage, to the strategic objectives of the circular economy.

It is also transformative for the circular economy because it acts as a transparency and traceability tool in organizing supply chains. Blockchain, which is in Industry 5.0, allows organizations to follow the products and materials' evolution at the different stages of the supply chain, from purchasing the raw materials or resources to their disposal as waste or recycling. By implementing transparency about the type of material to be used and the way the recyclable material would be processed, the company is well placed to tackle the problem of waste material while promoting recycling. In addition, blockchain technology contributes to the validation of the stated sustainability, letting the consumers and stakeholders be assured and confident that products produced and recycled undergo a circular economy system. Furthermore, it contributes to decentralization that lets industries work more cooperatively, sharing data in a secure way not only inside the industry but also with related ones and making all members of industrial chains contribute to building up a more sustainable environment for industries.

2.4 Smart Manufacturing and Sustainable Supply Chains

Industry 5.0 is a reflection of smart manufacturing that closely relates to the CE through the way products are built and resources are managed. Smart factories are computerized manufacturing facilities the basic components of which are integrated through AI and IoT that efficiently manage resource consumption cutting excess costs such as waste, while also improving product quality. Such systems enable organizations to manage manufacturing processes in real time thereby creating awareness of misuse of resources or energy and attending to them before they occur. Smart manufacturing combines the creative Components of human inputs with the technical accuracy of machine outputs thereby enabling a flexible manufacturing line where products can be made in such a way as to respond to certain sustainability parameters such as the generation of fewer carbon footprints, use of more sustainable raw materials and to other such like parameters.

Through sustainable consumption and production patterns, smart manufacturing supports the utilization of more circular resources and a reduction in the uptake of non-circular materials. It lets us devise items that are meant to be used, repaired, and recycled for as long as possible—that are solidly based on the principles of the circular economy. In addition, smart manufacturing promotes product-service systems where products are handed to consumers for use on a long-term rental basis, which can be recalled for repair, redesign, or recycling rather than being disposed of. This allows for minimal accumulation of raw materials, and discourages manufacturers from creating end products that cannot be recycled, repaired, or upgraded—highlighting the importance of the circular economy (Shah et al., 2024).

Another facet of Appell's worldview championed by Industry 5.0 is sustainable supply chains as part of the circular economy. Sustainable supply chain management

focuses on the reduction of the environmental cost of sustaining a supply chain by improving efficiency in the utilization of resources through the reduction of emissions, obtaining materials from sustainable sources, and ensuring that products are made in the right manner. In Industry 5.0, AI and blockchain are adopted to partially improve the performance and transparency of supply chains where it is possible to investigate how the materials are sourced, used, and recycled. This way companies can decide on the type of suppliers they select and have the confidence that the partner follows sustainable practices. Also, they have shown how AI can be used to substantially minimize fuel use and emissions by determining the best transit routes.

3 Synergies Between Circular Economy and Industry 5.0

Industry 5.0 is to change industries with circular economy models in the future, especially in manufacturing and waste recycling industries and supply chain businesses. Industry 5.0, in contrast to the previous industry versions, is based on the concept of people and automation, AI, and Deep learning mixed with the ethic of intelligence. This synergy in the manufacturing sector is revolutionizing production by successfully integrating sustainable design practices to minimize the use of resources and recycling. Industry 5.0 provides the opportunity for more personalized production hence eliminating the problem of overproduction and wastes. This is even more significant in circular economy strategies where the object is to delay or possibly eliminate as many outputs as possible. Integrating human beings more into the decision-making loop with intelligent machines will help manufacturers come up with devices that are more robust, easy to disassemble for repair and recycling, and meet the requirements of the closed-loop system (Barata & Kayser, 2024).

In the context of waste management Industry 5.0 offers innovation potential driven through the capability of sorting, collecting, and recycling processes. Symbiosis with machines in this sector may result in the creation of new methods for waste management and opportunities to extract the most amount of material possible back into the production cycle. The opportunities also include the ability to dynamically determine efficient routes for waste collection or predict the amount and types of waste produced to recommend the best steps for increasing the recycling ratio. It also creates an avenue through which industries can obtain recycled materials for use in their production, fulfilling some objectives of the circular economy as well. Moreover, blockchain can improve transparency in waste management through the use of a distributed ledger with records on where waste comes from, the processes it undergoes, and where it ends up.

Industry 5.0 will be greatly beneficial for both linear and circular economy systems and supply chains that are pivotal to both. The global supply of products tends to involve a long sequence and becomes very cumbersome to determine the origins of their production, processing, and usage, and this makes it very inefficient and wasteful. Suppliers connected in Industry 5.0 and utilizing AI and blockchain can track various materials from the extraction phase up to the disposal or recycling

phase. This level of transparency helps companies make the right decisions about the products that they source or produce by first embracing sustainable sourcing and production. It also increases the sharing and utilization of resources for greater overall productivity and minimizes time and stock constraints in the transportation of goods; hence minimizing waste. In this manner, Industry 5.0 adds a novel trying to schemes of circular economy and afford assurance that consistent environmental sustainability has been integrated into the constructions of furtherment completing schemes (Dwivedi et al., 2022).

3.1 Enhancing Recycling, Resource Efficiency, and Waste Management

The technologies of Industry 5.0, including digital twins as well as the use of AI in robotics are predicted to improve considerably the single steps of recycling, the usage of assets, and disposal. A digital twin is a virtual clone of a physical product through which the whole lifecycle of the product is easily monitored from design through utilization to disposal. That is why such an analysis is extremely important at this level for searching and evaluating the potential for increasing the efficiency of using resources and decreasing the number of undesirable outcomes. For example, using digital twins, manufacturers can test how a product will deteriorate or how various materials can be recycled, giving them the necessary information to decide how to build products that are easier to break down and recycle by the end of their life cycle. This also helps in cutting wastage while making sure that material that is used in production can be recycled for use by other industries in the future hence making it easier to follow the circular economy (Asibey et al., 2023).

Just like in manufacturing and assembly lines, Artificial Intelligence can help with recycling processes and it is augmented by robots. In single-stream recycling, in particular, items such as glass, plastic aluminum, and paper contaminate themselves and are hard to sort out. Nonetheless, AI-based robots can observe materials and sort the items with very high efficiency and prevent the loss of valuable items in the recycling process. Such robots also can have the ability to sort different types of plastics, metals, and other kinds of materials, thus making it easier and unlikely for contaminants to mix with other materials. This results in higher recycling rates and also assumes that the recycled materials come to the required quality required in manufacturing new products. Moreover, these AI systems can learn to evaluate a large amount of data relative to a company's recycling and disposals, allowing organizations to fine-tune their processes and cut back on waste productively.

In the aspect of Resource Wing, the Industry 5.0 technologies enable organizations to optimize the resources they already have instead of using raw materials. For instance, AI can be used in the determination of production schedules for the usage of resources so that the material is used properly. It can also assist in cutting down unnecessary production, preventing wastage, and recycling any used materials

as much as is feasibly possible. Furthermore, human input enables those creative loops to help advance these processes more resourcefully, for instance by coming up with new strategies for the use of resources like creating objects that are easy to disassemble and reuse or arriving at a more distinctive way of recycling the most intricate materials. Such a level of resource intensity is crucial for achieving the objective of the circular economy as it aims to keep the product, parts, components of the product, and materials in the economic and technological processes as long as possible without entering the level of waste.

3.2 Real-World Examples of Circular Economy Boosted by Industry 5.0

A few examples of Industry 5.0 technologies can show how those technologies help to improve circular economy activities, especially in the EU and Japan. The EU has been at the forefront of the advancement and implementation of the circular economy, and advancement in Industry 5.0 enhanced this. At the EU level, the Circular Economy Action Plan incorporates into supply chain digital technologies, including Artificial Intelligence and blockchain (Javaid et al., 2024). This has also facilitated the understanding of the life cycle of products purchased by some firms about procurement of sustainable materials to use in their production and disposing of those that are redundant in the market after their useful life has expired. Furthermore, some European corporations have adopted smart manufacturing industries that would allow them to create products that are easy to repair, durable, and can easily be recycled; objectives that are supported by circular economy.

The country, which is at the forefront of introducing smart factories and resource-saving technologies, has also successfully implemented Industry 5.0 into circular economy approaches. In their smart manufacturing, Japanese systems like Toyota and Panasonic have integrated AI and robot use to enhance their use of resources and avoid waste. They have also adopted product-based services, where consumers pay to use products as they have adopted the option of being able to make returns, rework, and recycle these products once their leases run out. This not only minimizes the need for procurement of new resources but also maximizes the utilization rates of resources for as long as is feasible.

Several companies in the United States have also adopted Industry 5.0 technology into their circular economy business ventures. For instance, the firm, Apple has incorporated the use of enhanced Robot technology in its recycling units for the intended purpose of repurposing various components of the retired gadgets. These robots can take apart these gadgets with a lot of efficiency; anything in particular such as rare metals will be reused in creating more of these devices. Such a level of accuracy is hard to provide by applying conventional recycling technologies; nevertheless, using AI-based robotic technologies Apple managed to enhance the corresponding recycling coefficients and, in turn, minimizing the negative impact on the surrounding environment (Kannan et al., 2024).

By depicting how these advanced technologies can be applied to facilitate a circular economy, these case studies provide a perspective on the potential types of uses for Industry 5.0 systems in making the industrial world cyclical to create sustainable end products, services, and processes. With time these technologies will have a broader role in contributing to the enhancement of sustainability practices so that developing industries do not compromise the sustainability of the environment.

4 Challenges and Risks of Industry 5.0 in Circular Economy

The major challenge, however, is the king's need to invest large capital to actualize Industry 5.0 for circular economy improvement, specifically in the two sub-disciplines of smart production and artificial intelligence. The technologies of Industry 5.0 like advanced robotics, digital twin technology, and AI-based analytics require a significant level of initial investment for the creation, deployment, and upgrade of facilities. These costs can be quite high, particularly for SMEs, which generally cannot afford to allocate the sort of capital that large companies can. The advantages of adopting these technologies for the enhancement of resource usage efficiencies, minimization of waste generation, and relatively lower operational costs are however apparent but the issue of high initial capital expenditure is a challenge to most businesses. For organizations in developing economies, it is even worse because they can barely access cheap funding and there is no incentive or government policies that seek to encourage the use of Industry 5.0 technologies shown in Fig. 3 (Hassan et al., 2024).

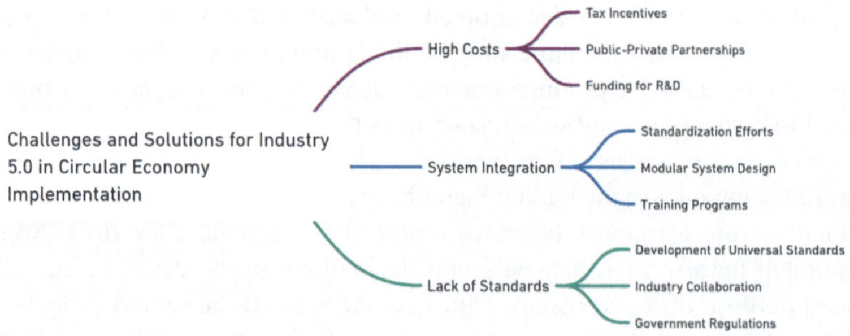

Fig. 3 Challenges and solutions for implementing Industry 5.0 in circular economy: Key challenges such as high costs, system integration problems, and lack of international standards can be addressed through targeted solutions like tax incentives, public-private partnerships, and training programs

However, apart from the cost of integrating the Industry 5.0 technologies, the business incurs costs of training employees and developing the skills needed for interaction with the advanced machines. This goes a long way in deepening the already existing financial costs on firms especially in such areas where market access for technical education and training programs could be a challenge. Furthermore, is the fact that technology evolves quite quickly, which requires firms to regularly update their systems, resulting in perpetual costs that can be expensive to the financial capacity of any entity. This poses chances that will see those firms who are well endowed financially reap big out of Industry 5.0 while firms in the developing regions or small firms will be left behind adding to the already existing disparities in the economy.

4.1 System Integration Problems and Technology Gaps

The application of Industry 5.0 technologies to the existing circular economy system also poses unique challenges mainly because of the compatibility of the different systems and the level of technology development. A lot of organizations including those operating in emerging markets are using relics that cannot communicate with several innovative technologies under Industry 5.0. The integration of AI, robotics, and IoT into these legacy systems is a challenging and costly endeavor that normally entails the complete redevelopment of the original system. In other cases where modern components happen to be in place in organizations, the integration of the sub-components of the Value Chain, the humans, the machines, and its components presents a challenge. Inadequate integration of old and new technologies in many organizations and industries can create problems such as efficiency losses and disconnected data, which nullify the value of Industry 5.0 (Negi et al., 2024).

Technology gaps are also a big challenge especially because in many developing countries, they may not have access to some of these sophisticated digital technologies. In most firms, the strategic direction, necessary skills, or resources to enable the leverage technology systems of Industry 5.0 are not available. This digital divide not only hinders the transformation of these economies to the more circular types of economies but also threatens to exacerbate the existing gap between the developed and the developing nations as far as technological and economic growth is concerned. However, there is a question that arises from Industry 5.0 and it relies on technologies that use data such as artificial Intelligence and digital twins and the question is the availability and reliability of the data used. When it comes to collecting data and managing data systems, it may prove challenging for regions and businesses to amass the necessary data to properly integrate an AI plan or product. With quality data missing in the equation, maximizing the promise that AI and the core concepts of Industry 5.0 have about optimizing resource use and addressing waste is difficult.

4.2 Lack of International Standards and Global Cooperation

The last major challenge regarded to use of Industry 5.0 for the support of the circular economy is the lack of standardized and coordinated policies across the countries. In the case of Industry 5.0, however, there is no globally agreed set of principles and norms, or established regulatory frameworks in many countries, including the European Union (EU), to support circular economy business models or to define rules regulating the application of digital technologies in industrial settings. Such lack of parallelism reduces policy certainty for cross-jurisdictional business entities because they will be in a precarious position of conducting business where policies may differ so greatly between countries. For example, a company may encounter different legal and regulatory expectations of data protection, AI application, or waste disposal, across its different markets, which can complicate the ability of the firm to execute a coherent Industry 5.0 strategy on a global level.

It also hinders progress in making circularity global where international collaboration is much needed to address circular economy challenges. Circuit business aspects are intrinsically transnational in their character as material, goods, and wastes move across borders in supply/value chains. The lack of standard rules makes it challenging to monitor these flows to ensure they are sustainable. Also, there is a lack of international collaboration on Industry 5.0 technologies which can cause possible problems of equitable access to the said tools. Although with the potentials and the legislation in place developed countries can move to Industry 5.0, the developing countries may be handicapped and would not engage fully in the emerging circular economy. Such disparities can slow or regress economic growth while deepening inequalities of Industry 5.0 and its impact on the ecology.

In addition, no internationally uniform standard can guarantee the ethical use of Industry 5.0 technologies. A major challenge of AI and automation systems is the fact that many are tied to algorithms that call for data; this consequently has raised privacy and data security as well as bias-related issues. Lacking internationally recognized standards for the ethical application of such technologies, companies may incur legal and image-related consequences in particular territories where regulations tend to be either permissive or too stringent. The current state of regulation provides increasing risk and uncertainty to industrial corporations adopting Industry 5.0 technologies, which could slow down the prospects of Industry 5.0 as the key driver toward a circular economy (Klein & Knight, 2005).

5 Policy Recommendations and Solutions

Perhaps one of the most influential strategies for promoting Industry 5.0 technologies in circular economy practice is the use of the tax credit. These incentives would make the costs of adopting the advanced technologies a bit lower to the businesses, especially the SMEs that may wish to adopt such technologies as AI, robotics, and digital twins, among others. This is true because, through tax credit policies where governments set incentives for additional spending on the research and development of these technologies, more people and businesses can be embraced toward embracing the industry 5.0 practices. This approach has been tried in several industries around the world where the government offers tax exemptions for investment in renewable energy or green infrastructure. Using the same concept in Industry 5.0 would not only encourage investment into efficient innovation technologies but also enhance the possibility for industries to adopt a circular economy.

These tax credit schemes could be extended further aimed tailored to support native innovations by addressing the peculiarities of resource utilization and waste management typical for developing countries, particularly India. For example, a sliding scale approach to tax credits could then be used, where those businesses assisted who can best prove innovation in recycling solutions or efficient use of resources in traditionally inefficient sectors including apparel or farming. This would improve the implementation of circular economy business practices and fund the relevant technology without undue economic pressure. Furthermore, tax credits could be provided to multinationals doing business in the developing world to encourage them to introduce their Industry 5.0 technologies to areas where they will be most effective (Garcia-Muiña et al., 2019).

5.1 Public-Private Partnerships in Technology Deployment

The transition to Industry 5.0 and, inter alia, the application of circular economy solutions is impossible without the cooperation of both sectors. There is great value in PPPs especially when it comes to implementing policies and creating technological solutions in the context of limited resources in many regions. There is an argument that governments have to provide the essential platforms for establishing advanced technologies, setting the policies, and standards, and providing the necessary financial incentives, while private sectors bring in the experience, innovation, and funds to implement those technologies. The interaction between these two sectors can then help to pick up the pace of the shift toward a circular economy.

That is why PPPs can influence the development of smart manufacturing sites and innovation centers related to Industry 5.0. Ministries can bring funding and construction, while corporations supply the innovation to design top-notch centers for integrating man and machine in recycling, waste management, and optimal manufacturing. These centers could act as a reference for other industries on how technologically advanced can be implemented in the circular economy. However, PPPs could also encourage the establishment of innovation grants or the co-financing mechanism whereby the government and private companies contribute toward the development of technologies designed to advance circular economy processes. In India, for instance, such collaboration might look toward improving the plurality of recycling for plastic wastes or boosting sustainable agriculture through the use of Artificial Intelligence-based precision farming methods.

Furthermore, PPPs help the governments support startups and small business entities that are unable to invest in Industry 5.0 technologies. The technical collaboration and funding by public-private partnerships ensure that these advanced technologies can be tested by small businesses without incurring higher risks. This cooperation approach makes certain that opportunities for Industry 5.0 and circular economy models are available not merely for multinational companies, which can afford a significant number of investments.

5.2 Literacy and Skill Development Programs for Workers

The enhancement of Industry 5.0 technologies in circular economy strategies implies an increased need to Bamboo Train raise the competency of workers, especially in developing countries. Since social cooperation between humans and machines is a key characteristic of Industry 5.0, people have to be prepared to work with AI, robotics, and other types of digital technologies. This calls for effective literacy, and skill development processes that allow the personnel to gain both theoretical and practical knowledge of circular economy.

These programs are particularly important in developing countries like India. Employees in manufacturing companies, farmers in the agricultural sector, and waste collectors in shown that they may not be in a position to exploit hi-tech Industry 5.0 technologies. Governments and private entities need to come up with training needs that may be suitable and affordable to such workers. This could involve providing supported government vocational programs that can provide certification in Industry 5.0 relevant competencies such as the operation of AI-based machinery or the running of smart supply systems. In addition, the principles of circular economy should become the foundation of changes in educational programs so that future employees create sustainable businesses from scratch (Poláková et al., 2023).

Besides the formal school-based education and training, there is a requirement for experience-based learning resulting from new technologies that are commonly in vogue. Every company that implements Industry 5.0 solutions must ensure their employees are trained in how to work with machines and in circular economy principles. Similar to incentives, government subsidies, or tax credits for companies willing to finance worker training programs, would encourage corporations to focus on it. When the workers are empowered with the necessary skills, it becomes possible for businesses to leverage the Industry 5.0 technologies together with the idea of the circular economy.

Besides, these measures for persons' skill development should not be restricted to technical training only. Employees are required to be aware of the underlying goals of CE in terms of the resource hub concept, resource-limited environment, and sustainable processes of production and using goods. The authors explain how increased knowledge of the circular economy can help workers to change the current industrial processes much more effectively. For example, training programs could be relevant to India and could indicate sectors such as textile and agriculture that have for a long time dominated the destructive usage of resources. For these reasons, the eighth pattern of Industry 5.0 involves teaching the workers in these sectors how to use technology to bring innovation to better practices and make a broader impact on the circular economy.

6 Future Outlook: The Role of Industry 5.0 in Achieving Global Sustainability

This stacked bar graph shown in Fig. 4 visualizes the anticipated growth in market size across various Industry 5.0 technologies. The chart highlights significant expansions in each technology sector, illustrating a compounded annual growth rate (CAGR) of 31.3%, and a projected increase from USD 50.8 billion in 2023 to USD 773.6 billion by 2033 (Zare et al., 2024).

Industry 5.0 will likely be instrumental in advancing circular economy initiatives around the world as industries start aiming for zero-waste manufacturing. In comparison to the four previous industrial revolutions, Industry 5.0 focuses on sustainability by capitalizing on the strengths of human creativity and adding AI, Robotics, and Digital Twins among other things to the process. This integration makes it possible for industries to work toward achieving a circular economy where products are made with design-for-recycling characteristics and made on-demand to cause minimal waste (Preud'homme, 2024). CO delivery through smart manufacturing systems of products eliminates the utilization of resources in the understanding and fabricating of products since products are made on-demand, right from the extraction stage to

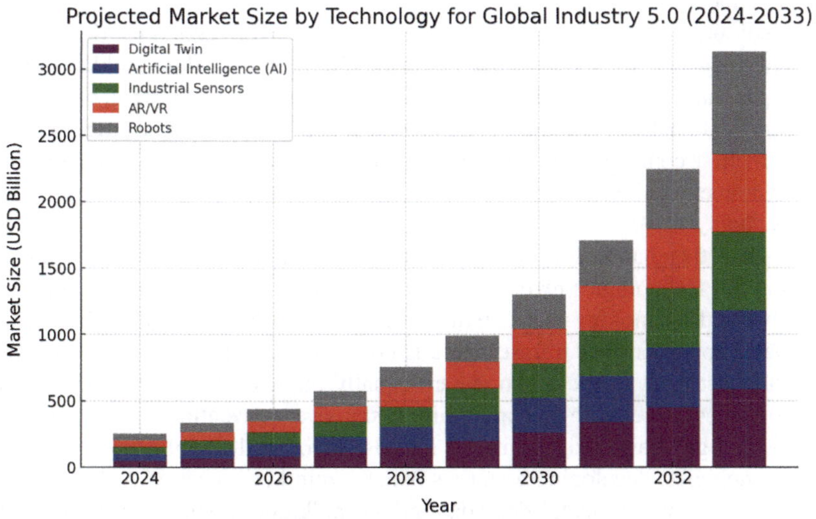

Fig. 4 Projected market size by technology for global Industry 5.0 (2024–2033)

the disposal stage of products. Industry 5.0 is built on human-machine collaboration and supports fair employment, sustainable procurement, and incorporates even the use of artificial intelligence into the pursuit of industrial development to meet global welfare and sustainability objectives.

Industry 5.0 enables an improvement in the circular economy, an approach for keeping resources in use, recovering, and regenerating them. Incorporating digital tools and artificial intelligence in industries can help these industries develop better recycling systems that are more efficient and more transparent to help avoid additional raw material production, and to minimize waste production. Industry 5.0 forward-thinking approach to decentralize human creativity also applies in societal problems, such as the power of inventive recycling methods and environmentally friendly materials, toward a circular economy on the global level. Since the drive toward the formulation of Industry 5.0 practices is gaining popularity among countries and companies globally, achieving the goal of a "zero waste" system collectively will be more easily achievable hence placing Industry 5.0 at the center of global sustainability efforts in the future decades (Ghobakhloo et al., 2022).

6.1 Long-Term Implications for Developed and Developing Nations

The consequences of Industry 5.0 in the long term vary depending on the country: developed countries and developing ones have different prospects and pain points, barriers, and opportunities. Industry 5.0 can be implemented more easily in developed countries because their industries have already established the necessary base for implementing new technologies such as AI, robotics, or IoT. Industry 5.0 can be mostly utilized by developed nations as a tool to foster economic development without a negative impact on the environment. For instance, industries like Germany and Japan that boast of smart manufacturing and green technology can easily adopt Industry 5.0 to balance resources, minimize wastage, and develop an efficient supply chain. In these contexts, Industry 5.0 may provide the avenue by which organizations will continue to enjoy competitive advantages while at the same time aligning with ENES goals.

On the other hand, developing nations may find it more difficult to join the industry 5.0 technologies revolution because, for the most part, they cannot access the advanced technology that is requisite for this revolution, they might not have adequate capital, or their infrastructure is not well developed. However, the potential gains accruing to these nations are equally impressive. Industry 5.0 presents developing countries with an opportunity to skip over conventional industrialization models and go straight to sustainable models. Through the application of Industry 5.0 technologies, developing countries benefit from key concerns such as scarcity of resources, environmental degradation, and pollution. Such applications as the application of AI and robotics in the farming sector could enhance the efficiency of the usage of resources and the prosecution of activities that are prejudicial to the carrying capacity of the environment, smart manufacturing could assist countries adopt industries that are sustainable economically and environmentally (Pachouri et al., 2023).

However, it is also certain that, the high diffusion of Industry 5.0 in developing nations will call for a significant push in the form of PPP, global compacts, and development on education and infrastructural development. Lack of these key supports would mean that there is a likelihood for the digital gap between the developed and developing nations to extend, thus leading to an increased gap of nations to join the global bandwagon toward the shift to the circular economy. For the policymakers and global organizations, the big question will be, how can Industry 5.0 fair take society forward, and promote sustainable development around the world? Widening the gap between the developed and the developing world based on economic and technological systems.

6.2 The Role of Innovation in Overcoming Future Challenges

As much as the future of sustainable Practice will depend on Industry 5.0, Innovation will always have a major role in ensuring that it meets the demands of the people. Companies strive toward achieving better alternatives, and the emerging breakthroughs in fields for instance AI, robotic engineering, and renewable resources are pivotal in attaining better solutions to current drawbacks. For example, the improvement of asset optimization subsectors might give rise to even better resource utilization systems, including control of materials and energy, which could not even be foreseen at the present stage. Likewise, technological advancements in renewable energy like increased efficiency of solar panels, or innovations in battery storage will decrease industries' dependence on carbon and other fossil fuels, and power the smart manufacturing of the new world.

Another area that needs further innovation is recycling. Modern recycling processes however are not effective; most of the material ends up in the loop without being recycled or is recycled in forms that reduce their quality. The next-generation materials science breakthroughs, like the discovery of new biodegradable plastics, or recyclable materials of a cyclical nature, would greatly cut down waste production. Robotics and AI shall also feature in the enhancement of recycling, as this means that materials can be sorted better and the outcomes are better. When these innovations appear, they will assist industries in enhancing manufacturing sustainability and minimizing wastage on an international level.

In addition, there is a need to enhance the international standards for Industry 5.0 and support circular economy practices. Nowadays, there is no single global integration as well as standard rules that could foster the advent of Industry 5.0 across industries. Thus, through the setting up of a global structure of guidelines that compel innovation while at the same time not harming the environment, international relations can be able to have a system that has easier means of sharing as well as adopting technological breakthroughs. This will particularly be important in an attempt to avail the needed technologies for the developing nations hence facilitating their complete integration in the circular economy systems (Foster & Yaoyuneyong, 2016).

7 Conclusion

Thus, the introduction of Industry 5.0 into circular economy models holds great prospects that will foster the discontinuity of the current linear economy model which is rigid to the environment and the economy. Compared to the previous industrial revolutions, Industry 5.0 improves resource efficiency, optimizing waste and recycling processes through the combination of humans and technologies such as AI, robotics, digital twins, and blockchain. These technologies make the manufacturing process more intelligent, the supply chain clearer and recycling methods more efficient which supports the global sustainable development initiative plan.

But it is also important to note that there are large challenges still, especially for the member countries in the developing world. The challenges that could further widen the developed countries and developing countries gap are; the cost of installing 5.0 industry technologies, compatibility issues with the new systems, and lack of standardization across the international market. To eliminate these barriers, policy efforts like tax credit initiatives, partnerships, and educating workers about sustainability are crucial for all developing nations to reap the sustainability solutions that define Industry 5.0.

Finally, despite all the challenges, Industry 5.0 offers the global marketplace the best prospects for transformation in the direction of a zero-waste circular economy. This current solution and focus on the development of new Industry 5.0 can serve as a foundation for sustainability in both developed and developing states as it may indeed be possible to invent a way for economic growth and for saving the environment together.

References

Arcot, S. (2023). Industry 5.0 and Society 5.0. In *Industrial Revolution and Metaverse: Industry 5.0 (Chapter 1).* QUING Publications. https://doi.org/10.54368/qpbc.2023.1.1

Asibey, M. O., Abdulai, A.-S. J., Dosu, B. J., &Tornyeviadzi, P. (2023). Waste management and resource recycling in the developing world (pp. 309–323). *In Waste management and resource recycling in the developing world.* Elsevier.

Barata, J., & Kayser, I. (2024). How will the digital twin shape the future of Industry 5.0? *Technovation, 134,* Article 103025. https://doi.org/10.1016/j.technovation.2024.103025

Dwivedi, A., Agrawal, D., & Jha, A. (2022). Studying the interactions among Industry 5.0 and circular supply chain: Towards attaining sustainable development. *Computers & Industrial Engineering, 176*, Article 108927. https://doi.org/10.1016/j.cie.2022.108927

Foster, J., & Yaoyuneyong, G. (2016). Teaching innovation: Equipping students to overcome real-world challenges. *Higher Education Pedagogies, 1*(1), 42–56.

Garcia-Muiña, F. E., González-Sánchez, R., Ferrari, A. M., Volpi, L., Pini, M., Siligardi, C., & Settembre-Blundo, D. (2019). Identifying the equilibrium point between sustainability goals and circular economy practices in an Industry 4.0 manufacturing context using eco-design. *Social Sciences, 8*(8), 241.

Ghobakhloo, M., Iranmanesh, M., Mubarak, M. F., Mubarik, M., Rejeb, A., & Nilashi, M. (2022). Identifying industry 5.0 contributions to sustainable development: A strategy roadmap for delivering sustainability values. *Sustainable Production and Consumption, 33*, 716–737. https://doi.org/10.1016/j.spc.2022.08.003

Hassan, M. A., Zardari, S., Farooq, M. U., Alansari, M. M., & Nagro, S. A. (2024). Systematic analysis of risks in Industry 5.0 architecture. *Applied Sciences, 14*(1466). https://doi.org/10.3390/app14041466

He, D., Zainol, R., & Azali, N. S. (2024). Navigating challenges in the sustainable development of urban brownfields: A PLS path modeling perspective. *Ain Shams Engineering Journal, 103*, Article 103002. https://doi.org/10.1016/j.asej.2024.103002

Javaid, M., Haleem, A., Khan, I. H., Singh, R. P., & Khan, A. A. (2024). Industry 4.0 and circular economy for bolstering healthcare sector: A comprehensive view on challenges, implementation, and futuristic aspects. *Biomedical Analysis, 1*(2), 174–198. https://doi.org/10.1016/j.bioana.2024.06.001

Kannan, D., Shamekhi Amiri, A., TarighiShaayesteh, M., Khalili Nasr, A., & Mina, H. (2024). Unveiling barriers to the integration of blockchain-based circular economy and Industry 5.0 in manufacturing industries: A strategic prioritization approach. *Business Strategy and the Environment*. https://doi.org/10.1002/bse.3886

Klein, K. J., & Knight, A. P. (2005). Innovation implementation: Overcoming the challenge. *Current Directions in Psychological Science, 14*(5), 243–246. https://doi.org/10.1111/j.0963-7214.2005.00373.x

Malik, S., Muhammad, K., & Waheed, Y. (2024). Artificial intelligence and industrial applications—A revolution in modern industries. *Ain Shams Engineering Journal, 15*(9), Article 102886. https://doi.org/10.1016/j.asej.2024.102886

Negi, P., Singh, R., Gehlot, A., et al. (2024). Specific soft computing strategies for the digitalization of infrastructure and its sustainability: A comprehensive analysis. *Archives of Computational Methods in Engineering, 31*, 1341–1362. https://doi.org/10.1007/s11831-023-10018-x

Pachouri, V., Kathuria, S., Gehlot, A., Negi, P., Thakur, G., & Chaudhary, M. (2023). Sustainable environment with technological intervention: Benefits and challenges. *2023 3rd International Conference on Pervasive Computing and Social Networking (ICPCSN)* (pp. 1609–1614), Salem, India. https://doi.org/10.1109/ICPCSN58827.2023.00269

Poláková, M., Horváthová Suleimanová, J., Madzík, P., Copuš, L., Molnárová, I., &Polednová, J. (2023). Soft skills and their importance in the labor market under the conditions of Industry 5.0. *Heliyon, 9*(8), Article e18670. https://doi.org/10.1016/j.heliyon.2023.e18670

Preud'homme, G. (2024, August 1). *Industry 5.0: A sustainable revolution at the heart of tomorrow's plants*. Proaction International Blog. https://blog.proactioninternational.com/en/industry-50-sustainability-revolutionizing-plants

Rame, R., Purwanto, P., & Sudarno, S. (2024). Industry 5.0 and sustainability: An overview of emerging trends and challenges for a green future. *Innovation and Green Development, 3*(4), Article 100173. https://doi.org/10.1016/j.igd.2024.100173

Sariatli, F. (2017). Linear economy versus circular economy: A comparative and analyzer study for optimization of economy for sustainability. *Visegrad Journal on Bioeconomy and Sustainable Development, 6*(1), 31–34.

Shah, S., Nautiyal, H., Gugliani, G., Kumar, A., Namboodri, T., & Singla, Y. K. (Eds.). (2024). *Sustainability in smart manufacturing: Trends, scope, and challenges* (1st ed.). CRC Press (Taylor & Francis). https://doi.org/10.1201/9781003467496

Song, J., Yuan, Y., Chang, K., Xu, B., Xuan, J., & Pang, W. (2024). Exploring public attention in the circular economy through topic modeling with twin hyperparameter optimization. *Energy and AI, 100*, Article 100433. https://doi.org/10.1016/j.egyai.2024.100433

Van Erp, T., Carvalho, N. G. P., Gerolamo, M. C., Gonçalves, R., Rytter, N. G. M., & Gladysz, B. (2024). Industry 5.0: A new strategy framework for sustainability management and beyond. *Journal of Cleaner Production, 461*, 142271.

Zare, L., Ben Ali, M., & Rauch, E. (2024). Industry 5.0 and SMEs future work competency fields: A literature review. In *International Symposium on Industrial Engineering and Automation* (pp. 200–208). Springer.

A Study on Impact of Social Media Influencers on Impulse Buying of Branded Cosmetic Products in India

Neha Gupta, Rita Prusty, Smita Dash, Sathya Swaroop Debasish, and Mohammad Helal Hossain

Abstract This research aims to test the consequences of endorsements from social media influencers (SMIs)/Internet Celebrities (IC) on consumer trust in marketing campaigns, and how this in turn, impacts impulse purchases of branded cosmetics SMIs. The research focuses on five key elements that encourage impulsive purchases: SMIs popularity, audience identification with the SMIs, admiration for the SMIs, perceived social distance, and the perceived compatibility between the cosmetics brand and the influencer. A survey of 97 internet celebrity or influencer followers in India was conducted, revealing that trust is essential for influencing impulse buying behavior. Both identification with the influencer and perceived fit greatly increases impulsive purchases. Conversely, a great deal of social distance can erode the relationship between trust and impulsive purchases. The study suggests that branded cosmetics should collaborate not only with popular but also relevant SMIs with whom people relate more. SMIs that project a modest and relatable image are more likely to build consumer trust, thereby amplifying the effectiveness of their endorsements. The primary findings of this research have been that trust plays a crucial role in impulse buying of branded cosmetic products. This research also demonstrates that identification and adoration with SMIS impacts impulse buying as more people go toward their own personality closer to SMIS and are ready to buy products when recommended by them.

Keywords Impulsive purchases · Social media influencer · Trust · Branded cosmetics · Consumer behavior

---

N. Gupta (✉) · R. Prusty
IBCS SOA Deemed to Be University, Bhubaneswar, Odisha, India
e-mail: nehagupta@soa.ac.in

S. Dash
School of Management, Driems University, Bhubaneswar, Odisha, India

S. S. Debasish
Department of Business Administration, Utkal University, Bhubaneswar, Odisha, India

M. H. Hossain
CEO & MD, Daystar HRM, Market Research Centre, Dhaka, Bangladesh

R. Kumar and S. N. Mohanty (eds.), *Sustainable Economy Models in the Age of Industry 5.0*, https://doi.org/10.1007/978-981-96-4104-8_16

1 Introduction

The sudden increase in popularity of social networking sites like Instagram, Facebook (FB), Twitter, and Pinterest recently changes customer buying behavior, enabling businesses adapt to support marketing and commercial activities on the said platforms, a phenomenon called social media shopping (Liang & Turban, 2011; Xiang et al., 2016). As such, because social media is capable of leveraging Web 3.0 and in order to encourage engagement with the material created by users on those, it helps online communities and marketplaces when consumers make decisions and actually make purchases (Huang & Benyoucef, 2013; Zhang et al., 2014). Social media's rise has changed how marketers use advertising since it allows for interactive, two-way engagement with customers, which increases the impact of peer endorsement. The social media influencer has become a new type of endorsement. By showcasing their life and abilities on social media and learning how to influence their fan base, some customers become into online opinion leaders (Chen et al., 2021). Influencers on social media are online producers who specialize in a certain market. They get to work with their favorite businesses, create and interact with an audience of followers, and publish stuff they are passionate about. A content creator is always producing and publishing content that will demonstrate competence in the area and, therefore help build a connection with the consumer. Two ways through which creators can "influence" followers to "try the new thing" include: production of insightful contents and significant engagement with audiences.

Even though, many academicians have focused on social commerce but still fewer researches are conducted on impulsive buying (Busalim & Hussin, 2016). The term "impulse buying" is defined as the intentional or unintentional buying action from the side of the customer, induced by extraneous stimulus. Impulsive buying is defined by outside advertising, unawareness, and expectation (Zhang et al., 2018). The emergence of social economy has been strongly influencing consumer's purchase behavior concerning the percentage of impulse buying. Decisions regarding purchases are generally intuitive. As a person goes about their routine, most of the time, they end up purchasing products impulsively. Impulse buying relates to impulsive customer behavior while making purchases. It is an impulsive, very strong, unplanned, and constant want to purchase anything right away.

The cosmetics business is flourishing and does not seem to be nearing a slowdown because customers everywhere now highly emphasize wellness and self-care. The market has greatly benefited from the increased popularity of most platforms for social media. Apart from the potential to dramatically affect specific demographics, these websites foster demand for beauty products and fill a gap that may not have existed for the makeup companies lately: taking the company directly to the end-user. For some customers, these social media websites are a means of becoming an online opinion leader, or social media influencer, guiding impulsive or deliberate buying behavior in followers. Their parasocial contact increases the consumer impulsive purchase inclination (Xiang et al., 2016). Moreover, SMIs possess more influence and credibility than the typical celebrities, (Djafarova & Rushworth, 2017). Because academics haven't given SMIs much consideration until lately, these superstars with high degrees of attraction and connection have the power to influence impulsive buying. For example, the influencers have gained publicity and have followers up to an enormous population because they are genuine as stated by (Pittman et al., 2016; Sheldon et al., 2016). It is yet unknown how consumers' impulsive purchasing habits relate to digital influencers on social media sites. Because of this, the study intends to help academicians along with practitioners understand how SMIs and contextual variables might lead to impulse purchase behavior. In light of the aforementioned gap, this study evaluated the extent to which SMIs promote impulsive purchases. Here, we narrowed down to the SMIs pages/groups where the users don't only view their own posts but also those of their peers' interactions. They are influencers who belong to an online community and have many followers. Besides that, businesses use SMIs for marketing their products to influence the purchase behavior of consumers (Pemberton, 2017). Based on the research problem above, this study attempted to answer a gap: whether and how contextual social connection and authentic posts by SMIs influence consumers to make impulse purchase of cosmetic when the buyer shows impulse traits.

We gathered data for hypothesis testing from India, a market that is expanding. India's social structure is changing quickly, and excessive spending is driving higher middle-class group income. In addition, social media superstars have surfaced, and companies encourage these celebrities to test out their products and write blog posts in an effort to draw in more clients (Bradri, 2019; Pemberton, 2017). This study is important in a number of ways. Initially, a study framework is put up to determine the effect of number of variables on the impulsive purchase of cosmetics. Next, investigate the most persuasive elements of SMI's endorsement as indicators of customers' propensity for impulsive purchases. Additionally, in this study a conceptual model is put to test that best captures the essence of SMI's endorsement and presents a realistic picture.

Therefore this study's main objective is to identify and examine the primary elements of SMI's support that influence consumer's impulsive purchase. The dimensions of SMI's support which is examined in this study are trust, SMIs' Popularity, Perceived Fit, Identification, Social Distance and Social Media Influencer Adoration and how they impact the impulsive purchase.

2 Literature Review

Using the latent state-trait theory, (Zafar et al., 2021), investigated the effects of polarization of feelings learning through observation, impulse buying propensity, and post-authenticity of celebrities. Outcome of the study showed that the suggested constructions had a major influence on the impulse to purchase. Furthermore, all correlations except negative feelings are moderated by the desire to buy on impulse. Surprisingly, there is little moderating influence from celebrities' honesty.

Fadillah et al. (2021), The research aimed to identify the variables influencing TikTok users' impulsive beauty product purchases. This research employs a hybrid technique (qualitative and quantitative methods) with a judgmental sampling. According to the study's findings, value, pleasure, and model shopping all have an impact on impulsive online purchasing behavior.

Zafar et al., (2021), in his study utilized the LST theory for investigating the effects of content authenticity of celebrities, emotional conflict, learning through observation, and impulsive purchasing tendency. For the data analysis technique, SPSS was used in this study along with the PLS-SEM. The proposed constructions had a major effect on the impulse to purchase. Further, all correlations except negative feelings are moderated by the desire to buy on impulse. Surprisingly, there is little moderating influence from celebrities' honesty. Moreover, the QCA results demonstrate the structures' resilience to the impulse to go for impulsive purchases.

The researchers of this study, (Saselah et al., 2024), conducted the study to see whether beauty influencers and hedonic shopping value have a relationship with impulsive purchases. From this study it was found out that both the independent variables are positively and significantly influencing on the impulsive purchase's variable. Specifically, the hedonic shopping value variable has significantly influenced the impulsive purchase variable, as has also influenced the online opinion leaders in beauty segment variable.

Wangshu et al. (2020) aim to study how influencer marketing activates the impulse buying behavior of the Chinese millennials customers using the research on the social media platform, Instagram. According to the study, when the Chinese millennials are engaging in impulsive buying behavior based on the influence of the influencers, the qualities of the customer, customer emotion, and self-perception all have a role to play but at different levels. Influence of influencers on customers' impulsive buying behavior can be seen in the sense that its characteristic induces deep feelings in customers, and to some extent, the consumer's self-concept plays a special role in managing these feelings.

A sea change has been seen in customer experience and behavior due to social media marketing according to (Pham et al., 2024). Therefore, when exposed to such an atmosphere, people frequently find themselves more prone to impulsive purchase behavior. Previous studies only showed how celebrities are becoming more and more influential on social media and how social engagement and the environment of impulsive purchase are related. The authors discover that the propensity for impulsive purchases is positively impacted because of faith customer places in celebrity and learning which takes place from their posts. Recognizing the significance of engaging and genuine postings from celebrities in shaping customers' buying decisions on social media, this study provides insightful information for those involved in digital celebrity marketing and communication.

Gita et al., (2023) stated in their research that the rise of social media has made celebrity endorsements fashionable and has an impact on business strategy. Even though the effects of impulsive purchase on the economy are widely recognized, a model is created to look at the connection between impulsive buying and online celebrity endorsement. The article investigates the potential relationship between e-WOM and purchasing behavior. The findings show a connection between impulsive purchasing and online celebrity endorsement. Propensity to buy skincare products is strongly associated with impulsive purchasing, with online word-of-mouth acting as a partial mediator. This study uses response theory, social identity, and planned behavior to investigate the effects of celebrity endorsements in cyberspace. It also demonstrates how to effectively find celebrity endorsers to boost customer purchasing intent, which encourages rash purchases.

3 Hypotheses Development

3.1 Impulse Buying

Impulse buying refers to a spontaneous, pleasure-driven purchasing behavior that is more emotional than rational, characterized by excitement rather than cognitive deliberation (Xiao & NSMIsholson, 2013). It often arises when the desire to purchase overwhelms an individual's capacity for self-control. This behavior is linked to the concept of self-discrepancy, where individuals seek to resolve differences between their actual and ideal selves. These individuals may be driven by materialistic values, believing that acquiring goods enhances their self-identity and provides satisfaction (Dittmar & Beattie, 1996). Given that luxury cosmetic products embody materialism and self-image satisfaction, this study focuses on impulse buying within the context of luxury cosmetic.

3.2 Trust

Despite the shift to digital commerce, establishing consumer trust remains a critical factor for success in buyer–seller relationships (Chen et al., 2019; Gawron & Strzelecki, 2021). Mayer et al. (1995), has defined trust as the willingness of one person to follow other person's actions and conduct, regardless of their control, on the assumption that the other would behave as expected. It encompasses beliefs about the integrity, benevolence, and competence of the trusted party (Jarvenpaa, 1998), and is strongly associated with consumer attitudes toward purchasing (Hsu et al., 2013).

Trust can act as a heuristic, lowering the amount of reasoning involved in decision-making, claim (McEvily et al., 2003). Given that impulse buying is driven primarily by emotions, trust is most likely to foster impulsive purchases. Particularly in the context of online transactions, where uncertainty is high, trust becomes even more critical (Gefen et al., 2003). Since impulse buying carries inherent risks, trust helps mitigate the perceived risk and stress associated with impulsive purchases

(Verplanken et al., 2011). Therefore, trust is expected to positively influence impulse purchases.

Hypothesis 1 (H1): Trust Has a favorable impact on impulsive purchase for branded cosmetics.

3.3 SMIs' Popularity

Influencers who have a greater fan base are able to spread positive word of mouth in a broader way as their presence in the social media is very effective (McQuarrie et al., 2013). Their centrality in social media amplifies their reputation, increasing their influence on others. Popular SMIs are often seen as more credible and trustworthy, partly because they stand to lose more if perceived as inauthentic. Numbers with respect to number of followers or the frequency of postings by them serve as basis for consumers to judge the competence and credibility of SMIs. Research by (Edwards et al., 2013), shows that higher social influence metric (e.g., Klout score) enhance perceptions of credibility (Huffaker et al., 2010).

The larger the SMIS's following, the greater their influence on the online community. This effect, known as the bandwagon heuristic, leads consumers to assume that an SMIS's popularity is based on expertise or other valued traits (Aydin et al., 2021, Marwick et al., 2015). The opinions of popular SMIs can thus shape consumer trust by reducing caution and facilitating quick decision-making. As (Wei et al., 2019), demonstrated, social influence has a neurological basis in shaping trust.

Hypothesis 2 (H2): The popularity of SMIs positively influences trust.

3.4 Perceived Fit

Influencers differ in their character attributes and content focus, attracting specific audiences based on these traits. In the context of branded cosmetic, where SMIs are seen as experts, the fit between the SMIS and the brand they endorse is crucial. It is believed that fit between a celebrity's image and the product being advertised leads to more effective messaging to followers (Kamins, 1990). When there is strong alignment between the SMIS's style and the brand's characteristic, this can boost the SMIS's credibility and make the communication more trustworthy.

Hypothesis 3 (H3): Perceived fit positively affects trust.

(Pradhan et al., 2014), showed that the collaboration between a celebrity and a brand positively influences consumer attitudes and reduces perceived risk. In luxury cosmetic, where high prices and exclusivity are key factors, the SMIS's attractiveness and expertise become even more important. If there is a lack of congruence between the SMIS and the brand, consumers may become more skeptical, which could hinder impulse buying behavior (Kapitan & Silvera, 2016).

Hypothesis 4 (H4): Perceived fit positively affects impulse buying in luxury cosmetics.

3.5 *Identification*

Identification refers to the emotional connection consumers feel when they perceive shared values, interests, or characteristic with others. This concept is particularly important in brand communities, where identification fosters trust and loyalty (Lopez et al., 2017) emphasized the role of identification in building trust in a brand community, which can extend to SMIs in social commerce contexts.

Hypothesis 5 (H5): Identification positively affects trust.

Identification also drives consumer behavior by encouraging value transformation, where consumers adopt the values or characteristic of SMIs they admire. This is tied to the social constructionist view that possessions are an extension of self-identity (Dittmar et al., 1996). Customers could be inspired to imitate in order to bridge the distance between their ideal and real selves the SMIs they identify with, particularly through purchasing luxury cosmetic products endorsed by those SMIs.

Hypothesis (H6): Identification positively affects impulse buying in luxury cosmetic.

3.6 *Social Distance*

Social distance refers to the perceived gap between individuals, whether emotional, cognitive, or decisions, influencing their self-control and impulse buying tendencies (Trope et al., 2010). When consumers perceive greater social distance from SMIs, they tend to engage in more abstract, rational thinking, which can lead to stronger self-control and reduced impulse buying (Trope et al., 2000).

Hypothesis 7 (H7): Social distance negatively affects trust.

Moreover, social distance is closely related to interpersonal similarity. Followers are more likely to believe and communicate with individuals they perceive resembles to themselves (Steffes et al., 2009). This similarity fosters trust and facilitates positive attitudes toward the SMIS and the products they endorse (Leonidou et al., 2006).

Hypothesis 8 (H8): Social distance negatively affects impulse buying in luxury cosmetic.

3.7 *Social Media Influencer Adoration*

SMIS adoration reflects a consumer's intense admiration or devotion toward influencers (Chiou et al., 2005). This concept parallels the idea of brand love, where consumers form passionate emotional attachments to a brand, influencing purchase intentions and loyalty (Batra et al., 2012; Palazon et al., 2019). SMIS adoration, driven by emotional attachment, is likely to similarly foster impulse buying behavior.

Consumers who adore SMIs may also exhibit a lower level of self-esteem, leading them to use impulsive purchasing as a coping strategy for negative emotions. Additionally, SMIs' entertaining and hedonic content aligns well with the emotional triggers of impulse buying (Yang et al., 2008). As a result, consumers with a higher propensity for SMIS adoration may be more likely to buy the products recommended by SMI's.

Hypothesis 9 (H9): SMIS adoration positively affects impulse buying in luxury cosmetic.

4 Methodology

4.1 Research Approach

The research constructs were based on previously validated scales from existing literature. A three-item scale whose Cronbach's alpha value is 0.75 and its composite reliability is 0.77 which is modified from (Chen et al., 2014) to gauge the popularity of SMIs. For Perceived fit, a scale adapted from Till and Busler and Ohanian, which consists of five items is focusing on three key aspects: expertise, attractiveness, and trustworthiness, it showed a Cronbach alpha score of 0.83 and composite reliability of 0.84. A four-item (McCroskey et al., 1975) scale with a reliability value of 0.89 and a CR of 0.55 was used to quantify social distance. Four items modified from Maltby et al. were used to measure SMIS devotion, and the Cronbach's alpha was 0.78. A scale from Schramm and Hartmann, which looked at identification of role through both internal and external characteristics and character success, was used to measure identification. Cronbach's alpha value of this scale 0.875 and 0.88 is its content validity.

Trust was adapted from a five-item scale based on Xiao and Benbasat's conceptualization of trust with reliability value of 0.85. A four-item scale was used to measure impulse buying, and a scale was created by Verhagen and Van Dolen, from which two items were taken, which had a composite reliability of 0.78.

A pre-test with 30 participants was conducted to establish the reliability and validity of the questionnaire. All constructs demonstrated acceptable Cronbach's alpha values, above 0.0.6. Additionally, the average variance extracted for all constructs was above the 0.5 threshold. Most respondents found questionnaire easy to understand and interpret. A seven-point Likert scale was used weights ranging from 1–7 where 1 refers to "strongly disagree" and 7 refers to "strongly agree."

4.2 Participants and Procedure

A total of 150 respondents were sent questionnaires who are also active on social media, out of which 120 responses were received. After removing incomplete submissions, 97 valid responses were retained for the final analysis. Convenience sampling method was used to select the respondents. Partial least squares–structural equation modeling (PLS–SEM) was used to analyze the data with the analysis conducted through the widely-used software SmartPLS®3.0.

5 Results

5.1 Analysis

As shown in Table 1, all constructs under the study are found to be consistent with each other as reliability as the Cronbach's alpha coefficient was greater than the 0.6 level along with the CR scores higher than the acceptable level of 0.6. Also, when compared to the AVE criterion at the 0.5 level, the convergent validity was satisfactory. With the exception of SMIS1 and IB1, all of the constructions' factor item loadings were more than 0.6.

A ratio of correlations between hetero- and mono-traits as indicated in Table 2, the HTMT test is used to determine discriminant validity and is based on Henseler et al. Every HTMT ratio is below the 0.85 most cautious cut-off (Tables 3 and 4).

5.2 Discussion of Analysis

An outline of the interlinkages between several predictors and impulse buying behavior is, therefore, availed by the path analysis results. The following, therefore, summarizes the findings of the hypothesis testing:

Trust and Impulse Buying: Analysis showed that there is a positive correlation of significant magnitude at the beta coefficient level of 0.168 between Trust and Impulse Buying, which is at the $p < 0.001$ level, thereby supporting H1. This therefore means that Trust is an essential mediator to facilitate impulsive purchasing decisions.

Perceived popularity was therefore a dimension of influencer credibility, and its relationship with Trust was such that it positively and significantly affected Trust ($\beta = 0.182$, $p < 0.001$), as indicated by H2. The perceived popularity of the influencers increases trust with the consumer an important variable in creating impulsive buying behavior.

Perceived Fit and Trust: As hypothesized, the findings were in opposition, and Perceived Fit failed to develop any significant effect on Trust ($\beta = 0.078$, $p = 0.106$), thereby achieving the rejection of H3. This suggests that compatibility or association

Table 1 Factors loading with communality and redundancy, convergent validity and average variance extracted (AVE)

Table 1 MM analysis				
Factors	Standardized loading	Cronbach's alpha	AVE	CR
SMI's Popularity (IP)	0.025	0.753	0.498	0.773
	0.736			
	0.711			
	0.539			
	0.702			
Perceived Fit (PF)	0.008	0.836	0.606	0.848
	0.816			
	0.816			
	0.783			
	0.655			
Identification (ID)	0.808	0.875	0.667	0.881
	0.856			
	0.782			
	0.8			
	0.836			
Social Distance (SD)	0.872	0.890	0.738	0.55
	0.943			
	0.815			
	0.799			
SMI Adoration (SMIA)	0.688	0.785	0.607	0.804
	0.793			
	0.812			
	0.815			
Trust (T)	0.7	0.852	0.632	0.858
	0.819			
	0.802			
	0.771			
	0.872			
Impulse Buying (IB)	0.121	0.692	0.491	0.785
	0.63			
	0.849			
	0.806			
	0.821			

Authors own calculation

Table 2 Discriminant validity (HTMT ratio), threshold: HTMT < 0.9

	SMI adoration	Identification	SMI' popularity	Perceived FIT	Trust	Social distance	Impulse buying
IC adoration				–	–		
Identification	0.5655				–		
ICS' popularity	0.052	0.534					
Perceived fit	0.148	0.187	0.479				
Trust	0.117	0.1479	0.652	0.202			
Social distance	0.564	0.369	0.781	0.518	0.746		
Impulse buying	0.624	0.587	0.591	0.795	0.735	0.653	

Authors own calculation

Table 3 Path coefficients of tested model & hypothesis testing and structural model evaluation

Table 4 Structural model analysis process					
		Path coefficients	Collinearity assessment		
Hypotheses	Hypothesized association	Hypothesized relationships	VIF	t-value	p-value
H1	T-IB	0.168 ***	1.761	4.286	0.000
H2	ICP-T	0.182 ***	1.934	3.58	0.000
H3	PF-T	0.078	1.912	1.23	0.106
H4	PF-IB	0.245 ***	1.243	5.143	0.000
H5	ID-T	0.567	1.101	1.103	0.147
H6	ID-IB	0.135 ***	2.61	6.34	0.000
H7	SD-T	-0.025***	1.149	3.104	0.000
H8	SD-IB	0.020 ***	1.112	1.745	0.000
H9	ICA-IB	0.232 ***	1.915	10.61	0.000

Authors own calculation *Significance level of path coefficients established at *** $p < 0.001$

Table 4 Structural model evaluation

	R2	Q2
T	0.449	0.347
IB	0.345	0.362

Authors own calculation

between the influencer's characteristics and the brand is not an important variable for trust building in this study.

Perceived Fit and Impulse Buying While it did not affect Trust at a significant level, Perceived Fit was positively highly correlated with Impulse purchase $\beta =$

0.245, $p < 0.001$, and hence supports H4. This means that in case proper alignment exists between influencers and brands, consumers may promptly adopt impulsive buying behavior regardless of the degree of trust.

Identification and Trust: H5 was not accepted because the relationship was not significant between Identification and Trust, $\beta = 0.567$, $p = 0.147$. This implies that the dimension through which the level of consumers' identification with influencers translates into formation of trust is not directly relevant to this context.

Identification and Impulse Buying: Identification, however was positively and highly influentially impacting on Impulsive purchase at $\beta = 0.135$, $p < 0.001$ thereby supporting H6. This outcome points to the fact that identification with influencers makes consumers reach impulsive buying decisions though the direct impact on trust was not found.

Social Distance and Trust: Social Distance was shown to negatively impact Trust (β = -0.025, $p < 0.001$). H7 is thus supported. The finding would imply that a perceived lack of social proximity between consumers and influencers results in reduced levels of trust; this has further implications for both the selection of influencers and audience targeting strategies.

Social Distance and Impulse Buying: Contrary to the initial hypothesis, Social Distance had a positive impact on Impulse Buying ($\beta = 0.020$, $p < 0.001$), and H8 was rejected. The counterintuitive finding shows that an increased sense of social distance may actually enhance the impulsive purchasing behavior of consumers, with this effect perhaps mediated by the consideration of an aspirational value from the distant influencer.

Influence Adoration and Impulse Buying: Last but not the least, influence adoration also had a positive relation with impulsive buying ($\beta = 0.232$, $p < 0.001$); thus, H9 was accepted. This states that the more adoration toward an influencer, the better the chance of inducing impulsive purchases in the individual, thus showing that the emotional attachment will play the crucial role for impulsive buying behavior. Figure 1 shows the result from smart PLS analysis.

6 Conclusion

The study proposes a model which can explain the perspective of consumer and social media influencer both on impulse buying behavior of luxury cosmetics products. The primary findings of this research have been that trust plays a crucial role in impulse buying of branded cosmetic products. Trust on Internet celebrity can lead directly to increase in impulse buying and trust on SMIS popularity also has direct positive influence on impulse buying of branded cosmetics Hill (Troshani, 2020), while greater social distance can prove to be negative on the aspect of trust on celebrity and hence can negatively affect impulse buying. This research also demonstrates that identification with SMIS impacts impulse buying as more people go toward their own personality closer to SMIS and are ready to buy products when recommended by them (Rook, 1987). Finally, this finding confirms earlier studies suggesting that

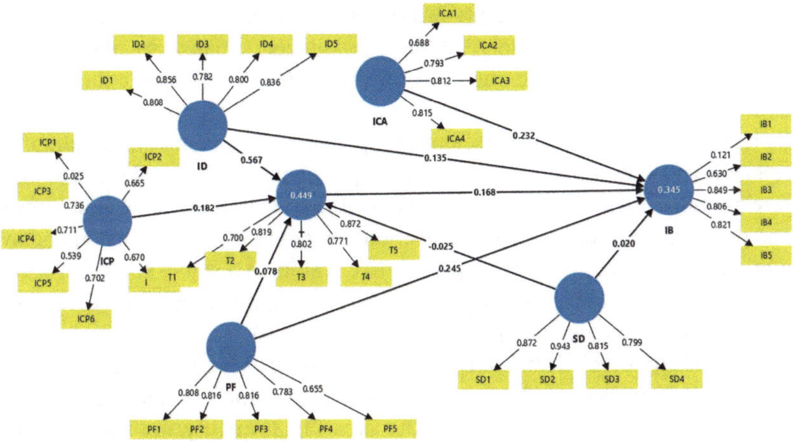

Fig. 1 Bootstrapped model

the celebrities whom the people admire will be of more consideration in forming the purchase intentions by Chiou (2005). This implies that self-centered motivation of people should play a significant role in determining influence on the shopping behavior. According to (Yang, 2004), SMIs' adoration can lead straight to impulsive purchase since the mental state involving emotional attachment may make it harder to resist impulse buying.

7 Managerial Implications for Sustainable Economic Development:

Based on our research we can suggest following managerial implications for sustainable economic development.

- **Consumer Trust and Sustainable Long-Term Relationships Customer** trust lies at the heart of sustainable economic results, as trusted brands are able to achieve higher loyalty and purchase repeat behavior from customers (Chen et al., 2021). For this trust to be developed, therefore, managers require influencers who are credible and aligned with the values of the brand, thereby encouraging sustained engagement from the customers with the sustainably positive growth implications (McQuarrie et al., 2013; Verhagen & Van Dolen, 2011).
- **Identification of Relevant Influencers in Sustainable Development**: In sustainable development, marketers need to concentrate on influencers who adhere to the values of the brand instead of metrics measuring the popularity of influencers. Influencers advocating value-based consumption instead of impulse consumption will enhance more mindful buying, which aligns with the principles of sustainable development (Pradhan et al., 2014; Tafesse & Wood, 2021).

- **Balancing Social Distance for Effective Engagement**: The perceived social distance between consumers and the influencer, created by interactive and approachable content (live-streamed reviews or Q&A), can make the influencers come across as closer to the customers. This feeling of closeness assists in the inclusiveness of an ethical conscious group of diverse consumers, including WOM referrals and sales impact, as noted by (López et al., 2019). In addition to impulse buying, other metrics of sustainable economic growth include what are inwardly and outwardly driven, such as satisfaction, trust, and loyalty by customers toward a brand (Liu et al., 2013). This perspective results in a stable source of revenues due to the eradication of short-term sales. It eventually aligns with the principles behind sustainable development (Smith et al., 2020).
- **Using Influencers' Influence for Educational Messages:** In the learning process, influencers can design a healthy demand among customers by offering them durable, long-lasting products that are of high-quality, not fast fashion, thus making them more responsibly ethical by a brand for sustainable messages that, in the long term, are positive to both society and the environment.

8 Limitations and Future Scope

The study has limitations regarding generalizability and scope as well as the findings. Applicability is limited to other cohorts and demographics since the sample is mainly comprised of socially networked consumers. This is also yet another limitation of the study: based on a single cultural context, it misses out the diversity of influencer impact across different cultural contexts. Moreover, using hypothetical scenarios as a means of measuring responses may often lack ecological validity since real-life contacts with influencers may elicit other forms of consumer behavior that are not represented in the hypothetical scenarios designed to measure responses. A cross-sectional design also very tightly constrains possible causal inferences and cannot provide insight into how influencer trust and impulsive buying evolves over time. In the end as the study focuses solely on impulse buying, it excludes other important consumer outcomes, like brand loyalty or satisfaction, that can add richness to the inferences regarding the impact of influencer marketing.

Funding: This research is not funded by any institute or organization.

Conflicts of Interest: The authors declare no conflict of interest.

References

Aydin, G., Uray, N., & Silahtaroglu, G. (2021). How to engage consumers through effective social media use—Guidelines for consumer goods companies from an emerging market. *Journal of Theoretical and Applied Electronic Commerce Research, 16*, 768–790.

Batra, R., Ahuvia, A., & Bagozzi, R. P. (2012). Brand love. *Journal of Marketing, 76*, 1–16.

Bogardus, E. S. (1933). A social distance scale. *Sociology and Social Research, 22*, 265–271.
Bradri. (2019). Pakistani influencers who are making the most of influencer marketing. Retrieved from https://bradri.com/blog/pakistani-influencers-who-are-making-the-most-of-influencer-marketing/
Busalim, A. H., & Hussin, A. R. C. (2016). Understanding social commerce: A systematic literature review and directions for further research. *International Journal of Information Management, 36*, 1075–1088.
Chen, Y., Lu, Y., Wang, B., & Pan, Z. (2019). How do product recommendations affect impulse buying? An empirical study on WeChat social commerce. *Information & Management, 56*, 236–248.
Chen, M., Xie, Z., Zhang, J., & Li, Y. (2021). Internet celebrities impact on luxury fashion impulse buying. *Journal of Theoretical and Applied Electronic Commerce Research, 16*(6), 2470–2489.
Chiou, J. S., Huang, C. Y., & Chuang, M. C. (2005). Antecedents of Taiwanese adolescents purchase intention toward the merchandise of a celebrity: The moderating effect of celebrity adoration. *Journal of Social Psychology, 145*, 317–334.
Dittmar, H., Beattie, J., & Friese, S. (1996). Objects, decision considerations, and self-image in men's and women's impulse purchases. *Acta Psychological, 93*, 187–206.
Edwards, C., Spence, P. R., Gentile, C. J., Edwards, A., & Edwards, A. (2013). How much Klout do you have? A test of system-generated cues on source credibility. *Computers in Human Behavior, 29*, A12–A16.
Fadillah, F., & Kusumawati, N. (2021). Factors affecting makeup products online impulsive buying behavior on TikTok. *Proceeding Book of The 6th ICMEM*, no. August, 11–13.
Gawron, M., & Strzelecki, A. (2021). Consumers adoption and use of e-currencies in virtual markets in the context of an online game. *Journal of Theoretical and Applied Electronic Commerce Research, 16*, 1266–1279.
Gefen, D., Karahanna, E., & Straub, D. W. (2003). Trust and TAM in online shopping: An integrated model. *MIS Quarterly, 27*, 51.
Gita, M. T., Satriawan, B., & Satyawisudarini, I. (2023). The influence of consumer impulse buying and electronic word of mouth on purchasing decisions. *International Journal of Advanced Multidisciplinary, 2*(2), 488–495.
Hill, S. R., Troshani, I., & Chandrasekar, D. (2020). Signalling effects of vlogger popularity on online consumers. *Journal of Computer Information Systems, 60*, 76–84.
Hsu, C. L., Lin, J. C. C., & Chiang, H. S. (2013). The effects of blogger recommendations on customers online shopping intentions. *Internet Research, 23*, 69–88.
Huang, Z., & Benyoucef, M. (2013). From e-commerce to social commerce: A close look at design features. *Electronic Commerce Research and Applications, 12*(4), 246–259.
Huffaker, D. (2010). Dimensions of leadership and social influence in online communities. *Human Communication Research, 36*, 593–617.
Hwang, K., & Zhang, Q. (2018). Influence of parasocial relationships between digital celebrities and their followers on followers purchase and eWOM intentions. *Computers in Human Behavior, 87*, 155–173.
Jarvenpaa, S. L., Knoll, K., & Leidner, D. E. (1998). Is anybody out there? Antecedents of trust in global virtual teams. *Journal of Management Information Systems, 14*, 29–64.
Kamins, M. A. (1990). An investigation into the "match-up" hypothesis in celebrity advertising: When beauty may be only skin deep. *Journal of Advertising, 19*, 4–13.
Kapitan, S., & Silvera, D. H. (2016). From digital media influencers to celebrity endorsers: Attributions drive endorser effectiveness. *Marketing Letters, 27*, 553–567.
Leonidou, L. C., Barnes, B. R., & Talias, M. A. (2006). Exporter–importer relationship quality: The inhibiting role of uncertainty, distance, and conflict. *Industrial Marketing Management, 35*, 576–588.
Liang, T.-P., & Turban, E. (2011). Introduction to the special issue social commerce: A research framework for social commerce. *International Journal of Electronic Commerce, 16*(2), 5–14.

Liu, Y., Li, H., & Hu, F. (2013). Website attributes in urging online impulse purchase: An empirical investigation on consumer perceptions. *Decision Support Systems, 55*(3), 829–837.

López, M., Sicilia, M., & Moyeda-Carabaza, A. (2019). The effect of social influence and self-presentation on consumers intentions to participate in user-generated content on social media. *Journal of Business Research, 96*, 300–310.

López, M., Sicilia, M., & Moyeda-Carabaza, A. A. (2017). Creating identification with brand communities on Twitter: The balance between need for affiliation and need for uniqueness. *Internet Research, 27*, 21–51.

Marwick, A. E. (2015). Instafame: Luxury selfies in the attention economy. *Public Culture, 27*, 137–160.

Mayer, R. C., Davis, J. H., & Schoorman, F. D. (1995). An integrative model of organizational trust. *Academy of Management Review, 20*, 709.

McEvily, B., Perrone, V., & Zaheer, A. (2003). Trust as an organizing principle. *Organization Science, 14*, 91–103.

McCroskey, J. C., Richmond, V. P., & Daly, J. A. (1975). The development of a measure of perceived homophily in interpersonal communication. *Human Communication Research, 1*(4), 323–332.

McQuarrie, E. F., Miller, J., & Phillips, B. J. (2013). The megaphone effect: Taste and audience in fashion blogging. *Journal of Consumer Research, 40*, 136–158.

Palazon, M., Delgado-Ballester, E., & Sicilia, M. (2019). Fostering brand love in Facebook brand pages. *Online Information Review, 43*, 710–727.

Pemberton, C. (2017). *Energize your influencer marketing - smarter with gartner*. Retrieved December 5, 2018, from https://www.gartner.com/smarterwithgartner/energize-your-influencer-marketing

Pham, A. D., Nguyen, H. N., Le, T. T., Nguyen, H. K., Khuat, H. T., Phan, H. T., & Vu, H. T. (2024). Young consumers' impulse buying tendency on social media: An empirical analysis in Vietnam in light of the LST theoretical perspective. Young Consumers.

Pradhan, D., Duraipandian, I., & Sethi, D. (2014). Celebrity endorsement: How celebrity–brand–user personality congruence affects brand attitude and purchase intention. *Journal of Marketing Communications, 89*, 1–18.

Rook, D. W. (1987). The buying impulse. *Journal of Consumer Research, 14*, 189.

Saselah, K. T., Lapian, S. J., & Gunawan, E. M. (2024). The effect of hedonic shopping value and beauty influencer in TikTok towards impulse buying among TikTok users in Manado. *Jurnal EMBA: Jurnal Riset Ekonomi, Manajemen, Bisnis Dan Akuntansi, 12*(01), 630–639.

Smith, M., Johnson, R., & Lee, K. (2020). The role of consumer trust in e-commerce: An examination of impulse buying behavior. *Journal of Marketing, 84*(2), 45–67.

Steffes, E. M., & Burgee, L. E. (2009). Social ties and online word of mouth. *Internet Research, 19*, 42–59.

Tafesse, W., & Wood, B. P. (2021). Followers engagement with Instagram influencers: The role of influencers content and engagement strategy. *Journal of Retailing and Consumer Services, 58*, 102303.

Trope, Y., & Fishbach, A. (2000). Counteractive self-control in overcoming temptation. *Journal of Personality and Social Psychology, 79*, 493–506.

Trope, Y., & Liberman, N. (2010). Construal-level theory of psychological distance. *Psychological Review, 117*, 440–463.

Verplanken, B., & Sato, A. (2011). The psychology of impulse buying: An integrative self-regulation approach. *Journal of Consumer Policy, 34*, 197–210.

Verhagen, T., & Van Dolen, W. (2011). The influence of online store beliefs on consumer online impulse buying: A model and empirical application. *Information & Management, 48*(8), 320–327.

Wangshu, G., & Guanhua, W. (2020). How influencers marketing motivates consumers' buying behaviour: A focus group investigation of the impulse buying behaviour via Chinese millennials' lens.

Wei, Z., Zhao, Z., & Zheng, Y. (2019). Following the majority: Social influence in trusting behavior. *Frontiers in Neuroscience, 13*, 1–8.

Xiang, L., Zheng, X., Lee, M. K. O. O., & Zhao, D. (2016). Exploring consumers impulse buying behavior on social commerce platform: The role of parasocial interaction. *International Journal of Information Management, 36*(3), 333–347. https://doi.org/10.1016/j.ijinfomgt.2015.11.002

Xiao, S. H., & Nicholson, M. (2013). A multidisciplinary cognitive behavioural framework of impulse buying: A systematic review of the literature. *International Journal of Management Reviews, 15*, 333–356.

Yang, C., Wang, Y., & Niu, H. J. (2008). The effects of idolatry and personality traits on impulse buying: An empirical study. *International Journal of Management, 25*, 633–641.

Zafar, A. U., Qiu, J., Li, Y., Wang, J., & Shahzad, M. (2021). The impact of social media celebrities posts and contextual interactions on impulse buying in social commerce. *Computers in Human Behavior, 115*, 106178.

Zhang, K., Xu, H., Zhao, S., & Yu, Y. (2018). Online reviews and impulse buying behavior: The role of browsing and impulsiveness. *Internet Research, 28*(3), 522–543.

Zhang, H., Lu, Y., Gupta, S., & Zhao, L. (2014). What motivates customers to participate in social commerce? The impact of technological environments and virtual customer experiences. *Information & Management,* 51(8), 1017–1030. https://doi.org/10.1016/j.im.2014.07.005

Sustainable Market Dynamics and Supply Chain Management: Examining the Availability of Hybrid Paddy Seeds in Different Regions of Odisha

Sumit Gouraba Patra and Manmath Nath Samantaray

Abstract Hybrid paddy seeds were developed in order to enhance production, pest control and adapt to new conditions of the environment. However, the use of these seeds has not been implemented effectively in Odisha. This research aims at analyzing the market structure and supply chain of hybrid paddy seeds in Odisha, India and its sustainable dimensions by exploring the distribution, usage and issues faced by the concerned stakeholders in different parts of the State. The research employs both quantitative questionnaires and qualitative interviews with actual and potential farmers, seed distribution companies, seed retailers and government officials involved in the policymaking process. Some of the emerging constraints for the delivery of improved seeds include the locations and accessibility for the private seed companies, prices for small-holder farmers and distribution of the seeds across the targeted areas. It also draws out environmental sustainability issues surrounding hybrid paddy production and called for integrated and sustainable practices in agriculture. The study highlights the role of synergy between government organizations at national and state levels, private stakeholders and farmers and rural people in increasing the availability and utilization of hybrid paddy seeds in Odisha.

Keywords Hybrid paddy seeds · Supply chain management · Sustainable market dynamics · Odisha agriculture · Seed adoption

S. G. Patra (✉) · M. N. Samantaray
Research Scholar, ASBM School of Business, ASBM University, Bhubaneswar, India
e-mail: sumit.patraphd2022@asbm.ac.in

M. N. Samantaray
e-mail: manmath.samantaray@asbm.ac.in

R. Kumar and S. N. Mohanty (eds.), *Sustainable Economy Models in the Age of Industry 5.0*, https://doi.org/10.1007/978-981-96-4104-8_17

1 Introduction

Odisha is an agricultural state and its around 70% of population is directly or indirectly involved into agriculture business (Government of Odisha, 2022). Rice is the most common crop, taking up a large area of the state's irrigated agriculture. The farming practices in Odisha remain agrarian, and monsoon based, making the sector sensitive to climate change and other related shocks (Nayak & Behera, 2018).

One of the most significant innovations in the crop production is hybrid paddy seeds that have brought improved yields, quality and productivity in numerous states including Odisha—the leading producer of paddy in the country. Paddy is not only the crop that has been cultivated from time immemorial but also the crop that generates primary income for tens of millions of farmers in India. Studying the effect of the hybrid seeds on agriculture productivity in Odisha is justified as this state is divided into twelve agro-climatic regions, which makes it possible to analyze the problem in detail and from different perspectives. Biotic seeds have been developed to increase yield potential, improve resistance to pests and diseases and perform well under different environment this makes them a key tool in the quest for food security and sustainable agriculture (Spielman et al., 2013).

It is consequently important to foster the creation of long-term sustainable market systems for the hybrid paddy seeds to help feed the growing global population particularly Odisha state and meet the increasing rice demands. Hybrid paddy seeds which are expected to yield better results, increase resistance to pests and diseases as well as withstand different climate conditions have now become vital tools for farmers seeking to increase production and therefore income (Dorairaj & Govender., 2023). Sustainable market dynamics in this regard encompass establishing a solid system of providing such seeds without compromising the physical, fiscal and social sustainability (Johns et al., 2013). This can be quite a challenging undertaking as it needs to be backed by a sound supply chain management (SCM) arrangement that seeks to encompass seed producers and distributors, policymakers and farmers among others (Sharma et al., 2023).

Sustainability in supply chain management for hybrid paddy seeds addresses the ability to deliver seeds to different regions of Odisha especially during planting seasons. An effective supply chain enhances this accessibility by improving the flow of materials and information, strategic planning and proper management of resources and transportation systems. However, supply of hybrid paddy seeds depends on certain constraints such as market forces, changes in weather conditions and changes in policy (Dar et al., 2021). Therefore, there are disparities for the seeds distribution in the region, some areas in Odisha continue to experience a challenge of seeds shortage while others are well supplied. Such disparities have informed the call for targeted approaches in directing seed distribution in the various regions of world to increase access and coverage of seeds (Pradhan et al., 2015).

Policies exist to try to regulate the supply of hybrid seeds but there are key challenges that affect availability in Odisha. The problems are: High costs that seed producers face in developing and putting seed hybridity on the market, in this process, quality controls must be set up to guarantee the seed's reliability and efficiency (Afzal et al., 2019). The production process is resource demanding and requires capital investment; this makes it difficult for producers to produce a large number of seeds for dissemination (Chartres et al., 2015). As for distributors, problems are associated mainly with transport and storage, especially in the countryside where freezing storage and transport infrastructure are not well developed. Seed quality has to be well maintained, because high temperatures and poor storage can actually deteriorate seeds, thus complicating their functionality, even though they reach the farmers at last (Kameswara et al., 2017).

On the same note, the primary beneficiaries are the farmers and they also experience some factors hindering the acquisition of hybrid paddy seeds. High initial costs of investment keep it away due to financial constraints and credit crunch that is experienced by most farmers in Odisha who cannot afford to invest on hybrid seeds. Furthermore, it shows that seeds in local markets are also seasonal and the information given to farmers on where they can obtain improved seeds is inaccurate; therefore, they use traditional seeds that require more land, yield low and are resistant to pests (Smale et al., 2009). Awareness barriers also exist; some farmers may have poor attitude toward hybrid seeds or are ignorant of the advantages that come with the use of such seeds hence impaired adoption rates will in one way or the other deter agricultural productivity (Dethier & Effenberger, 2012).

The adoption of hybrid paddy seeds in Odisha can be improved through coordinated effort with the major stakeholders including policymakers, agricultural institution and private players to improve the distribution channel and stabilizing market structures. Government interventions in the form of incentives to fertilization of hybrid seeds, provision of better credit facility for farmers and encouraging the existing outlets at localized level can possibly solve the existing challenges (Langyintuo et al., 2010). Also, there is need for educational programs that could educate farmers on the importance of using hybrid seeds to boost production since programs with such information can improve on adoption and productivity rates, respectively (Matuschke et al., 2007).

The significance of this study arises from its specificity of the subject region for instance Odisha where there is an effort to conduct empirical research on the subject of availability of hybrid paddy seeds and challenges posed to stakeholders. Consequently, taking into consideration the outcomes of the given research, and looking at the existing studies, it is possible to assert that appreciable insights into the nature of sustainable market dynamics and supply chain management can be derived if the specific area under consideration is Odisha.

2 Literature Review

In terms of the distribution of infrastructural facilities, markets and the capacities of farming communities to embrace new innovations agriculture in Odisha is still highly regionalized. These areas include the coastal plains, hilly terrains and the interior tracts all of which afford different prospects and difficulties in the production of paddy (Thapa et al., 2018). A good example is the use of hybrid paddy seeds which has proven most useful in regions that have borne the full brunt of climate change, land degradation and much more recently, pest infections. However, comparing them and checking the quality of seeds that were adopted and not across the state shows the variety of sustainable market elements, more effective supply chain, and governmental support (Dash & Singhi, 2019).

Therefore, the supply chain of hybrid paddy seeds is important in influencing it availability and its usage by the farmers. A good supply chain means that seeds are produced, stocked and dispensed in the market and in the farmer's fields at the right time especially during planting seasons. The intermediate supply chain for hybrid paddy seeds is composed of actors such as seed producers and processors, distributors, retailers and government actors in the Odisha region. This supply chain efficiency, however, differs significantly across various geographical zones of the state (Singh & Sahoo, 2021).

The result also showed that farmers in the coastal areas where there is good infrastructure and market have adopted the use of hybrid seeds more than other areas. This can be attributed to improved transport systems, tighter bond with distributors and extensive government support. For instance, wherever the government has launched certain programs to support the use of hybrid seed, the farmers have shown more preference toward hybrid seed and in turn they have increased the yield in agriculture (Shravanthi, 2022). On the other hand, in the remote areas that often lack development in the Odisha's tribal areas, the hybrid paddy seeds have not been embraced several reasons which includes high costs, farmer's low awareness level and lack of structures.

Another challenge to the utilization of hybrid paddy seeds in Odisha is the fluctuation of the prices and cost factors that are easily accessible or preferably cheap for the small-scale farmers grow. Hybrid seeds are usually costly than conventional seeds, and the initial capital that is needed to buy them may be large for most farmers especially those with little capital (Rutsaert et al., 2021). This price sensitivity is especially troublesome regarding geographical locations that either lack subsidies or have those poorly performed. Therefore, while hybrid seeds bring long-term rewards—such as increased yields and generally lower input—the initial costs discourage many farmers from using them for their planting (Pandey et al., 2020).

Another important activity in delivering the hybrid paddy seeds is the involvement of private sector, particularly in the distribution of such seeds. Breeders practicing in private seed companies have played a crucial role in deposing and commercializing of these seeds more so in the developed markets of these regions. However, their exposure is generally limited to the rural areas, where the overall market might be less, and the barriers to distribution higher. This has further created a situation where different regions of Odisha have positive changes in availability and choice of seeds, for instance, coastal plain regions are able to access better varieties and choices compared to interior and tribal regions (Kumar et al., 2023).

It is crucial to look through the lens of government policies and interventions, majorly because the latter substantially influence the sustainable market patterns of the hybrid paddy seeds. Hybrid seeds have been promoted through programs implemented by the Government of India in collaboration with the state government of Odisha. These have included subsidies, training and credit facilities targeted at making farmers adopt the Hybrid varieties. Yet, the impact of these policies has been progressive meaning that regions with decent local governance and functional agricultural extension systems benchmarks higher results than regions with weak implementation (Airee et al., 2020).

However, factors that contribute to the regularity of the growth of hybrid paddy seeds are economics and policy-related factors besides the awareness and education of farmers on the need to use the hybrids. Out of these, in many parts of Odisha especially the tribal and remote areas, the farmers have not got much information about the hybrid variety seeds and therefore they still practice the traditional seeds. This informative deficit is further compounded by the celebrated issue of restricted access to agricultural extension and the problem of informing farmers in the rural tracts. There is a need therefore to develop an effective outreach and education programs to help all farmers to capitalize on the potential benefits by embracing hybrid paddy seeds (Gautam et al., 2018).

Environmental sustainability as a final factor that influences the use of the hybrid paddy seeds is worthy of consideration. Though these seeds are technically developed to be less sensitive to environmental conditions, they demand more water, fertilizers and pesticide to grow. Paddy intercropping has also now been considered as a threat to the flora and fauna, soil condition and water conservation systems that have been initiated for long-term effect (Goel & Bhaskaran, 2006). These pressures further point to the importance of enhancing integrated pest management and organic farming as well as planting of hybrid seeds to avert these risks (Devkota et al., 2020).

Other factors which also create sustainable market dynamics for the hybrid paddy seeds include changes in demand and supplies, strategies of pricing of the seeds and market competition among seed firms. Hybrid seeds, of course, have also found their strong acceptance by farmers when the market is competitive and demands have been established. But the price of seed in areas that have a relatively Small populated market structure coupled by low competition is relatively higher and farmers have a limited number of choices. These market dynamics are a key factor leading to this differential level of adoption of hybrid seeds between regions: where some parts of Odisha are witnessing a positive impact and others are falling behind (Kumar & Barker, 2023).

Another facet of social science is reflected in the Odisha investigation of hybrid paddy seeds as well. The gains arising from other forms of agricultural improvement including hybrid seeds are never equally spread across the entire society, especially the powerless groups of peoples inclusive of the small scale and tribal farmers. The flow of these benefits to such communities needs proper interventionary measures in terms of financial assistance, and awareness, as well as infrastructural facilities appropriate to the type of these special groups (Dalei & Dutta, 2015).

3 Research Objectives

Based on the inputs from the numerous literature studies, a research study was conducted with the following objectives:

a. To understand the sustainable market dynamics of hybrid paddy seeds in the regions of Odisha.
b. To examine supply chain efficiency and availability of hybrid paddy seeds in Odisha.

4 Methodology

The method used for this research was systematic to capture appropriate information on sustainable market aspects and supply chain of hybrid paddy seeds in various zones of Odisha. Both quantitative and qualitative data collection tools were used in an effort to provide adequate results about the pending changes.

4.1 Study Area

The study was conducted in five distinct regions of Odisha: coastal areas, central, northern regions, southern and western zones. These regions were chosen because they presented various agro-climatic environment and varying market accessibility that was thought would capture the true nature of the hybrid paddy seeds.

4.2 Study Design

This study incorporated a cross-sectional concurrent triangulation research design, with data collection being done at a one point in time of the different regions of Odisha. The research model was therefore developed to cover the supply chain logistics of the hybrid paddy seeds, and the prevailing market trends within this crucial sub-sector, in addition to capturing the actual experiences and real-life difficulties bearing on the farmers and other key players interested in this business.

4.3 Study Setting

The study was conducted in rural and semiarid settings of the chosen areas of operation as most of the population depends on agriculture. Informal interviews were conducted in farmers' houses, local markets, seed distribution points and government offices. The setting was chosen in such a way so as to try capture as diverse opinions as possible from the key stakeholders in the production, distribution and consumption of hybrid paddy seeds.

4.4 Study Duration

The data collection in the study was done in three months from March to May 2024. This period was selected depending on the agricultural calendar because it corresponds to the time still waiting for the kharif season, which is the most important period in Odisha for the planting of the paddy.

4.5 Participants—Inclusion and Exclusion Criteria

The target population in this study comprised of farmers, seed distributors and retailers as well as government representatives. These famers were chosen because most of them had used hybrid paddy seeds during at least any of the two planting seasons. Paddy seed distributors and retailers participating in the study who engaged in the distribution or sale of hybrid paddy seeds. The relevant authorities: the officials who were in charge of agricultural policy, seed supply for the study regions of the identified countries were considered for the study. But small-holder farmers who did not adopt hybrid paddy seeds were treated out of sample as well as those farmers whose knowledge could not help to explain market and supply factors.

4.6 Study Sampling

The target population of the study was students studying in higher education institutions in Odisha state; hence, participants were selected from each of the five regions: coastal, central, northern, southern and western Odisha. This way made certain that the sample selected covered a wide range of the scenarios in Odisha. Participants included within each region were sampled based on their activities of supply chain of hybrid paddy seeds.

- **Farmers:** A total of 150 farmers (30 from each region) were selected to participate in the survey.
- **Seed Distributors:** 20 distributors (4 from each region) were selected for in-depth interviews.
- **Retailers:** 30 retailers (6 from each region) were interviewed.

- **Government Officials:** 10 officials (2 from each region) were included in the study.

4.7 Study Sample Size

The number of respondents in the study was arrived at through the use of probability sampling technique based on the population of farmers and stakeholders in the study regions in order to give the study an improved statistical significance and reliability of the results. The study's sample included 150 farmers, 20 seed distributors, 30 retailers and 10 policymakers for a total of 210 samples.

4.8 Study Groups

Participants were categorized into four groups: farmers, seed distributors, retailers and government officials. This grouping facilitated the analysis of data from different perspectives within the supply chain, allowing for a more comprehensive understanding of the sustainable market dynamics.

4.9 Study Parameters

The study focused on the following parameters:

- **Market Access:** Availability of hybrid paddy seeds in different regions.
- **Distribution Efficiency:** Time taken for seeds to reach farmers from distributors.
- **Pricing Dynamics:** Price fluctuations and affordability of hybrid seeds.
- **Challenges:** Issues faced by stakeholders in the supply chain.

4.9.1 Study Procedure

The data collection process involved the following steps:

- **Survey Administration:** In the course of the study the researchers developed structured questionnaire to be used for the study which was administered to 150 farmers. The questionnaire confined to the thoughts about the availability of seed, buying pattern, problem encounter and satisfaction level with the seed supply chain.
- **In-depth Interviews:** Focus was conducted in seed distributors of 20, retailers 30 and government officials of 10. These interviews focused on providing specific overviews of the supply chain processes and sustainable market conditions and issues.

- **Field Observations:** Field studies were carried out at seed distribution centers and local markets to capture the flow of supply chain throughout the study area and bargaining power of buyers and suppliers.
- **Data Validation:** The findings from the surveyed and interviewed subjects were triangulated with the results of field observations and other secondary data from government and market research publications.

4.9.2 Data Analysis

In this study, both qualitative and quantitative research techniques were adopted throughout the data analysis process. Self-administered questionnaires completed by the farmers were keyed into a statistical package and analyzed using descriptive statistics including means, frequencies and cross tabulations. For qualitative analysis, interview texts and notes recorded during research were analyzed through the theme. Four key themes emerged, and the collected qualitative data were then coded into such categories. These were particularly complemented through triangulation with the quantitative data to provide a more holistic analysis.

4.9.3 Ethical Considerations

The research proposal for the study was approved by the institutional review board for ethics in research. Written desires to participate in the study and being informed of the aim, purpose and process of the study were obtained from all actors involved, before data collection was conducted. Participant data was kept anonymous and aggregate data was used to reduce one's individuality making confidentiality intact.

5 Results and Discussion

This section seeks to highlight the findings and recommendations of the research work and/or analyses in this study. The results are presented in relation with the research questions and goals and comprises sustainable market structures, SCM practices, hybrid paddy seed access and issues confronted.

5.1 Sustainable Market Dynamics

Regional differences in the Odisha market dynamics for hybrid paddy seeds were identified and highlighted based on five analyzed regions. Different factors such as the infrastructures, government and regional economic contributed to the ability, price and the extent of penetration of hybrids seeds (Table 1).

Table 1 Sustainable market dynamics of hybrid paddy seeds across different regions of Odisha

Region	Adoption rate (%)	Average price (₹ per kg)	Market penetration (Scale: 1–10)	Government support (Yes/No)
Coastal	85	250	9	Yes
Central	78	240	8	Yes
Northern	60	260	6	No
Southern	70	255	7	No
Western	55	270	5	No

Source Primary data

Table 1 summarizes the sustainable market factors of hybrid paddy seeds in five regions. The hybrid seeds of paddy was found to have high adoption at coastal and central zones due to better infrastructures and transportation facilities and increased government support. The market penetration was high as was observed in the coastal region due to good market networks in distributing seeds. Nonetheless, the north and the west had lower standards of market penetration of the sites due to poor infrastructure and lack of political backing. Higher average price in these regions also limited farmers' access and use due to low affordability. However, the use of adoption was moderate in the southern region while some constraints like the escalating and intermittent prices besides logistical difficulties hindered market diffusion.

5.2 Supply Chain Management

In this study it was found that the effectiveness of the management of supply chain of hybrid paddy seeds was unevenly distributed among regions. Other areas of focus in the supply chain analysis were transport structure, arranged distribution channels and role of private actors (Table 2).

Table 2 Supply chain efficiency metrics across different regions

Region	Average delivery time (Days)	Distributor network strength (Scale: 1–10)	Private player involvement (%)	Stock-out frequency (Times per Season)
Coastal	5	9	75	1
Central	6	8	70	2
Northern	12	5	40	4
Southern	10	6	50	3
Western	14	4	30	5

Source Primary data

Table 2 highlighted some core trends: coastal and central regions had shorter average delivery time and less stock-out frequency due to its proper distributor network and more involvement of private players. In contrast, the distribution time and high stock-out frequency was observed in northern and western region due to less actor network and less involvement of the private players which affected the efficiency of supply chain in a very worst manner. The distributor network in the southern region was moderate but problems like long delivery time due to transport problems and moderate involvement of Private players led to occasional stock-out.

5.3 *Availability of Hybrid Paddy Seeds*

Hybrid paddy seeds were observed to be available in limited quantities and distribution in the regions reviewed, due to the factors such as supply chain, demand and price (Table 3).

According to Table 3, high availability indexes in coastal and central regions were related to supply chain management, and market demand. These farmers indicated very high satisfaction with the availability of and quality of hybrid paddy seeds. However, an availability index in the north and west region was few because of poor stocks and unlamentable supply chain. Though, consumers in these regions showed dissatisfaction in the accessibility of these hybrid seeds, alleging the effort of farmers in acquiring seeds on time as a challenge. Some concerns with stock levels and access were observed regarding hybrid paddy seeds in the southern region and they were moderate. Farmers' satisfaction was also moderate as some at times complained of difficulty in accessing seeds.

Table 3 Availability of hybrid paddy seeds across different regions

Region	Availability index (Scale: 1–10)	Average stock availability (kg per Distributor)	Farmer satisfaction (Scale: 1–10)
Coastal	9	500	8
Central	8	450	7
Northern	5	300	5
Southern	6	350	6
Western	4	250	4

Source Primary data

5.4 Challenges Faced by Stakeholders

Different supply chain participants encountered various problems such as supply chain management problems, price instability problems and lack of sufficient government support in some areas (Table 4).

High logistics, relatively unstable food prices and limited government support were key problems in the north-western regions. The responses received had low satisfaction levels from the various stakeholders and they commonly complained about the challenges that they encounter in ensuring that seeds are available and affordable at all times. Thus, southern region had moderate problems regarding logistics and prices and the level of satisfaction among the stakeholders was higher than among the north and west region due to better availability of seeds and private sector support. Coastal and central region stakeholders indicated that they were less affected by the logistics of implementing the project and were generally content with the level of support they received from the government. Price stability led to enhanced satisfaction levels, as showed in Table 4.

6 Discussion

The future market model and management of hybrid paddy seeds in Odisha are more diversified and generalized, which is a promising as well as difficult path in agricultural advancement in Odisha. This discussion is an integration analysis of the study findings, with implications for the factors that influence the uptake of hybrid paddy seeds, supply chain effectiveness, regional imbalance and other issues affecting the stakeholders. Through the consideration of these aspects, the discussion interests in identifying the mechanism need to be improved and provide suggestion to facilitate the distribution and availability of hybrid paddy seeds in Odisha effectively.

The analysis exposed major regional differences in the sustainable market characteristics that are appropriate for hybrids paddy seeds, as the coastal and central regions had a higher rate of appropriate adoption in contrast to the northern, southern

Table 4 Challenges faced by stakeholders in different regions

Region	Logistical challenges (Scale: 1–10)	Price fluctuations (Yes/No)	Government support (Adequate/ Inadequate)	Stakeholder satisfaction (Scale: 1–10)
Coastal	3	No	Adequate	8
Central	4	No	Adequate	7
Northern	8	Yes	Inadequate	5
Southern	6	Yes	Inadequate	6
Western	9	Yes	Inadequate	4

Source Primary data

and western sectors. Hybrid seed adoption level was significantly high at 85% in the coastal region due to better physical infrastructure, government support and effective market. On the other hand, the western region recorded poorer adoption at 55% basing on factors such as poor infrastructure, high seed price and limited governmental support.

These differential adoption rates can therefore be explained for a number of reasons. First, the coastal and central regions enjoy favorable germination conditions and easy access to urban centers and other means of transporting seeds to the plantations on time. This has also be through incentives such as subsidies and training's that these governments have provided to farmers for them to use the hybrid seeds. On the other hand, the northern and western state, which are relatively less developed and located far away from the growth centers, lack access to these facilities. The prices for hybrid seeds in these regions remain relatively high and since there are no mechanisms for price support, all these increases the cost of hybrid seeds beyond the reach of the small-holder farmer.

The level of market penetration of hybrid seeds also reveals these disparities in the case of the coastal region with the score of 9 out of 10 in contrast to the score of only 5 of western region. This gap has been identified as an area of focus that requires major efforts to be directed toward enhancing market access in under developed markets. The analysis recommends that promotion of infrastructure facilities, decrease in seed prices and increase in the scope of subsidies for less developing region can lead to a four-fold increase of hybrid paddy seeds in the Odisha.

This research highlights Supply Chain Management as an influential component in provision of hybrid paddy seeds to farmers. This work revealed the following observation on the efficiency of the supply chain: a marked geographical dispersion; lower lead time and high distributor's strength in the coastal and central zones; a greater involvement of the regional play actors in the supply chain. For instance, the delivery time was only 5 days in the coast region and was 14 days in the west region. This difference effectively put into perspective the need to adopt an efficient supply chain network that will ensure prompt delivery of the seeds especially for farmers who rely on specific planting cycles.

This work identified distributor network as a major supply chain performance driver based on the strength of its network. The coastal and central regions also had stronger network coverage and hence experienced fewer stock-outs of seeds in any one year. Whereas the northern and western networks with relatively weaker social networks reported many stock-out incidences and delayed seed delivery. The role of private players was also another important factor in improving the supply chain functions. Consequently, benefiting from a higher density of private sector involvement (75% in the coastal region), the supply chain for food and other essentials was more sensitive to market requirements and capable of confronting logistics complexities.

However, the study also came across various on supply chain management issues. Areas of low network development, for instance the northern and the western zones experienced delays in transportation as a major bottleneck to delivery time and cost. This was especially compounded by the fact that few private players had ventured into these areas which was evident by the fact that the public sector was left to meet the demand for hybrid seeds. To address these problems, solutions have to be made on several fronts: provision of transport infrastructure, establishment of incentives for private sector participation and enhancements of supply chain integration.

The percentage of adoption depends with availability of hybrid paddy seeds. The results of this study suggest that, seed supply follows supply chain management and sound market systems. A cross comparison established that farmers in regions with efficient supply chain and strong market interface including the coastal and central administrative regions achieved higher availability indexes and high level of satisfaction. On the other hand, the supply chain in the early development stage, particularly in the northern and western region, posed a great challenge to ensure the right stock level.

Table 3 shows the availability index that was obtained from the key informants and ranged from 9 in the coastal region to 4 in the western region indicating poor access to seeds in Odisha. Better infrastructure and more pronounced distributor networks and higher private sector development can better satisfy and maintain consistent stock of the higher available stock in the coastal and central regions. These two regions demonstrated high level of satisfaction from the availability and quality of the hybrid paddy seeds hence higher level of adoption.

However, the supplies finality in the north and west states underscores the need to increase seed distribution initiatives in those zones. The study also revealed that many of the farmers could not access seeds of their required qualities due to delayed arrivals, which left them to use local seeds or be denied planting time if they had to wait longer for the improved seeds to arrive. Enhancements in these issues will thus not only entail supply side improvements such as better seed supply chains but also better strategies to supply hybrid seeds to these regions.

6.1 Challenges Faced by Stakeholders

The analysis of the case study therefore revealed that the identified stakeholders in the hybrid paddy seed supply chain experienced diverse and complex logistical, financial and policy-related difficulties. Other physical constraints for example transportation hitches and poor infrastructure were most felt in the northern and western zones. These challenges meant that delivery of stock took longer time, they were expensive and was accompanied by more instances of stock-out eliminating efficiency within the supply chain.

Another problem area that was clearly recognized in the study was the problem of price instability especially in regions with low state subsidies. These respondents in the northern, southern and western parts of the country expressed very serious problems in this area stating that prices vary sharply and that this makes it hard for farmers to anticipate price changes hence fail to afford hybrid seeds. Uncertainty regarding the prices to customers in these regions was also a major problem, compounded by the absence of price stabilization mechanisms; thus implying a lowered RAC and decreased market penetration.

One key factor was the levels of public support and or its lack, becoming a key determiner of stakeholder satisfaction. Relatively in regions which government subvention is fairly evident such as the coastal and central regions, the stakeholders' perceived satisfaction was higher than the perceived challenges. On the other hand, stakeholder in area that government support was deemed insufficient in the north and west complained about lack of investment in infrastructures, price support, and training. Based on these ideas, it is possible to conclude that the enhancement of governmental involvement can be useful in underdeveloped areas to solve some of the problems discussed in the present study.

6.2 Conclusion

In conclusion, study suggests extended outlook toward more sustainable market conditions in distinct sectors of Odisha along with an adequate supply chain system focused toward availability of those particular hybrid paddy seed. Coastal areas and central region have however experienced a strong base in infrastructures, government support and order supplies of hybrid seeds while cross the northern, southern and western countries of the world still experience the following challenges. Mitigating these challenges demands additional investment in infrastructure, supply chain enhancement, private sectors engagement and government's expansion of support in underdeveloped zones. If these problems are solved, Odisha may improve distribution of the hybrid paddy seeds and address the issues of poor accessibility hence increasing productivity of agriculture and better improvement of the farmers' lives in the state.

Acknowledgements I am sincerely grateful to Dr. Manmath Nath Samantaray, Professor at ASBM School of Business, for his invaluable guidance, support and encouragement throughout this research. His insights into sustainable market dynamics and supply chain management have been instrumental in shaping this study. Thank you for your mentorship and inspiration.

Funding The authors did not receive support from any organization for the submitted work.

Conflict of Interest The authors have no significant financial, professional and personal interest that might have influenced presentation of words in the manuscript.

Ethics Approval The research proposal for the study was approved by the institutional review board for ethics in research. Written desires to participate in the study and being informed of the aim, purpose and process of the study were obtained from all actors involved, before data collection was conducted. Participant data was kept anonymous and aggregate data was used to reduce one's individuality making confidentiality intact.

References

Afzal, I., Shabir, R., & Rauf, S. (2019). Seed production technologies of some major field crops. *Agronomic Crops: Volume 1: Production Technologies*, 655–678.

Airee, S., Ojha, B., Ojha, A., & Bhandari, A. (2020). Supply Chain Analysis of Rice (Oryza sativa. L.) Sub-Sector in Kanchanpur District, Nepal. *International Journal of Social Sciences and Management*.

Chartres, C. J., & Noble, A. (2015). Sustainable intensification: Overcoming land and water constraints on food production. *Food Security, 7*, 235–245.

Dalei, N. N., & Dutta, P. (2015). Sharecropping system and its supply chain management: A case study from coastal belt of Odisha. *International Journal of Applied Research*.

Dar, M. H., Bano, D. A., Waza, S. A., Zaidi, N. W., Majid, A., Shikari, A. B., & Singh, U. S. (2021). Abiotic stress tolerance-progress and pathways of sustainable rice production. *Sustainability, 13*(4), 2078.

Dash, S. K., & Singhi, R. (2019). Seed production: A better business prospect for farmers of Odisha. *Journal of Emerging Technologies and Innovative Research*.

Dethier, J. J., & Effenberger, A. (2012). Agriculture and development: A brief review of the literature. *Economic Systems, 36*(2), 175–205.

Devkota, K., Yadav, S., Khanda, C., Beebout, S. E. J., Mohapatra, B., Singleton, G., & Puskur, R. (2020). Assessing alternative crop establishment methods with a sustainability lens in rice production systems of Eastern India. *Journal of Cleaner Production*.

Dorairaj, D., & Govender, N. T. (2023). Rice and paddy industry in Malaysia: Governance and policies, research trends, technology adoption and resilience. *Frontiers in Sustainable Food Systems, 7*, 1093605.

Gautam, D., Magar, D., & Gautam, S. (2018). Rice seed production and marketing practices in Nepal. *Journal of Agriculture and Environment*.

Goel, V., & Bhaskaran, S. (2006). Marketing practices and distribution system of rice in Punjab, India. *Journal of International Food & Agribusiness Marketing*.

Johns, T., Powell, B., Maundu, P., & Eyzaguirre, P. B. (2013). Agricultural biodiversity as a link between traditional food systems and contemporary development, social integrity and ecological health. *Journal of the Science of Food and Agriculture, 93*(14), 3433–3442.

Kameswara Rao, N., Dulloo, M. E., & Engels, J. M. (2017). A review of factors that influence the production of quality seed for long-term conservation in genebanks. *Genetic Resources and Crop Evolution, 64*, 1061–1074.

Kumar, N., & Barker, N. (2023). A study on brand promotion of hybrid paddy in Koderma District of Jharkhand. *Asian Journal of Agricultural Extension, Economics & Sociology*.

Kumar, R., Stephen, A. J., & Kumar, A. (2023). Marketing of hybrid paddy seeds in Saharsa district of Bihar, India. *Asian Journal of Agricultural Extension, Economics & Sociology*.

Langyintuo, A. S., Mwangi, W., Diallo, A. O., MacRobert, J., Dixon, J., & Bänziger, M. (2010). Challenges of the maize seed industry in eastern and southern Africa: A compelling case for private–public intervention to promote growth. *Food Policy, 35*(4), 323–331.

Matuschke, I., Mishra, R. R., & Qaim, M. (2007). Adoption and impact of hybrid wheat in India. *World Development, 35*(8), 1422–1435.

Pandey, R., Vinayagam, S., Krishnan, M., & Akhila, K. (2020). Factors influencing the buying behavior of hybrid paddy seed growers. *Asian Journal of Agricultural Extension, Economics and Sociology*.

Pradhan, P., Fischer, G., Van Velthuizen, H., Reusser, D. E., & Kropp, J. P. (2015). Closing yield gaps: How sustainable can we be? *PLoS ONE, 10*(6), e0129487.

Rutsaert, P., Donovan, J., & Kimenju, S. (2021). Demand-side challenges to increase sales of new maize hybrids in Kenya. *Technology in Society*.

Shravanthi, A. R. (2022). Market potential and promotional strategies for VSPL hybrid paddy seeds in Koraput District of Odisha. *Indian Research Journal of Extension Education*.

Singh, N. P., & Sahoo, S. (2021). Supply chain of paddy and wheat seeds production in Tarai development corporation and private seed firms of Uttarakhand. *International Journal of Current Microbiology and Applied Sciences*.

Smale, M., Cohen, M. J., & Nagarajan, L. (2009). Local markets, local varieties: Rising food prices and small farmers' access to seed.

Spielman, D., Kolady, D., & Ward, P. S. (2013). The prospects for hybrid rice in India. *Food Security*.

Thapa et al. (2018). Supply Chain analysis of rice sub-sector in Dang District, Nepal. *International Journal of Applied Sciences and Biotechnology*.

Impact of Goods and Services Tax on the Performance of MSMEs in Uttarakhand State

Jyotika Kaur, Rupa Khanna, and Tanupriya Choudhury

Abstract GST has been explained as "Good & Simple Tax" by our Honorable Prime Minister Shri Narendra Modi as the reform seeks to achieve simplicity, transparency, and ease in 17 Central and State Indirect Taxes and 23 Cesses were subsumed. The present work was undertaken to study the impact of GST on the performance of MSMEs in the State of Uttarakhand. In view of studying the said problem, a questionnaire schedule was prepared to collect the desired data. The study has been conducted in the districts of Uttarakhand. In this context, statistical tools such as confirmatory factor analysis, reliability and validity test, and regression analysis were used for the purpose of analysis. The findings have therefore established that certain variables such as compliance, cost, market conditions, innovation and learning, and operating activity are some of the factors that determine performance by MSME in the Uttarakhand State. Further the study was limited to the state of Uttarakhand with an emphasis only on the MSME sector. Future studies can be conducted in the larger states which have densely populated MSMEs along with studying impact of the reform on the consumers.

Keywords Good and simple tax · Simplicity · Transparency · Questionnaire · Performance

1 Introduction

The new taxation was implemented "all over India (including the State of Jammu & Kashmir) with the assent of the President of India" (Government of Jammu and Kashmir, 2017). It is publicly acknowledged as the "Constitution (One Hundred & First Amendment) Bill", 2017 (CBIC, Government of India, n.d.). Goods & Services Tax has replaced "17 Central & State Indirect Taxes and 23 Cess" leading to a harmonized structure of indirect taxation in the nation.

J. Kaur (✉) · R. Khanna · T. Choudhury
Graphic Era Deemed to Be University, Dehradun, Uttarakhand, India
e-mail: kaur.jyotika@gmail.com

R. Kumar and S. N. Mohanty (eds.), *Sustainable Economy Models in the Age of Industry 5.0*, https://doi.org/10.1007/978-981-96-4104-8_18

The GST is levied and charged concurrently by Central and the State Government thus creating federalism in the country. The powers related to amendments in GST is vested in a body, i.e., GST Council which is constituted to take decisions related to the new taxation regime. The taxation possess uniformity in the tax rates across the nation. The foundation of the GST lies in the input tax credit mechanism which eradicates cascading effect resulting in lower prices of the goods and services. The regime is completely based on Information Technology for its operations. A mechanism termed as Anti Profiteering Mechanism has been formulated which looks upon that the reduction in prices of the goods and services are passed on to the consumers.

GST holds with it the concept of "One Tax, One Market, One Nation" (Jaitley, 2017) which aims at replacing "multiple indirect taxes and creation of one tax".

The new indirect taxation regime is designed to bring about radical changes in the economy of India by removing the lacunae present in the prior taxation regime. The same happens with the GST, which has uniform tax rates all over the country, a digitalized compliance process, an e-way bill system that removes waiting time at the check posts to ensure unrestricted flow of goods and services across Indian States.

2 Review of Literature

The execution of the new taxation regime in the nation marks a substantial transformation in the country's indirect tax regime. For the MSMEs in Uttarakhand, which forms a great proportion of the state's industrial base, comprehension of the GST regime's nuances is very important. The MSMEs in Uttarakhand represent diversity, from producers of traditional handicraft items to units which are modern manufacturing units, and each of them has problems and prospects facing the GST regime. The implications of GST on MSMEs in Uttarakhand requires an understanding of the tax system's intricacies and the specific characteristics of MSMEs in the state for which the study was undertaken (Bhalla et al., 2023) encompass the tax restructuring factors, which influence a firm's features and demographic structures. Using regression, SEM, exploratory, and CFA tools to access the impact are upon ROE and ROI as a measure of performance, it comes up with findings that firm size, turnover, and DuPont analysis drive upsurge positive influence on ROI and ROE. The result of has implications for policymakers, government, investors, as well as businesses in the understanding of factors boost returns and lead to realizing the sustainable development goals (Mohanty et al., 2024). It uses a structured questionnaire that elicits responses from registered MSMEs in India. It is based on the stratified sampling technique. "Then, EFA and CFA were performed to deduce the latent variables and assess the model's fit by indicating that the model had a good fit". From the findings, it is well noted that overall reforms in tax systems have positive effects on the financials of the enterprises and how firm characteristics and factors that influence financial performance, including reliability regarding various drivers of this performance.

Deloitte Touche Tohmatsu India LLP (2023), the survey tried to get industry views on the GST journey and focused on areas of ease of doing business (EoDB) and tax technology adoption. The methodology was online surveys with senior leaders across industries, comprising 28 questions on the organization's tax experience and the impact of recent compliance initiatives. 612 responses were collated from six industries. The study snapshot gave that 94% of respondents across industries showed an affirmative sentiment toward GST. The survey findings revealed the significance of digital laboratories to enhance the capabilities of the tax authorities in conducting risk assessments and profiling the complete spectrum of taxpayers and in capturing real-time data.

Press Information Bureau (2023), elaborates on the numerous benefits of GST, which are positive to consumers, reduction of the tax burden on common usage items, and facilitation of compliance for MSMEs through measures such as the composition scheme & technological advancements. It also stresses that technology, at the core of the GST Facilitation Revolution, enables increased revenue collection, improved compliance, and smooth processes. Furthermore, the study outlines recognition of compliant taxpayers from different sectors of industries, and the focus was laid on MSMEs.

Kankariya et al. (2021) laid the study where "the objectives of the study was to seek the effect of tax on prices of goods supplied by the traders, compare pre & post GST impact on compliance costs of the trader, understand the effect of GSTN on the online tax Compliance of Traders concerning the Technology Acceptance Model, evaluate the perception of traders of goods towards the Quarterly Return Monthly Payment Scheme under GST & assess the impact of Planned Behaviour of Traders on their GST Tax Compliance Intention concerning the Theory of Planned Behaviour." The study uses descriptive research design and both primary and secondary data. This study's conclusions are found to be stating that GST has no influential effect on the food & essential items' consumer price index. Further it was found that post GST implementation the prices of goods did not become astronomic but price increased due to COVID-19 in the month of June 2021. The impact of GST on traders of Jalgaon district is negligible while compliance cost rises in post GST regime. Traders also complain that due to GST regime they were facing issues in depositing their returns. Therefore, it was concluded with the improvements in the GST structure so that the issues noticed within the traders are removed.

3 Methodology

3.1 Objective and Hypotheses

For study, the research objective that was designed was:

1. To assess the impact of the GST on the performance of MSMEs in Uttarakhand State.

To fulfill the objective, the primary study was conducted to evaluate the GST. scenario on the MSMEs.

The hypotheses framed is as follows:

Hypothesis 1: Compliance and Performance

H0: Compliance does not have a significant impact on Performance.

H1: Compliance has a significant positive impact on Performance.

Hypothesis 2: Cost and Performance

H0: Cost does not have a significant impact on Performance.

H1: Cost has a significant positive impact on Performance.

Hypothesis 3: Innovation & Learning and Performance

H0: Innovation & Learning does not have a significant impact on Performance.

H1: Innovation & Learning has a significant positive impact on Performance.

Hypothesis 4: Market Conditions and Performance

H0: Market Conditions do not have a significant impact on Performance.

H1: Market Conditions have a significant positive impact on Performance.

Hypothesis 5: Operating Activity and Performance

H0: Operating Activity does not have a significant impact on Performance.

H1: Operating Activity has a significant positive impact on Performance.

3.2 Research Design

The study design refers to the layout for conducting the research study. It outlines the details of procedures for carrying out the research and solving the research problem.

The study is exploratory in nature in recognizing the factors in the context of the Indian markets, the scope is limited only to micro, small and medium enterprises.

3.3 Sample Area

The study was conducted in Uttarakhand state which is divided into two regions: "Garhwal Region & Kumaon Region. Under these two regions falls thirteen districts of the State; seven districts in the Garhwal region while six districts in the Kumaon region".

For the purpose of the study out of 13 districts, such districts were selected which are hub of MSME, i.e., Dehradun, Haridwar, Uttarkashi & Tehri Garhwal from Garhwal region while Udham Singh Nagar & Nainital from Kumaon region.

Data was collected by the MSME registered as well as unregistered in the above districts through personal visits & online via mail (google form).

3.4 Sample Size and Data Collection

For the requisite of data collection, the respondents are MSMEs located in the Garhwal region & Kumaon region. The questionnaire was designed in two parts: Part A & Part B. Part A consists of demographic profile of the respondents while Part B consists of statements in the form of variables based on the seven point Likert scale for the purpose of conducting a thorough study.

The questionnaire was circulated through personal visits & via google form. 500 questionnaires circulated across the respondents, 310 of them were suitable for the purpose of this study.

4 Data Analysis and Interpretation

The data obtained through questionnaire was analyzed as follows:

4.1 Cross-Loadings

Cross-loadings are one of the important aspects of confirmatory factor analyses, especially as "indicators of convergent and discriminant validity of constructs within a measurement model". In the study, what we find is that almost every indicator loads mainly on the corresponding construct; this is a good sign of convergent validity.

Cross-loadings describe the correlation of the latent variables (such as Compliance, Cost, Innovation & Learning, Market Conditions, Operating Activity, Performance) and their respective measured indicators, for example, CO1, CO2, CPM1, etc. as mentioned in the Table 1.

Interpretation

1. Cost—CO

CO1 to CO5: These indicators load significantly on the Cost construct, with high cross-loading values ranging from 0.775 to 0.815. This suggests that these indicators effectively capture different aspects of compliance within the model.

2. Compliance—CPM

CPM1 to CPM8: Similarly, indicators for Compliance exhibit strong cross-loadings with the Compliance construct, ranging from 0.737 to 0.847. This indicates that these indicators are robust measures of compliance-related factors.

3. Innovation & Learning—INL

INL1 to INL3: The fit indices from the Innovation & Learning construct range from 0.702 to 0.823, indicating how very effective the measurements are in capturing the dimensions of innovation and learning within a model.

Table 1 Cross-loadings of the constructs

	CPM	CO	INL	MC	OA	PF
CO1		0.775				
CO2		0.797				
CO3		0.789				
CO4		0.815				
CO5		0.788				
CPM1	0.756					
CPM2	0.742					
CPM3	0.838					
CPM4	0.846					
CPM5	0.837					
CPM6	0.737					
CPM7	0.847					
CPM8	0.745					
INL1			0.823			
INL2			0.733			
INL3			0.702			
MC1				0.842		
MC2				0.808		
MC3				0.740		
MC4				0.752		
MC5				0.845		
OPA1					0.800	
OPA2					0.892	
OPA3					0.876	
OPA4					0.851	
PF1						0.870
PF2						0.869
PF3						0.909

4. Market Conditions—MC

MC1 to MC5: Indicators for Market Conditions have strong cross-loadings with the Market Conditions construct: from 0.740 to 0.845. This indicates that indicators are good measurements of the factors related to the market.

5. Operational Activity—OA

OA1 to OA4: Indicators for Operating Activity have significant cross-loadings with the Operating Activity construct, ranging from 0.800 to 0.892, indicating the relationship between operational performance and the measures of the indicators.

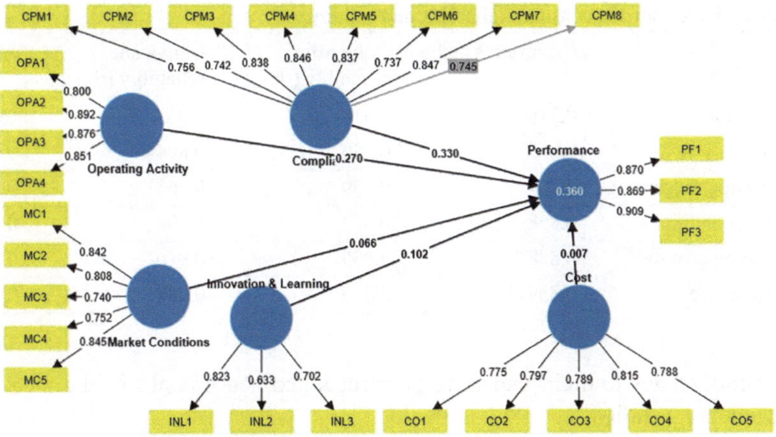

Fig. 1 Graphical representation of the cross-loadings

6. Performance—PF

PF1 to PF3: These indicators have strong cross-loadings with the Performance construct, ranging from 0.869 to 0.909. These should indicate that these indicators, taken together, are good at capturing different dimensions of organizational performance within the model. Figure 1 showing Graphical Representation of the Cross-loadings.

Strong cross-loadings indicated that the measured indicators captured their respective latent constructs properly.

The given model can be assured for reliability and validity, thus validating further analysis and interpretations in terms of relationships between constructs and indicators.

4.2 *Analysis*

Reliability Analysis—It assesses the consistency and stability of measurement instruments. One of the most commonly used measures of internal consistency is Cronbach's alpha, which measures how closely related a set of items is as a group. These values were high, ranging from 0.704 to 0.916 as stated in Table 2, thus proving that all constructs have a strong internal consistency in the study, and thereby showing the confidence in the reliability of measurement instruments.

Validity Analysis—It evaluates whether measurement instruments have the intended constructs or not. Convergent validity, as calculated with indicators including Average Variance Extracted, demonstrates what extent items in a particular construct are correlated. With such AVE values in this study ranging from 0.523 to 0.780 as stated in Table 2, this indicates that a sufficient amount of variance in the

Table 2 Table showing construct reliability and validity

	Cronbach's alpha	Composite reliability (rho_a)	Composite reliability (rho_c)	AVE
Compliance	0.916	0.919	0.932	0.632
Cost	0.853	0.856	0.894	0.629
Innovation & learning	0.704	0.759	0.765	0.523
Market conditions	0.858	0.873	0.898	0.638
Operating activity	0.878	0.885	0.916	0.732
Performance	0.859	0.865	0.914	0.780

indicators relates to their respective constructs and hence confirms the measuring instruments' validity.

Interpretation

(1) Compliance:

Reliability:

The computed Cronbach's alpha of 0.916 is indicative that the test items for Compliance are highly related, thus having high evidence of internal consistency. This means that respondents consistently answer the items in a similar manner.

Composite reliability measures also prove a high degree of internal homogeneity: rho_a = 0.919, rho_c = 0.932. Both are far higher than the conventional minimum acceptable value of 0.7, indicating good reliability.

Validity:

The AVE value to be 0.632, this shows that about 63.2% of the variance in the indicators—in this case, items—is accounted for by the Compliance construct.

(2) Cost:

Reliability:

The values of Cronbach's alpha of 0.853, the items measuring Cost validate high internal consistency.

Both composite reliability measures (rho_a = 0.856, rho_c = 0.894) also represents strong reliability.

Validity:

The value of 0.629 for AVE indicates that about 62.9% of the indicators' total variation was brought forth by the Cost construct, hence acceptable convergent validity.

(3) Innovation & Learning:

(a) **Reliability:**

While the Cronbach's alpha of 0.704 might be considered a little low compared to other constructs, it still exhibits acceptable internal consistency.

The composite reliability measures are higher, thus more robust: rho_a = 0.759 and rho_c = 0.765, indicating more robust internal consistency.

Validity:

With a lower AVE value of 0.523, it infers that around 52.3% of the indicators' variation is explained by the construct of Innovation & Learning. Compared to the other constructs, this is slightly lower; still, it is within the threshold that marks it as acceptable for convergent validity.

(4) Market Conditions:
(a) **Reliability**

The value of 0.85 of Cronbach alpha shows that the items measuring Market conditions demonstrate high internal consistency.

Both composite reliability measures (rho_a = 0.873 & rho_c = 0.898) further confirm strong reliability.

An AVE value of 0.638, most of the variance in the indicators is explained by the construct and, thus, it has good convergent validity.

(5) Operating Activity
(a) **Reliability**

With a Cronbach's alpha of 0.878, the items measuring Operating Activity demonstrate high internal consistency.

Both composite reliability measures (rho_a = 0.885 & rho_c = 0.916) further confirm strong reliability.

Validity

The Average Variance Extracted (AVE) value of 0.732 indicates that approximately 73.2% of the variance in the indicators (items) is attributed to the construct of Operating Activity. This suggests good convergent validity, as a substantial portion of the variance in the indicators is explained by the construct.

(6) Performance
(a) **Reliability**

The Cronbach's alpha value of 0.859, shows that the items measuring Operating Activity demonstrate high internal consistency.

The measures for composite reliability further confirm strong reliability: rho_a = 0.865, rho_c = 0.914.

Validity

The AVE value of 0.780 indicates that the construct of performance attributes around 73.2% of the variance in the items. This suggests a good convergent validity, whereby a substantial portion of the indicators' variances are explained by the construct.

Moreover, the measurement model finds its roots to be robust and reliable with acceptable thresholds for all the instruments. Although Innovation & Learning shows a little lower reliability compared to the others, it still meets acceptable thresholds for both reliability and validity. These results give confidence in the accuracy of measurement of the constructs.

4.3 Heterotrait-Monotrait Ratio (HTMT)

It estimates discriminant validity by comparing the between-construct correlations with the within-construct correlations of a research model. A high HTMT ratio could indicate that two constructs are not distinct from one another and, therefore, overlap in ideas they try to conceptualize.

Interpretation of the Table 3

- **Compliance**

No values are provided to compare Compliance with other constructs, this means that this construct is not compared directly in the analysis. This probably means that it is thought of as a unique and standalone construct within the model.

- **Cost**

HTMT Values:

- With Compliance: 7.076
- With Innovation & Learning: 0.333
- With Market Conditions: 0.240
- With Operating Activity: 0.487
- With Performance: 0.558

Interpretation:

The exceptionally high HTMT value of 7.076 with Compliance raises concerns about discriminant validity between Cost and Compliance. Such a high value suggests that these two constructs might not be sufficiently distinct from each other within the model.

Conversely, the HTMT values with other constructs (Innovation & Learning, Market Conditions, Operating Activity, and Performance) are all below 1, indicating acceptable discriminant validity.

- **Innovation & Learning**

HTMT Values:

- With Cost: 0.333

Table 3 Table showing Heterotrait-monotrait ratio (HTMT)—Matrix

	CPM	CO	INL	MC	OA	PF
Compliance						
Cost	7.076					
Innovation & learning	0.333	0.338				
Market conditions	0.240	0.261	0.596			
Operating activity	0.487	0.489	0.649	0.395		
Performance	0.558	0.550	0.471	0.321	0.566	

- With Market Conditions: 0.338
- With Operating Activity: 0.489
- With Performance: 0.550

Interpretation:
The HTMT values with all other constructs are less than 1, so there is satisfying discriminant validity. This would mean that Innovation and Learning are quite distinguishable from other constructs in the model.

- **Market Conditions**

HTMT Values:

- With Cost: 0.240
- With Innovation & Learning: 0.338
- With Operating Activity: 0.596
- With Performance: 0.471

Interpretation:
It shows relatively high HTMT value of 0.596 with Operating Activity in terms of discriminant validity between Market Conditions and Operating Activity.

- **Operating Activity:**

HTMT Values:

- With Cost: 0.487
- With Innovation & Learning: 0.489
- With Market Conditions: 0.596
- With Performance: 0.321

Interpretation:
The value of HTMT, 0.596 with Market Conditions, shows weak discriminant validity with Operating Activity. This suggests that these two constructs may not be distinct enough within the model.

- **Performance**

HTMT Values:

- With Cost: 0.558
- With Innovation & Learning: 0.550
- With Market Conditions: 0.471
- With Operating Activity: 0.321

Interpretation:
The HTMT values for Performance are below 1 indicating satisfactory discriminant validity with the other constructs in the model.

4.4 *Fornell-Larcker Criterion:*

It estimates discriminant validity by comparing the between-construct correlations with the within-construct correlations of a research model. A high HTMT ratio indicates that two constructs are not distinct from one another and, therefore, may overlap in the underlying ideas they conceptualize.

Interpretation of the values stated in the Table 4

1. **Compliance:**

   - Square root of AVE: 0.795

Interpretation: The AVE value 0.795 indicates that the variation that is captured by the construct itself with its measured variables.

2. **Cost:**

   - Square root of AVE: 0.949
   - Correlations:
   - With Innovation & Learning: 0.305
   - With Market Conditions: 0.221
   - With Operating Activity: 0.437
   - With Performance: 0.501

Interpretation: The AVE for Cost value is 0.949, which shows that the Cost has a substantial amount of variance in its measured variables. This advocates that Cost indicates discriminant validity; it is more correlated with its own measures than with other constructs of the model.

3. **Innovation & Learning:**

   - Square root of AVE: 0.724
   - Correlations:
   - With Market Conditions: 0.472
   - With Operating Activity: 0.549
   - With Performance: 0.384

Table 4 Fornell-larcker criterion

	CPM	CO	INL	MC	OA	PF
Compliance	0.795					
Cost	0.949	0.793				
Innovation & learning	0.305	0.295	0.724			
Market conditions	0.221	0.231	0.472	0.799		
Operating activity	0.437	0.424	0.549	0.348	0.856	
Performance	0.501	0.480	0.384	0.283	0.496	0.883

Interpretation: The values of square root of the AVE for Innovation & Learning was 0.724, indicating that it captures a significant portion of variance in its measured variables. Comparing this with its correlations with other constructs, it is found that it exceeds these correlations. Thus, Innovation & Learning demonstrates discriminant validity.

4. **Market Conditions:**

   - Square root of AVE: 0.799
   - Correlations:
   - With Operating Activity: 0.348
   - With Performance: 0.283

Interpretation: The values of square root of the AVE for Market Conditions is 0.799, indicating a substantial amount of variance captured by this construct. When comparing this with its correlations with other constructs, it surpasses these correlations. Thus, Market Conditions also exhibit discriminant validity.

5. **Operating Activity:**

   - Square root of AVE: 0.856
   - Correlations:
   - With Performance: 0.496

Interpretation: The values of square root of the AVE for Operating Activity is 0.856, indicating that it captures a significant portion of variance in its measured variables. Comparing this with its correlation with Performance, it is found that it exceeds this correlation. Hence, Operating Activity demonstrates discriminant validity.

6. Performance:

   - Square root of AVE: 0.883

Interpretation: Discriminant validity for Performance cannot be fully assessed without knowing its correlations with remaining constructs. The square root of the AVE is 0.883; hence, Performance itself has a large amount of variance relative to its measured variables.

All constructs in this study meet the Fornell-Larcker criterion and thus cannot be rejected as having good discriminant validity, hence supporting the validity of the measurement model.

4.5 Variance Inflation Factor (VIF) & R Square

Variance Inflation Factor (VIF) and R-squared (R^2) are important metrics used in regression analysis to assess multicollinearity and the explanatory power of the model, respectively.

4.5.1 Variance Inflation Factor

VIF thus describes the extent of multicollinearity among predictor variables in a regression model. A high value of VIF will indicate that a predictor variable is highly correlated with other variables in the model, hence leading to unstable estimates. It enumerates how many times the variance of an estimated regression coefficient has increased because of multicollinearity in the model.

Interpretation of the Tables 5 and 6

1. Compliance - > Performance:

   Vif: 2.567

   Interpretation: The VIF value of 2.567 for the relationship between Compliance and Performance suggests that there is relatively low multicollinearity between these two variables. This indicates that Compliance does not strongly correlate with other predictors in the model to a degree that would significantly inflate the variance of its regression coefficient when predicting Performance.

2. Cost - > Performance:

   Vif: 3.450

   Interpretation: The VIF value of 3.450 for the relationship between Cost and Performance indicates a moderate level of multicollinearity. While not extremely

Table 5 Variance inflation factor

	VIF
Compliance - > performance	2.567
Cost - > performance	3.450
Innovation & learning - > performance	1.649
Market conditions - > performance	1.315
Operating activity - > performance	1.632

Table 6 R square

	R-square	R-square adjusted
Performance	0.360	0.353

Table 7 Path coefficients

	Beta	STDEV	t value	P values
Compliance - > Performance	0.138	0.142	0.976	0.031
Cost - > Performance	0.206	0.130	1.590	0.021
Innovation & Learning - > Performance	0.159	0.150	1.058	0.031
Market Conditions - > Performance	0.029	0.082	0.352	0.012
Operating Activity - > Performance	0.274	0.107	2.553	0.011

high, it suggests that there may be some correlation between Cost and other predictors in the model, potentially leading to an increase in the variance of the regression coefficient for Cost when predicting Performance.

3. Innovation & Learning - > Performance:

   Vif: 1.649

   Interpretation: The VIF value of 1.649 for the relationship between Innovation & Learning and Performance suggests relatively low multicollinearity between these two variables. This indicates that Innovation & Learning does not strongly correlate with other predictors in the model to a degree that would significantly inflate the variance of its regression coefficient when predicting Performance.

4. Market Conditions - > Performance:

   Vif: 1.315

   Interpretation: The VIF value of 1.315 related to Market Conditions and Performance shows very low multicollinearity between the two variables. This can mean that market conditions have very little correlation with other predictors in the model and, therefore, minimally inflate the variance of the regression coefficient for this predictor when the dependent variable of performance is being predicted.

5. Operating Activity - > Performance:

   Vif: 1.632

   Interpretation: The value of VIF in the relation between Operating Activity and Performance is 1.632, showing relatively low multicollinearity between these two variables. Similar to other cases, this indicates Operating Activity does not strongly correlate with other predictors in the model to a degree that would inflate the variance of its regression coefficient when predicting Performance in a way that is of undue concern.

VIF values indicate that while there may be some level of multicollinearity present between Cost and Performance, the relationships between Compliance, Innovation & Learning, Market Conditions, and Operating Activity with Performance exhibit low to moderate multicollinearity, suggesting that these predictors can be considered fairly independent when predicting Performance in the regression model.

4.5.2 R Square

R-squared (R^2) is an "estimate to measure that how much the independent variables explain the variability of the dependent variable in a regression model. It signifies the proportion of variance in the dependent variable that is predictable from the independent variables". In the study, the R-squared values indicate the amount of variance in Performance explained by Compliance, Cost, Innovation & Learning, Market Conditions, and Operating Activity.

R-squared (R^2):

The R-squared value of 0.360 indicates that approximately 36.0% of the total variance in the dependent variable, Performance, is explained by the independent variables (predictors) included in the regression model. In other words, around 36.0% of the variability observed in Performance can be attributed to the variables included in the model. However, it also suggests that there are other factors not captured by the model that contribute to the remaining 64.0% of the variability in Performance.

Adjusted R-squared:

The values of adjusted R-squared 0.353 is somewhat lower than the R-squared value. The adjusted R-squared accords for the number of predictors in the model and the sample size.

In this case, the adjusted R-squared being slightly lower than the R-squared value suggests that the inclusion of additional predictors in the model may have led to a slight overestimation of the model's explanatory power when considering the sample size and number of predictors.

Interpretation:

While the values of R-squared and adjusted R-squared indicate that the regression model explains a moderate amount of the variability in Performance, there is still a significant portion of unexplained variability.

This implies that there are likely other factors influencing Performance that are not accounted for in the current model. It could be additional variables that were not included or other complex interactions between the included variables.

4.6 Path Coefficients

Path coefficients indicate the "strength and direction of the relationships between constructs in a structural equation model". In the study, the path coefficients show the impact of Compliance, Cost, Innovation & Learning, Market Conditions, and Operating Activity on Performance.

Interpretation of the Table 7

(1) Compliance - > Performance:
  - Beta (β): 0.138

This indicates that for every one-unit increase in Compliance, there is, on average, a 0.138 unit increase in Performance, holding other variables constant.

– Standard Deviation (STDEV): 0.142

It provides a measure of the variability or dispersion of the path coefficient estimate.

– t value: 0.976

The t value assesses the statistical significance of the path coefficient.

– p-value: 0.031

A p-value below the significance level (commonly 0.05) suggests that the relationship is statistically significant, i.e., rejecting the null hypothesis and accepting the alternate hypothesis.

Interpretation:

The path coefficient suggests a positive relationship between Compliance and Performance. A statistically significant p-value (0.031) indicates that this association is uncertain to have occurred by chance alone. Therefore, an increase in Compliance tends to be associated with an increase in Performance.

(2) Cost - > Performance:
- Beta (β): 0.206
- STDEV: 0.130
- t value: 1.590
- p-value: 0.021

Interpretation:

The path coefficient of 0.206 means that, on average, an increase in the Cost variable by one unit will result in a cumulative increase of 0.206 units in the Performance variable, keeping all other variables at a constant level. The relationship is statistically significant because the p-value is 0.021, indicating cost significantly affects performance, i.e., rejecting the null hypothesis and accepting the alternate hypothesis.

(3) Innovation & Learning - > Performance:
- Beta (β): 0.159
- STDEV: 0.150
- t value: 1.058
- p-value: 0.031

Interpretation:

This means that for a one-unit increase in Innovation & Learning, there is an average increase of 0.159 units in Performance, other variables being constant. The result shows that the relationship is statistically significant at a p-value of 0.031, hence Innovation & Learning is significant to Performance, i.e., rejecting the null hypothesis and accepting the alternate hypothesis.

(4) Market Conditions - > Performance:
- Beta (β): 0.029
- STDEV: 0.082
- t value: 0.352
- p-value: 0.012

Interpretation:

The path coefficient indicates a positive relationship between Market Conditions and Performance, but it is relatively weak compared to other predictors. However, it is statistically significant (p-value = 0.012), suggesting that even small changes in Market Conditions can influence Performance, i.e., rejecting the null hypothesis and accepting the alternate hypothesis.

(5) Operating Activity - > Performance:
- Beta (β): 0.274
- STDEV: 0.107
- t value: 2.553
- p-value: 0.011

Interpretation:

The path coefficient indicates that, other things being equal, for one unit of Operating Activity, an average increase of 0.274 units is observed in Performance. It is significant statistically with a p-value equal to 0.011; it means that Operating Activity contributes greatly to Performance, i.e., rejecting the null hypothesis and accepting the alternate hypothesis.

In a nutshell, the path coefficients demonstrate important information on the relationships between Compliance, Cost, Innovation & Learning, Market Conditions, Operating Activity, and Performance in your model. Therefore, the presented statistical significance means that the strength of the relationships may signal the effect of each construct on the performance outcome variable in your study.

Conclusively, the results above reflect the fact that the model has good reliability and validity, with proper relationships between constructs. These findings give insight into the factors that affect performance in the context of Compliance, Cost, Innovation & Learning, Market Conditions, and Operating Activity.

5 Findings, Limitations, and Recommendations for Future Research

5.1 Findings

- The study assessed the impact of the Goods and Services Tax (GST) on the performance of Micro, Small, and Medium Enterprises (MSMEs) where different hypotheses formulated for assessing whether factors, such as compliance, cost, innovation & learning, market conditions, and operating activity do exert significant influences on MSME performance.
- Results Major Findings: All the tested variables—compliance, cost, innovation & learning, market conditions, and operating activity—showed statistically significant effects on MSME performance (p-values < 0.05). However, the most significant influence was that of Operating Activity, $\beta = 0.274$, $p = 0.011$, which further implies that any positive performance. The cost had a positive impact, but it was a mild influence. Innovation & Learning had a positive impact. Compliance and Market Conditions bore weaker relationships in size, but were statistically significant. Null hypotheses were rejected because indeed these factors affect MSME performance under GST.

5.2 Limitations and Recommendations for Future Research

- The study was confined to the MSME sector only. Future researches can be done on other sectors such as Companies, i.e., Private & Public Companies operating nationally & internationally.
- The study was limited to the state of Uttarakhand which is small state. The study can be done in some MNC region to get more wider view of the reform.
- The study was conducted on the businessmen only. The future study can be conducted to study the impact on consumers specifically consumers buying luxuries goods where GST levy is 28% and cess which has made the consumer postpone or forgo their purchase for such goods.

6 Conclusion

GST has brought with it opportunities and challenges to MSMEs' performance. On the positive side, it has brought benefits like simplification, better compliance, and market access, while the implementation was characterized by hurdles and adjustment pains to MSMEs. Therefore, it is crucial for the future that policy interventions address issues with regard to MSMEs, coupled with capacity-building measures and technological support, to maximize the benefits of GST for their growth and sustainability.

Targeted policy interventions by policymakers in developing and implementing policies can work toward supporting the MSME sectors adapt effectively to the GST regime. Hence, the evidence-based policy implications will be instrumental in fostering the growth of MSMEs, enhancing compliance, and the promotion of economic development in the GST landscape.

References

Adhia, Dr. H., & Priya, S. S. (2018). Minutes of the 29th GST Council Meeting held on 04th August 2018. In GST Council, Goods and Services Tax Network (GSTN), Niti Aayog, Planning Commission, Federation of Indian Micro, Small and Medium Enterprises (FISME), Central Board of Indirect Taxes and Customs (CBIC), & Law Review Committee (LRC), *GST Council Meeting* (pp. 1–42).

Aggarwal, A. (2023). Indian Tax System: Situation before and after Goods and Service Tax (GST) [Report]. https://ssrn.com/abstract=4536080

Bhalla, N., Kaur, I., & Sharma, R. K. (2022). Examining the effect of tax reform determinants, firms' characteristics and demographic factors on the financial performance of small and micro enterprises. *Sustainability, 14*(14), 8270. https://doi.org/10.3390/su14148270

Bhalla, N., Sharma, R. K., & Kaur, I. (2023). Effect of goods and service tax system on business performance of micro, small and medium enterprises. *SAGE Open, 13*(2). https://doi.org/10.1177/21582440231177210

Breen, J. P., Bergin-Seers, S., Roberts, I., & Sims, R. (2002). The impact of the introduction of the GST on small business in Australia. *Asian Review of Accounting, 10*(1), 89–104. https://doi.org/10.1108/eb060751

CBIC. (2017, July 8). *Jammu and Kashmir joins GST* [Press release]. Goods and Services Tax. https://cbic-gst.gov.in/pdf/press-release/jammu-kashmir-joins-gst-08.07.2017.pdf. Central Government, State Government, Union territory, local authority. (n.d.). Rate of GST on Services. In nil rate services.

Council, G. (2017a). Minutes of the 20th GST Council Meeting held on 5th August, 2017. In A. Jaitley, Minute Book.

CRISIL. (2017). GST to create business efficiency in the medium term. In *CRISIL Impact Note.*

Daley, J., Wood, D., Grattan Institute, Parsonage, H., & Coates, B. (2015). *A GST reform package.*

Dalvi, S. & KPMG. (2016). Analysis of the GST Law: Is your business prepared for the change?

Datta Chaudhuri, D., & Dafria, M. (2022). The impact of goods and services tax on small businesses in India (WP/01/2021–22/ECO).

Deloitte Touche Tohmatsu India LLP. (2023). *GST @ 6 An insight into the sixth year of GST.*

Deshmukh, A. K., Mohan, A., & Mohan, I. (2022). Goods and Services Tax (GST) implementation in India: A SAP–LAP–Twitter Analytic perspective. *Global Journal of Flexible Systems Management/global Journal of Flexible Systems Management, 23*(2), 165–183. https://doi.org/10.1007/s40171-021-00297-3

EY. (2015). Worldwide VAT, GST and Sales Tax guide.

FLYERS, G. (n.d.). Input Tax Credit Mechanism in GST. In GST Flyers.

From Our Online Archive, & From Our Online Archive. (2017, July 1). GST will help in lowering inflation, propelling GDP: Finance Minister Arun Jaitley. The New Indian Express. https://www.newindianexpress.com/business/2017/Jul/01/gst-will-help-in-lowering-inflation-propelling-gdp-finance-minister-arun-jaitley-1622787.html

Garg, S., Priyanka, P., Narwal, K. P., & Kumar, S. (2023). Goods and Service Tax and its implications on revenue efficiency of sub-national governments in India: An empirical analysis. *American Journal of Business, 38*(4), 193–210. https://doi.org/10.1108/ajb-09-2022-0144

Goods & Service Tax, CBIC, Government of India :: 101st Constitution Amendment Act, 2016. (n.d.). https://cbic-gst.gov.in/hindi/constitution-amendment-act.html

Government of India. (n.d.-a). [A brief note on impact of GST on MSMEs].

Government of India. (n.d.-b). Rate of GST on goods. in *Schedule I: Llist of goods at nil rate.*

GOVERNMENT OF INDIA. (2023). What are Micro, Small & Medium Enterprises? In https://www.dcmsme.gov.in/ssiindia/defination_msme.htm.https://www.dcmsme.gov.in/ssiindia/defination_msme.htm

Government of India, M. of I. and B. (2017). *One Nation-One Tax Goods and Services tax.*

Government of India & Ministry of Finance. (2019). Major Decisions taken by the GST Council in its 32nd Meeting held today under the Chairmanship of the Union Minister of Finance. In *Press Information Bureau* (pp. 1–2). https://pib.nic.in/newsite/PrintRelease.aspx?relid=187403

Government of India, Ministry of Finance, & Department of Revenue. (2017). Notification No.1/2017-Compensation Cess (Rate). In *The Gazette of India* [Press-release].

Government of India & Ministry of Micro, Small and Medium Enterprises. (2021). *Annual Report202122.*https://msme.gov.in/sites/default/files/msmeenglishannualreport2021-22.pdf

Government of India & Ministry of Micro, Small and Medium Enterprises. (2022). *Annual Report 2022 23.*https://msme.gov.in/sites/default/files/MSMEAnnualReport202223English.pdf

Government of Jammu and Kashmir. (2017). THE JAMMU AND KASHMIR GOODS AND SERVICES TAX ACT, 2017.

Government of Uttarakhand. (2023, September 6). *Government of Uttarakhand- Micro,Small & Medium Enterprises Section No VII-3–2023/04(01)-MSME -2023.* https://investuttarakhand.uk.gov.in. Retrieved January 3, 2024, from https://investuttarakhand.uk.gov.in/themes/backend/acts/act_english1694183094.pdf

Grant Thornton. (n.d.). Getting up to speed with VAT/GST in emerging markets. *Grant Thornton.*

GST Council. (2017). Minutes of the 17th GST Council Meeting held on 18 June 2017. In GST Council, *GST Council* (pp. 1–46).

GST Council. (2023). Newsletter Issue-48, March, 2023. In *GST Council*. https://www.gstcouncil.gov.in

Behera, Shri R. B. Subba, & Shri Niranjan Pujari. (2019). *Minutes of 35th GST Council Meeting held on 21st June, 2019* (pp. 1–76).

(CBEC), Sarna, V. N., & Goyal, A. (2017). Minutes of 22nd GST Council Meeting held on 6 October 2017.

GST Council, Goods and Services Tax Network (GSTN), & Government of Assam. (2017). Minutes of 23rd GST Council Meeting held on 10 November 2017. In *MINUTE BOOK*.

GST Council, Goods and Services Tax Network (GSTN), Infosys, & Adhia, Dr. H. (2018). *Minutes of 25th GST Council Meeting held on 18 January 2018.*

GST Council, GST Implementation Committee, Adhia, H., & Pandey, R. (2018). *Minutes of 27th GST Council Meeting held on 04th May, 2018.*

GST Council Secretariat. (2021a). Agenda for 45th GST Council Meeting. In *GST Council Secretariat* (Vol. 1).

GST Council Secretariat. (2021b). *Confidential agenda for 46th GST Council meeting*.GST Council, Sitharaman, N., Singh, Y. J., Badal, M. S., Uniyal, S., Chaudhary, S. R., Singh, [38] S. R., Cheema, H. S., & Agarwal, P. C. (2022a). *Minutes of the 47th Meeting of GST Council held on 28th & 29th June 2022* (pp. 1–155).

GST Council, Vaghela, P. D., Tiwari, R. K., & Gupta, U. (2017). *Minutes of the 12th GST Council Meeting held on 16th March 2017* (pp. 1–31).

GST, L. (n.d.). Registration under GST Law. In *GST Flyers*.

GST Reforms AND Intergovernmental Considerations In India. (2009). In *Asia Research Centre Working Paper 26*.

Harishekar, L., & Manoj, G. (2021). GST and its impact on small and medium scale enterprises - A study of Peenya industrial area in Bangalore, Karnataka. *Studies in Business and Economics, 16*(1), 81–94. https://doi.org/10.2478/sbe-2021-0007

Indian Institute of Management Calcutta, Mitra, S., Palil, Ibrahim, Gober, Burns, Gold, Hakim, Nayyar, Singh, Poddar, Ahmad, Vasanthagopal, Dash, & Agarwal. (n.d.-a). *GST and Governance: Unintended Implications on the Poor and MSMEs.*

Indian Institute of Management Calcutta, Mitra, S., Palil, Ibrahim, Gober, Burns, Gold, Hakim, Nayyar, Singh, Poddar, Ahmad, Vasanthagopal, Dash, & Agarwal. (n.d.-b). *GST and Governance: Unintended Implications on the Poor and MSMEs.*

Ironkwe, U., & Peter, G. T. (2015). Value added tax and the financial performance of quoted Agribusinesses in Nigeria. In Academy of Business and Retail Management (ABRM), University of Port Harcourt, & Rivers State University of Science and Technology, International Journal of Business and Economic Development. https://www.ijbed.org

Jain, P., 3DUWQHU & PwC. (2018). Over 200 days of GST: the road ahead. In PwC (pp. 3–5) [Book].

Jaiswal, A., & Krishna, R. B. (2017). Mother of all tax reforms: India's goods and services tax 2017. *ResearchGate*.https://www.researchgate.net/publication/318820446_MOTHER_OF_ALL_TAX_REFORMS_INDIA'S_GOODS_AND_SERVICES_TAX_2017

Jaitley, A., Bommai, B., Rao, T. H., Khanna, S., Singh, B., Rathore, B. S., Kaushik, M., & Agarwal, R. (2019). Minutes of 37th GST Council Meeting held on 20th September, 2019. In N. Sitharaman, GST Council Meeting (pp. 1–118).

Jaitley, A., Gangwar, S. K., Narayanasamy, V., Mein, C., & Sisodia, M. (2016). Minutes of the 1st GST Council Meeting (22nd and 23rd September 2016). In GST Council Meeting (pp. 1–16).

Kankariya, R. A., Suryadatta Group of Institute Pune, Dongre, A., & North Maharashtra University. (2021). An analysis of impact of Goods & Services Tax (GST) on traders of goods in Jalgaon District [PhD dissertation]. In Kavayitri Bahinabai Chaudhari North Maharashtra University, Jalgaon.

Lim, B. P., Teoh, W. C., Yap, W. W., & Yong, K. H. (2014). GST in malaysia: An ugly truth or a beautiful lie? in bachelor of economics (hons) financial economics [Thesis; Paper]. Universiti Tunku abdul rahman. Retrieved April 16, 2024, from http://eprints.utar.edu.my/1416/

Loo, E. C., Md Taib, M. S. B., Faculty of Accountancy Universiti Teknologi Mara Melaka, & Account Division, Accountant General's Department of Malaysia. (2017). Goods and Services Tax (GST): challenges faced by business operators in malaysia. in SHS Web of Conferences (Vol. 36, pp. 1–2) [Journal-article]. https://doi.org/10.1051/00027

Mbugua, S., Kamunge, Njeru, A., & Tirimba, I., Ondabu. (2014). Factors affecting the performance of small and micro enterprises in Limuru Town market of Kiambu County, Kenya. *International Journal of Scientific and Research Publications, 4*(12).

Ministry of Finance, Government of India. (2019a). About GST.

Ministry of Finance, Government of India. (2019b). Brief History of GST.

Ministry of Information and Broadcasting, Government of India. (2017). One Nation-One Tax Goods and Services tax.

Ministry of Micro, Small & Medium Enterprises. (2023). Ease of doing business initiative. In Ministry of Micro, Small & Medium Enterprises. https://my.msme.gov.in/MyMsme/Reg/Msme_Background.aspx

Morris, S., Pandey, A., Agarwalla, S., & Agarwalla, A. (2017). Impact of the proposed GST on the consumer price index in India. In Indian Institute of Management Ahmedabad (pp. 1–4) [Report].

Mukherjee, S. & National Institute of Public Finance and Policy. (2019). Inter-Governmental fiscal transfers in the presence of revenue uncertainty: the case of Goods and Services tax (GST) in India. In NIPFP Working Paper Series (No. 255; pp. 1–3). https://www.nipfp.org.in/publications/working-papers/1850/

Mukherjee, S. & National Institute of Public Finance and Policy. (2023). Revenue Performance Assessment of Indian GST. In NIPFP Working Paper Series (No. 392; pp) https://www.nipfp.org.in/publications/working-papers/1992/

Ocheni, S. (2015). Impact analysis of tax policy and the performance of small and medium scale enterprises in Nigerian economy. *Strategic Management Quarterly, 3*(1). https://doi.org/10.15640/smq.v3n1a3

PDFs/03GS_120517578066848D9F4143AFF5FD9992150C0C.PDF

Shukla, S. P. (2018). GST compensation to states. In Ministry of Finance, Ministry of Finance. https://pib.nic.in/PressReleaseIframePage.aspx?PRID=1542747

Singh, N. P., Bisen, J., Venkatesh, P., & Aditya, K. (2018). GST in India: Reflections from food and agriculture. *Agricultural Economics Research Review/agricultural Economics Research Review, 31*(2), 175. https://doi.org/10.5958/0974-0279.2018.00035.6

Sinniah, R. (2015). Perception of manufacturers in Perak State towards Goods and Services Tax (GST) Implementation. In Master of business administration (corporate management) [thesis]. universiti tunku abdul rahman.

Sitharaman, N. (2023a). Draft Minutes of the 49th Meeting of the GST Council held on 18th February 2023.

Sitharaman, N. (2023b). Draft Minutes of the 49th meeting of the GST council held on 18th February 2023.

Sitharaman, N. & GST Council. (2023a). Minutes of the 50th Meeting of the GST Council held on 11th July, 2023 (By Vigyan Bhawan).

The Institute of Cost Accountants of India. (2023). A Significant Step towards making india economically competitive goods and services tax (GST). In *Journal Of* (Vol. 58, Issue 07, pp. 124–100). https://www.icmai.in

Mohanty, R., Samal, A., & Kumar, R. (2024). Green finance supports sustainable environment: Empirical approach towards applicability in tourism sector. *Theoretical and Practical Research in Economic Fields, 15*(1), 114–126.

Higher Education Institutes and Sustainable Development Progression Through the Lens of Industry 5.0: The Journey of Quest International University, Malaysia

Prashanth Beleya, Tilagavati Subramaniam, and Tanaraj Krishna

Abstract The contribution of Higher Education Institutes (HEI) to generate towards the progress of any nation is second to none. HEIs are responsible for cultivating the knowledge that enables students to contribute to sustainable development and prepare job-ready graduates for Industry 5.0. Nevertheless, as the world evolves seamlessly at lightning speed, HEIs have been tasked to be at the forefront to spearhead the necessary changes towards Industry 5.0 rapidly. Sustainable Development (SD) and the need to focus on Industry 5.0 have been identified as pivotal shifts. Thus, industries and nations are beginning to align policies and strategies by integrating the Sustainable Development Goals (SDGs) into their practices. This chapter focuses on 3 particular SDGs, namely quality education (SDG 4), Industry, Innovation, and Infrastructure (SDG 9) and Partnerships for the Goals (SDG 17) which have been integrated into the business policy and planning together with Industry 5.0 elements of Quest International University (QIU), Malaysia. Additionally, the chapter will dive into the benefits gained by QIU and also discuss the current challenges and future aspirations of QIU in further incorporating subsequent SDGs into the higher education business model.

Keywords Higher Education Institutes (HEIs) · Sustainable Development Goals (SDGs) · Industry 5.0 · Quest international university · Malaysia

T. Subramaniam · T. Krishna
Quest International University, Ipoh, Malaysia
e-mail: tilagavati.subramaniam@qiu.edu.my

T. Krishna
e-mail: tanaraj.krishna@qiu.edu.my

P. Beleya (✉)
INTI International University, Nilai, Malaysia
e-mail: prashanth.beleya@newinti.edu.my

R. Kumar and S. N. Mohanty (eds.), *Sustainable Economy Models in the Age of Industry 5.0*, https://doi.org/10.1007/978-981-96-4104-8_19

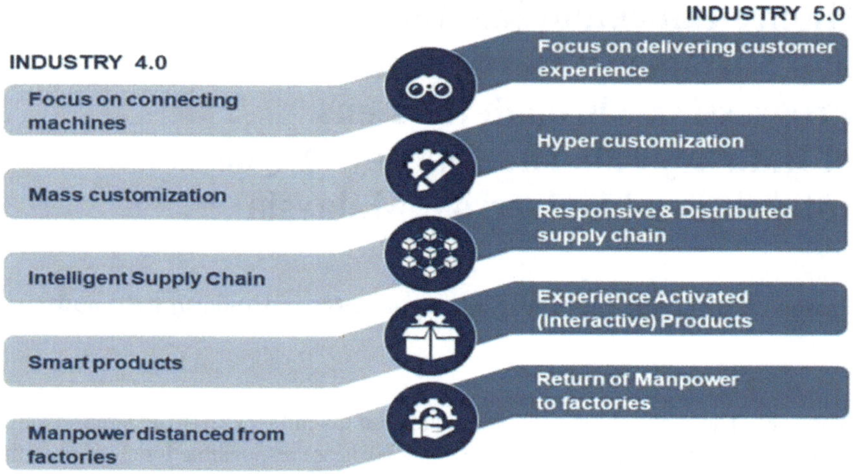

Fig. 1. Transformation from IR 4.0 to IR 5.0

Industry 5.0 (IR 5.0) signifies a dramatic transition in which human ingenuity integrates with modern technologies to emphasize sustainability, efficiency, and social accountability. In contrast to Industry 4.0, which emphasized automation, **IR 5.0** prioritizes human involvement, promoting collaboration between individuals and intelligent systems (Cronin & Doyle-Kent, 2022; Golovianko et al., 2023; Rame et al., 2024). Figure 1 illustrates the disparity between different approaches, highlighting the transition to a more human-centred paradigm.

Higher Education Institutions (HEIs) are significantly influenced by this transformation, particularly as the globe confronts pressing issues such as climate change, social inequality, and swift technological progress (United Nations, 2015). By incorporating the United Nations' Sustainable Development Goals (SDGs) into their fundamental plans, HEIs prepare students with technical competencies and a sustainability-oriented attitude crucial for Industry 5.0 and the intricate global environment (Žalėnienė & Pereira, 2021).

By adopting Industry 5.0, Higher Education Institutions integrate technology with ethical principles, promoting innovation consistent with the **Sustainable Development Goals**. This transition fosters a balanced strategy for efficiency and accountability, tackling social and environmental issues while cultivating graduates who exemplify responsibility (Mohamed Hashim et al., 2024). Higher Education Institutions persist as society's "immune system" for knowledge, fostering new ideas and enabling students to formulate innovative, ethical answers to global challenges (Mohamed Ali, 2021; Carayannis & Morawska-Jancelewicz, 2022; Gürdür Broo et al., 2022; Hashim et al., 2024). They cultivate future leaders proficient in change management, sophisticated problem-solving, and the promotion of sustainable development objectives (Patel et al., 2024).

In recent years, Higher Education Institutions (HEIs) have progressively adopted sustainability as a core component of their purpose, incorporating it into courses, research, and community collaborations (Abo-Khalil, 2024). Higher Education Institutions (HEIs) are equipping students to confront real-world concerns by providing courses in sustainable development, environmental science, and social entrepreneurship, alongside conducting research that addresses environmental and social issues (Tomasella et al., 2022). Partnerships with businesses, governments, and nonprofits augment students' education by linking theoretical knowledge to practical application, fostering a sense of responsibility and community involvement (Jain, 2023).

Higher Education Institutions (HEIs) acknowledge their crucial role in promoting a sustainable future, leading many to embrace the concepts of Industry 5.0, which emphasize human-centred technological integration and innovation. By aligning with Industry 5.0, institutions are harmonizing their educational objectives with global Sustainable Development Goals (SDGs). This alignment bolsters essential Sustainable Development Goals (SDGs) for Higher Education Institutions (HEIs), including **SDG 4 (Quality Education), SDG 9 (Industry, Innovation, and Infrastructure), and SDG 17** (Partnerships for the Goals). Higher Education Institutions (HEIs) can stimulate innovation, promote ethical technology utilization, and cultivate social responsibility in students through curricula centred on sustainability, research initiatives, and community engagement.

In Malaysia, Higher Education Institutions are essential for national development and economic growth, facilitating the country's ambition for innovation-driven progress. **The Malaysian Higher Education Blueprint 2025** delineates a strategy for preparing graduates for the global labour market, enhancing research initiatives, and integrating technology-facilitated education within a sustainability paradigm. The Blueprint underscores the necessity of collaboration between Higher Education Institutions and industry to improve research, establish pertinent curricula, and strengthen public-private partnerships, thereby connecting educational objectives with Malaysia's overarching economic and developmental aspirations.

Quest International University (QIU) embodies this dedication. Through the integration of Industry 5.0 and sustainability, QIU has achieved a top-300 global position for Zero Hunger and SDG collaborations in the Times Higher Education Impact Rankings 2021. QIU, as one of only 19 Malaysian universities and one of 7 private institutions in these rankings, exemplifies how education aligned with Industry 5.0 and sustainability fosters national development and global educational achievement. QI Group's Social Impact and Sustainability (SIS) Committee directs QIU's efforts towards achieving Sustainable Development Goals (SDGs) through projects aimed at empowering individuals, improving communities, and protecting the environment, thereby underscoring its dedication to sustainability and social impact.

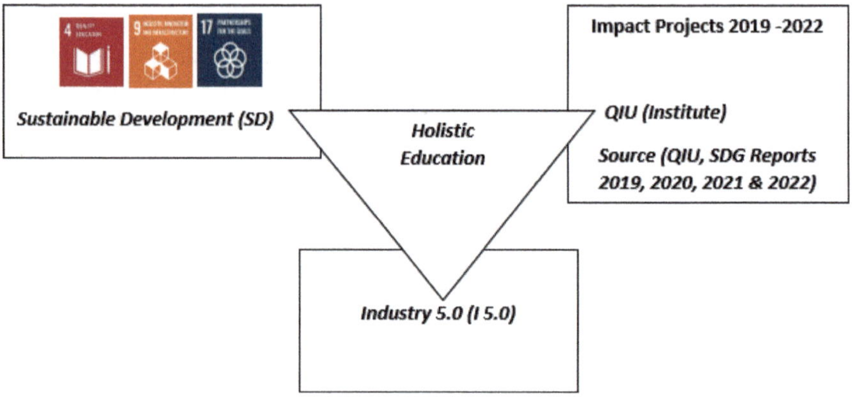

Fig. 2 Framework of integration of sustainable development and moving towards industry 5.0 in Quest International University (QIU)

1 Framework of Integration of Sustainable Development and Moving Towards Industry 5.0 in Quest International University (QIU)

Higher Education Institutes (HEIs) are responsible for cultivating the knowledge that enables students to contribute to sustainable development and prepare job-ready graduates for Industry 5.0. HEIs have been tasked to be at the forefront to spearhead the necessary changes towards Industry 5.0 rapidly. Sustainable Development (SD) and the need to focus on Industry 5.0 have been identified as pivotal shifts. Therefore, Quest International University (QIU) has taken drastic moves to achieve sustainable development and attain Industry 5.0. In line with this, QIU plays an important role in driving impactful projects aligned with sustainability goals.

This chapter deliberates on SD impact projects at QIU which focuses on Sustainable Development Goals 4, 9, and 17. Figure 2 exhibits the integration of Sustainable Development Goals 4, 9, and 17 and QIU impact Projects moving towards Industry (5.0).

2 The Contribution of QIU to the Attainment of the SDGs

According to Roy and Ateeq (2024), HEIs play a crucial role in educating and shaping future leaders who will spearhead the pursuit of attaining and implementing agendas aimed at sustainable development. As an emerging Higher Education Institution, Quest International University (QIU) is dedicated to capitalizing on significant opportunities to contribute to advancing Sustainable Development Goals. In provision to achieve QIU's Sustainable Development Goals, the university has established four clusters comprising Education, Research, Operations, and Leadership. These clusters are dedicated to working on impactful projects. As an educational institution, QIU prioritizes is SDG 4, which interconnected with all other Sustainable Development Goals. QIU believe education is a vital tool for empowering people, fostering economic growth, developing skills, and enhancing techno-entrepreneurship, ultimately contributing to the achievement of all SDGs. While focusing on SDG 4, QIU also contributes to other SDGs, such as SDG 9 and SDG 17. These contributions towards SDG 9 and SDG 17 will drive infrastructural development and foster positive partnerships, aligning with the principles of Industry 5.0.

3 Impact Projects at QIU (2019-2022)

QIU has up front to initiate impact projects aligned with the Sustainable Development Goals (SDGs), route towards Industry 5.0. These initiatives cover a wider range from community development to techno innovation. Despite the hiccups faced during the Coronavirus pandemic, QIU has robustly continued to push boundaries, to create a sustainable future (Kumar. R, et.al., 2024). All the clusters have contributed their parts to produce effective sustainable development projects.

Figure 3 shows the QIU's impact projects related to SDG 4, 9, and 17 from 2019 to 2022. In 2019 QIU initiated 8 impact projects underscore SDG 4, 9, and 17. However, due to the Coronavirus pandemic from 2020 to 2021 the number of impact projects was decreased, and it struck back in 2022 with 11 projects (Source: QIU Impact Projects Report 2019, 2020, 2021 & 2022).

Figure 4 provides a structured overview of QIU projects related to education, infrastructure, and partnerships. The limelight was projects under SDG 4 such as inclusive education and health literacy relevant to the pandemic circumstances. Hence, projects under SDG 9 championed environmental sustainability and industry innovation emphasizing on human-centred approach by considering both environmental and industry impacts. Additionally, QIU's collaborative initiatives under SDG 17 showcase its enthusiasm to invest in the local community through skills development and economic empowerment. This firmly establishes QIU as a community-focused institution committed to sustainable and equitable growth. Instantly, QIU's projects lay out a clear roadmap for how institutions can advance towards Industry 5.0 while tackling crucial environmental and social challenges. Indirectly, this infuses wealthy learning experiences for the graduates and prepares them for the future job market.

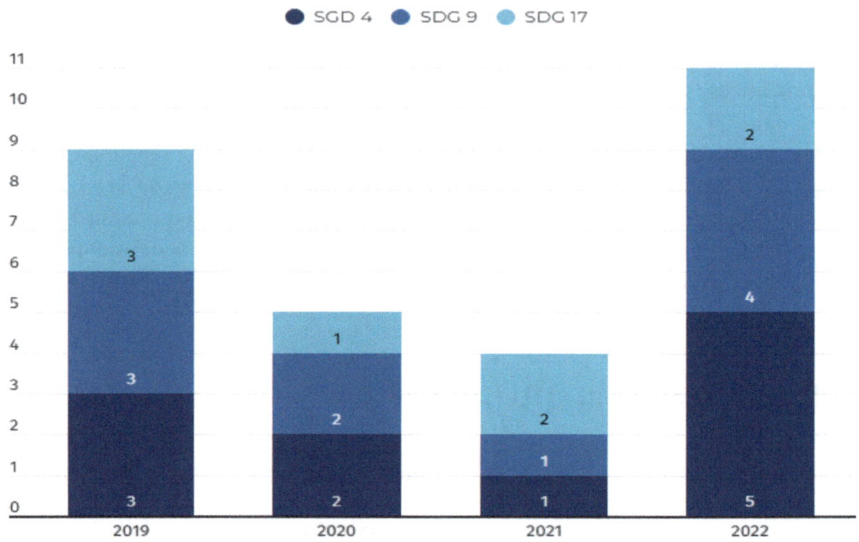

Fig. 3. QIU impact projects related SDG4, 9, 17 from 2019 to 2022

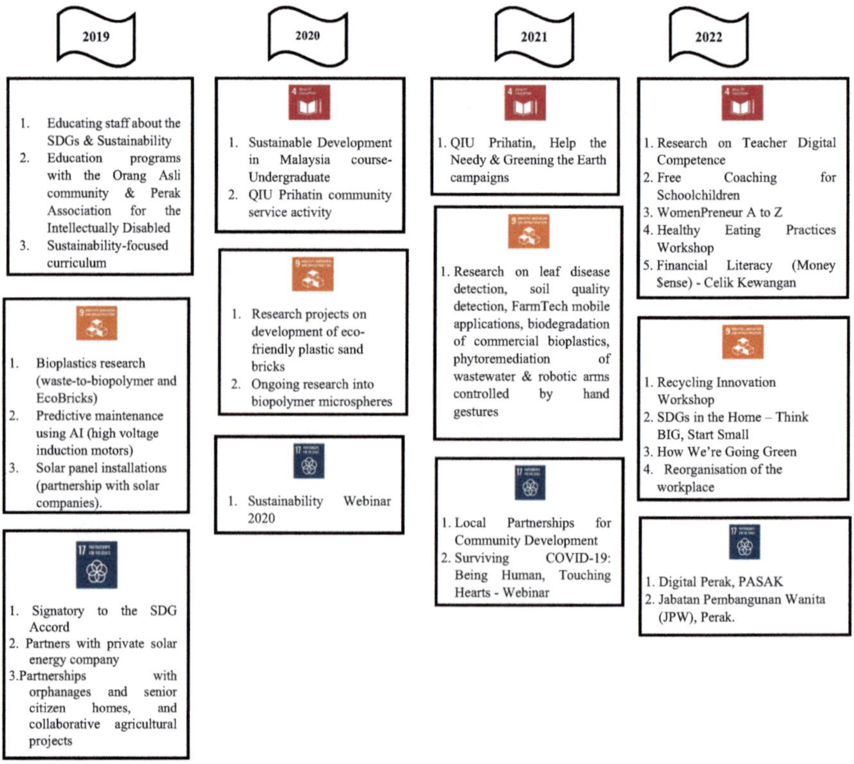

Fig. 4. Overview of projects in education, infrastructure, and partnerships

3.1 Opportunities in SDGs for QIU

Higher education is increasingly pressured to include the Sustainable Development Goals (SDGs) into their strategy, research, and curriculum. There must be an understanding to break silos and take a more holistic approach to education by embracing sustainability, and adopting an interdisciplinary approach that can tackle the real-life challenges of our times and fulfil students' expectations to be prepared for those challenges. Higher Education Institutions (HEIs) are a crucial pillar of the pursuit of the SDGs. They provide the knowledge, skills, and mindsets essential for solving the world's sustainable development challenges and the meeting the SDGs.

The Sustainable Development Goals Report 2022 painted a worrying picture of progress across almost all areas. It said that "cascading and interlinked crises are putting the 2030 Agenda for Sustainable Development in grave danger, along with humanity's very own survival". The urgency of the situation demands that all sectors of society galvanize to secure greater participation and leadership, more resources and improved, game-changing solutions to address increasing inequality and the escalating climate emergency. Achieving the SDGs is a big task, and considering the vital roles that universities play in supporting and implementing them, the sector must strive for advancement.

Quest International University (QIU) has been at the forefront of change where SDG is considered and various plans are in place to further elevate this progress. As such, the opportunity to unlock potentials through SDGs and incorporating Industry 5.0 elements become ever so important. QIU has already taken the necessary steps to be a pioneer in developing and embracing certain if not all SDGs available. This is evident from the rankings available that list QIU in the top 10 of SDG embracing universities in Malaysia. Through the incorporation of SDGs, QIU is steadfast in its commitment of progress and this bounds to open more opportunities.

Incorporating the Sustainable Development Goals (SDGs) into QIU offers multiple benefits that enhance the university's role as an academic institution and as a responsible, forward-looking community member. Aligning with the SDGs strengthens QIU's reputation attracts socially conscious students and staff, boosts research opportunities, and drives positive social, environmental, and economic impacts. The benefits are elaborated below.

- Academic Excellence and Attractiveness

Integrating the SDGs into QIU's curriculum raises the university's academic profile by offering students a contemporary, socially relevant education that resonates with global priorities. Taught programmes that address issues like clean energy (SDG 7), reduced inequalities (SDG 10), and sustainable cities (SDG 11) will provide students with cutting-edge knowledge and skills that are increasingly valued in today's job market and globalization. By aligning educational offerings with SDGs and industrial revolution 5.0 (IR 5.0) components, QIU would naturally become the focal attraction to students passionate about making a positive impact, thus drawing a motivated and diverse student body. This will also help attract staff and faculty who prioritize social responsibility, enhancing QIU's reputation as an institution that values both academic rigour and ethical considerations. As the global population diversifies towards generation Alpha, this would open abundant opportunities for QIU to grow their business strategically.

- Research and Partnerships

Aligning research priorities with SDGs creates pathways for collaboration with national and international bodies focused on sustainability. By prioritizing research areas like renewable energy (Goal 7), industry innovation (Goal 9), and climate action (Goal 13), QIU can attract research funding from not just Malaysian government agencies and private companies, but also international organizations committed to sustainable development. These partnerships not only would increase the impact of QIU's research capabilities, but also provide real-world experience for students and faculty members through collaborative projects and internships. Additionally, research on SDG-related topics has a high potential for publication, which will improve QIU's visibility and academic standing in the global research community in the long run. Additionally, such collaborations have the potential to develop commercial values for projects that could generate income for QIU.

- Community Social Responsibilities

Incorporating SDGs into community engagement initiatives would enable QIU to have a meaningful impact on local communities. This has been well defined in the previous areas. For instance, through health education programmes aligned with Good Health and Well-being (Goal 3), QIU can address public health issues that directly affect nearby populations. Education and training programmes that promote Quality Education (Goal 4) and Decent Work and Economic Growth (Goal 8) help uplift communities by creating economic and social opportunities, particularly for underserved populations. Such initiatives would build strong bonds between QIU and the community, demonstrating the university's commitment to being an active participant in social development. This engagement also raises QIU's visibility and influence as a positive force in society, creating a ripple effect that benefits the university's reputation and encourages community support. A university with positive word of mouth encourages students and organizations to increase their exposure towards such establishments.

- Promoting Sustainability and Operational Efficiency

By integrating SDG principles into campus operations, QIU can lead by example in sustainability practices. Reducing waste, conserving water, implementing renewable energy, and encouraging sustainable consumption contribute to a smaller environmental footprint, which is aligned with Climate Action (Goal 13) and Responsible Consumption and Production (Goal 12). Not only does this make QIU a model for sustainable behaviour, but it also leads to cost savings on resources like energy and water, making campus operations more efficient. These sustainability practices further attract students, staff, and faculty who prioritize working and studying in an environmentally responsible setting.

- Preparing Students for Global Challenges and Employability

Through SDG-oriented education, students at QIU are better equipped to understand and address global challenges, making them highly adaptable in a dynamic job market. Exposure to pressing issues like inequality, environmental degradation, and healthcare access creates graduates who are not only skilled but also socially aware and proactive. Such an environment would deem QIU graduates attractive to employers who at present and the future value sustainability and corporate responsibility. With global demand growing for professionals who can contribute to sustainable development, QIU's emphasis on the SDGs ensures that graduates are well-positioned for roles in public policy, international organizations, and socially responsible companies. Incorporating the SDGs positions QIU as a leader in sustainable education, setting an example for other institutions in Malaysia and the region. QIU's commitment to sustainable practices and social responsibility attracts recognition and opportunities for leadership within academic and policy circles, strengthening its brand as a progressive, ethical institution. By embracing this global agenda, QIU demonstrates that it is committed not only to academic excellence but also to contributing to a better world.

3.2 Anticipated Challenges in SDGs for QIU

The funding disparity between public and private universities in Malaysia presents substantial obstacles for institutions such as QIU when they embark on SDG-related initiatives. Substantial government support is provided to public universities, which are frequently regarded as central actors in the advancement of national education and development objectives. This support includes grants and subsidies that are specifically designed to support sustainability initiatives. This public funding allows them to establish collaborations with international organizations, conduct impactful research, and pursue long-term sustainability initiatives. The financial support affords public institutions a strategic advantage in advancing the SDG agenda, as they can utilize these resources for both research and campus-wide sustainable practices.

In contrast, private institutions such as QIU primarily rely on self-generated funds, including tuition fees, fees from consultancy services, private contributions, and limited corporate grants. QIU is at a disadvantage when competing for the same SDG funding as public institutions as a result of this funding model. QIU's long-term contributions to Malaysia's sustainability landscape may be impacted by its frequent prioritization of core academic requirements over sustainability projects due to its limited financial resources. Although government grants and subsidies are occasionally available to private institutions, they are subject to rigorous eligibility requirements that frequently favour established public institutions with substantial research infrastructure. Despite the fact that private universities are eligible, they are subject to intense competition for these restricted funds, which results in a significant administrative burden and a strain on resources.

It is imperative to resolve these disparities as Malaysia advances towards IR5.0, which prioritizes sustainable growth that is human-centred. The vision of IR5.0 necessitates that institutions such as QIU incorporate sustainable practices with advanced technologies. Private institutions may encounter difficulties in completely supporting Malaysia's IR5.0 and SDG objectives in the absence of more inclusive funding policies. QIU's sustainability initiatives were primarily self-funded or supported by private contributions between 2019 and 2022, demonstrating the university's dedication to sustainability in the face of limited resources. The SDG report emphasizes that QIU effectively executed a multitude of impactful activities with minimal funding, showcasing strong community engagement and resourcefulness. This track record indicates that QIU has the potential to increase its capacity to implement more extensive, transformative initiatives that would have a broader social and environmental impact if it were granted access to more substantial funding. Increased funding would allow QIU to expand its SDG initiatives, support more comprehensive research, and cultivate impactful partnerships, thereby making a substantial contribution to Malaysia's sustainability objectives.

Although QIU's SDG initiatives are beneficial to its community, they frequently fail to garner the same level of recognition as those of more established universities in Malaysia. For instance, QIU has implemented numerous environmental conservation initiatives, including tree-planting programmes and local clean-up campaigns that have effectively engaged students and local stakeholders. Nevertheless, these endeavours, despite their significance, are not extended beyond the immediate campus and adjacent communities as a result of QIU's limited media presence and brand recognition. Conversely, larger universities with established reputations can effortlessly amplify their impact and awareness by utilizing their media networks to promote comparable activities on a national or even international scale.

Another example is QIU's Health and Well-being Campaign, which offers complimentary health screenings and wellness talks to the local community. Despite their significant impact, these activities are frequently restricted to the immediate region as a result of budgetary and outreach constraints. Larger universities with more established networks and more resources are frequently able to expand similar campaigns across multiple regions, securing partnerships with national healthcare organizations that further increase visibility. The capacity of QIU to secure industry partnerships is also impacted by the lack of widespread recognition for its SDG work. For example, although QIU has organized sustainable development forums and seminars, these events are more likely to garner local rather than national recognition, which restricts the potential for engagement with key stakeholders who could offer sponsorships or collaboration opportunities. In contrast, large universities have the ability to secure government endorsements and corporate sponsors for comparable events, thereby broadening the scope and influence of their SDG initiatives.

These initiatives could be substantially expanded if QIU had more extensive media networks and partnerships. QIU's ability to make a more significant contribution to Malaysia's SDG and IR5.0 aspirations is contingent upon its visibility, which is essential not only for impact but also for attracting the necessary funding and resources for sustainable development initiatives.

The absence of a comprehensive mechanism to measure the true impact of its SDG initiatives on target groups is a critical challenge for QIU, as it is for many institutions. Although QIU has implemented numerous advantageous initiatives, including a financial literacy workshop for women in the B40 demographic, there is inadequate follow-up to evaluate whether these initiatives have accomplished their intended goals. For instance, the workshop's objective was to equip B40 women with financial skills to enhance their economic resilience. However, the absence of a structured measurement framework renders it uncertain whether the participants experienced substantial enhancements in their quality of life, income management, or financial literacy as a result of the workshop.

There are numerous obstacles associated with this absence of impact measurement. To begin with, it is challenging for QIU to demonstrate the effectiveness of its programmes or accountability to potential partners, donors, and stakeholders in the absence of evidence regarding the workshop's outcomes. These days, corporate sponsors and funding authorities are becoming more concerned with quantifying outcomes to guarantee that their contributions generate genuine, enduring transformations. Furthermore, the QIU is unable to accurately evaluate the programmes most beneficial or need for improvement in the absence of data that indicates whether the workshop authentically impacted financial habits or long-term economic empowerment. Due to this uncertainty, QIU is unable to replicate or expand successful initiatives, which could lead to the misallocation of resources to less effective activities.

Additionally, QIU's capacity to strategically plan for future SDG objectives is impeded by the absence of measurable outcomes. The university may overlook the valuable insights that programmes such as the financial literacy workshop could provide into the needs and challenges of the B40 community, thereby missing out on critical data that could steer future projects, if follow-up assessments are not conducted.

Implementing a structured impact assessment mechanism would enable QIU to collect meaningful data on participant outcomes, behavioural shifts, and community benefits. Such a system would strengthen QIU's contribution to Malaysia's SDG objectives, enhance its ability to secure further support, and ensure that its programmes genuinely address the needs of underserved groups in a way that aligns with IR5.0's human-centred sustainability principles.

3.3 Conclusion

QIU's commitment to Sustainable Development Goals (SDGs) 4, 9, and 17 forms a strong foundation for its alignment with Industry 5.0, aiming to create a future where technology serves humanity, drives sustainable growth, and strengthens cross-sector partnerships. Through SDG 4, the university emphasizes quality education, equipping students with a unique blend of technical expertise and a sustainability-oriented mindset that prepares them to advance Industry 5.0's human-centred objectives. By fostering a curriculum and research opportunities that prioritize ethical, sustainable practices, QIU ensures that its graduates are ready to address contemporary challenges while maintaining a strong commitment to social responsibility. Through SDG 9, QIU promotes innovation, resilient infrastructure, and research that enhance industrial capabilities while prioritizing environmental and social impact.

Looking to the future, QIU recognizes the importance of expanding its efforts beyond just SDGs 4, 9, and 17, striving to address the full spectrum of the SDGs in its educational approach. At the same time, QIU aims to integrate a holistic focus on additional SDGs, such as climate action (SDG 13), responsible consumption and production (SDG 12), and decent work and economic growth (SDG 8). The university envisions expanding its research and programmes to foster more sustainable industrial practices, while also promoting social equity, environmental sustainability, and ethical innovation. By strengthening partnerships across academia, industry, and society (SDG 17), QIU seeks to create a collaborative ecosystem where students, researchers, and industry leaders work together to solve global challenges. This forward-thinking approach ensures that QIU's graduates are not only equipped to meet the technical demands of Industry 5.0 but are empowered to drive a future focused on ethical progress, sustainable development, and inclusive innovation.

References

Abo-Khalil, A. G. (2024). Integrating sustainability into higher education challenges and opportunities for universities worldwide. *Heliyon, 10*(9), e29946. https://doi.org/10.1016/j.heliyon.2024.e29946

Ali, M. (2021). Vocational students' perception and readiness in facing globalization, industry revolution 4.0 and society 5.0. *Journal of Physics: Conference Series, 1833*(1), 012050. https://doi.org/10.1088/1742-6596/1833/1/012050

Carayannis, E. G., & Morawska-Jancelewicz, J. (2022). The futures of Europe: Society 5.0 and industry 5.0 As driving forces of future universities. *Journal of the Knowledge Economy, 13*(4), 3445–3471. https://doi.org/10.1007/s13132-021-00854-2

Cronin, M., & Doyle-Kent, M. (2022). Creating value with environmental, social, governance (ESG) in Irish manufacturing smes': A focus on disclosure of climate change risks and opportunities. *IFAC-PapersOnLine, 55*(39), 48–53. https://doi.org/10.1016/j.ifacol.2022.12.009

Golovianko, M., Terziyan, V., Branytskyi, V., & Malyk, D. (2023, January 13). *Industry 4.0 Vs. industry 5.0: Co-existence, transition, or a hybrid*. Elsevier BV. Retrieved from https://www.researchgate.net/publication/367094997_Industry_40_vs_Industry_50_Co-existence_Transition_or_a_Hybrid

Gürdür Broo, D., Kaynak, O., & Sait, S. M. (2022). Rethinking engineering education at the age of industry 5.0. *Journal of Industrial Information Integration, 25*, 100311. https://doi.org/10.1016/j.jii.2021.100311

Jain, V. (2023, June 15). *Unveiling the crucial role of stakeholders in higher education: Key insights and strategies*. Unknown. Retrieved from https://www.researchgate.net/publication/371580869_Unveiling_the_Crucial_Role_of_Stakeholders_in_Higher_Education_Key_Insights_and_Strategies

Mohamed Hashim, M. A., Tlemsani, I., Mason-Jones, R., Matthews, R., & Ndrecaj, V. (2024). Higher education via the lens of industry 5.0: Strategy and perspective. *Social Sciences & Humanities Open, 9*, 100828. https://doi.org/10.1016/j.ssaho.2024.100828

Patel, N. S., Puah, S., & Kok, X.-F. K. (2024). Shaping future-ready graduates with mindset shifts: Studying the impact of integrating critical and design thinking in design innovation education. *Front. Educ. 9,* 1358431. https://doi.org/10.3389/feduc.2024.1358431

Rame, R., Purwanto, P., & Sudarno, S. (2024). Industry 5.0 and sustainability: An overview of emerging trends and challenges for a green future. *Innovation and Green Development, 3*(4), 100173. https://doi.org/10.1016/j.igd.2024.100173

Mohanty, R., Samal, A., & Kumar, R. (2024). Green finance supports sustainable environment: Empirical approach towards applicability in tourism sector. *Theoretical and Practical Research in Economic Fields, 15*(1), 114–126.

Kumar, R., Goel, R., Singh, T., Mohanty, S. M., Gupta, D., Alkhayyat, A., & Khanna, R. (2024). Sustainable finance factors in indian economy: analysis on policy of climate change and energy sector. *Fluctuation and Noise Letters, 23*(2), 2440004–339.

Tomasella, B., Wylie, A., & Gill, D. (2022). The role of higher education institutions (HEIs) in educating future leaders with social impact contributing to the sustainable development goals. *Social Enterprise Journal, 19*(4), 329–346. https://doi.org/10.1108/sej-03-2022-0027

United Nations. (2021). The sustainable development goals report. Retrieved from https://unstats.un.org/sdgs/report/2021/The-Sustainable-Development-Goals-Report-2021.pdf

Žalėnienė, I., & Pereira, P. (2021). Higher education for sustainability: A global perspective. *Geography and Sustainability, 2*(2), 99–106. https://doi.org/10.1016/j.geosus.2021.05.001

Roy, R., & Ateeq, A. (2024). Moving towards industry 5.0: Opportunities and challenges in Bahrain higher education. *The AI Revolution: Driving Business Innovation and Research, 1*, 425–435.

GPSR Compliance

The European Union's (EU) General Product Safety Regulation (GPSR) is a set of rules that requires consumer products to be safe and our obligations to ensure this.

If you have any concerns about our products, you can contact us on ProductSafety@springernature.com

In case Publisher is established outside the EU, the EU authorized representative is:

Springer Nature Customer Service Center GmbH
Europaplatz 3
69115 Heidelberg, Germany

Batch number: 10704035

Printed by Printforce, the Netherlands